Lecture Notes in Statistics

Volume 227

Lecture Notes in Statistics (LNS) includes research work on topics that are more specialized than volumes in Springer Series in Statistics (SSS).

The series editors are currently Peter Bühlmann, Peter Diggle, Ursula Gather, and Scott Zeger. Peter Bickel, Ingram Olkin, and Stephen Fienberg were editors of the series for many years.

Preface

Machine learning as a whole and, in particular, statistical learning have become a major tool in engineering research and industry during the last two decades, first on a small scale and in the recent past in more and more settings. It allows to use the large amount of information from the construction of production plants and from measurements of process parameters to achieve considerable improvements in product quality and in production time and costs.

This book is dedicated to statistical learning in engineering applications. Its main purpose is to introduce engineers to this exciting new area without requiring more than the most basic training in mathematics and statistics, which usually is part of the engineering education. On this level, it is also of interest for students of data science or related disciplines as a first primer on statistical learning with a focus on ideas and not on technical details.

In the first part, a short leisurely introduction to statistical learning as a part of machine learning, particularly important for industrial users, is given. Here, the focus is on the main ideas and concepts as well as on the prerequisites and the limitations of the methods. We also provide a review of common models and algorithms, but we do not discuss the computational problems of implementing learning methods as we expect engineering users to apply ready-made software packages, which are now widely available, respectively, to hire specialists in more complicated situations. In the latter case, good communication between data scientists and engineers is crucial for success, and the first part of the book will provide a common basis for this interchange.

The second part of the book presents case studies in applications of statistical learning from projects with industry from the *Fraunhofer Institute of Industrial Mathematics* (ITWM). They illustrate the complexity of real-world problems occurring in industry which differ in several respects from the famous standard applications of machine learning in the recent past, where the main difference is often the relative scarcity of data on the one hand, but the presence of sound engineering expertise on the other hand.

The Fraunhofer ITWM (www.itwm.fraunhofer.de) in Kaiserslautern, Germany, is a large applied mathematical research institute with about 400 researchers from mathematics, engineering, computer science, physics, and other disciplines. Almost half of the ITWM's annual operating budget of about 40 million euro is generated from contracts with industry, and the rest includes many public research grants. The institute has a decades-long history of successful innovative projects with large, medium-sized, and small companies and with organizations from a large spectrum of sections of industry: process engineering and mechanical/plant engineering, mobility and automotive industry and suppliers, medicine and medical technology, energy and raw materials industry, technical textiles, information technology, and the finance industry.

The ITWM consists of six departments (Image Processing; Financial Mathematics; Material Characterization and Testing; Flow and Material Simulation; System Analysis, Prognosis and Control; Transport Processes) and of three larger divisions (High Performance Computing; Mathematics for Vehicle Engineering; Optimization). Some of these groups used machine learning in industrial projects already 25 years ago, but in the last decade it has become a ubiquitous tool in all kinds of modeling, simulation, and optimization of processes in industry. Consequently, almost all groups of the ITWM have contributed to the second part of the book, showing the broad spectrum of applications of statistical learning.

Kaiserslautern, Germany
September 2023

Jürgen Franke
Anita Schöbel

Contents

An Introduction of Statistical Learning for Engineers

Jürgen Franke

Abstract Statistical learning is a branch of machine learning that is concerned with the statistical aspects of data analysis by means of machine learning algorithms. In the following, we present a leisurely introduction to this topic, as it is of particular interest for applications in engineering research and practice.

Statistics is one of the roots of modern machine learning, in particular the discipline of nonparametric statistics, which more than 40 years ago developed a lot of concepts and ideas that are still relevant nowadays, e.g., tree-based learning (Sect. 7), local smoothing (Sect. 10), or resampling techniques like crossvalidation (Sect. 3.4) or bootstrap (Sect. 9.1). Using the advances of computational technology at that time, it was possible to leave the restricted framework of classical statistics, which mainly is based on models for the data-generating process, which are completely specified up to a manageable number of parameters. The latter could be estimated from the small- and medium-sized datasets that were typical in the past. However, researchers always were aware that those models only were an approximation to truth, in particular in realistic situations apart from ideal laboratory conditions, and that this model misspecification may be the cause of suboptimal or even false results from a data analysis. Nonparametric statistics allowed to adopt a much broader view on the data-generating process (compare Sect. 2.6), which is also the basis of modern machine learning methods.

1 Introduction

Statistical learning is a branch of machine learning that is concerned with the statistical aspects of data analysis by means of machine learning algorithms. In the

J. Franke (✉)
Department of Mathematics, University of Kaiserslautern-Landau (RPTU), Kaiserslautern, Germany
e-mail: franke@mathematik.uni-kl.de

J. Franke, A. Schöbel (eds.), *Statistical Machine Learning for Engineering with Applications*, Lecture Notes in Statistics 227,
https://doi.org/10.1007/978-3-031-66253-9_1

following, we present a leisurely introduction to this topic, as it is of particular interest for applications in engineering research and practice.

Statistics is one of the roots of modern machine learning, in particular the discipline of nonparametric statistics, which more than 40 years ago developed a lot of concepts and ideas that are still relevant nowadays, e.g., tree-based learning (Sect. 7), local smoothing (Sect. 10), or resampling techniques like crossvalidation (Sect. 3.4) or bootstrap (Sect. 9.1). Using the advances of computational technology at that time, it was possible to leave the restricted framework of classical statistics, which mainly is based on models for the data-generating process, which are completely specified up to a manageable number of parameters. The latter could be estimated from the small- and medium-sized datasets that were typical in the past. However, researchers always were aware that those models only were an approximation to truth, in particular in realistic situations apart from ideal laboratory conditions, and that this model misspecification may be the cause of suboptimal or even false results from a data analysis. Nonparametric statistics allowed to adopt a much broader view on the data-generating process (compare Sect. 2.6), which is also the basis of modern machine learning methods.

The considerable progress of statistical learning during the last three decades was due the cooperation of computer science and statistics together with the rapid advances of computational facilities for collecting, organizing, and processing data. Apart from the necessary hardware, a data analysis based on machine learning requires the following ingredients:

(1) preparation of suitable data,
(2) specification of a flexible model for solving the task of interest, e.g., a neural network depending on a large number of unknown parameters,
(3) learning, i.e., the estimation of the model parameters from the data, which usually requires solving a complex optimization problem for which an efficient training or estimation algorithm, adapted to the hardware at hand, is needed,
(4) validation and interpretation of the results.

Getting these tasks done requires a joint effort of experts from the respective field of science or engineering, from computer science, statistics, and, in particular with respect to (3), from optimization and numerical analysis.

In industrial applications, it is only rarely necessary to develop a machine learning (ML) procedure from scratch. Usually, one chooses a suitable one from the widely used model classes and then uses publicly available apps or ML packages for fitting it to the data without worrying about the computational details. Therefore, with few exceptions, we do not discuss the training algorithms required in step (3), which, together with the models, are at the heart of machine learning, but not so important for the user turning to ready-made solutions.

For engineering applications, it is more important to understand the scope and limitations of the models that we review in the following chapters without going into too much details. It is also essential to understand how to collect and process data before they are fed into the learning algorithm, how to tune the algorithms by suitable choice of so-called hyperparameters such that they provide reasonable outputs, and how to retrospectively decide if the trained ML system does what it should. Finally, engineers frequently do not only want to have a result, i.e., an estimate of an unknown quantity or a classification of a given object, but they also would like to have some information about its reliability and precision. For example, a face recognition algorithm used in deciding if a person belongs to the group of people allowed to enter a building results in a "yes" or "no" answer. In developing and selling such a system, the probabilities that an authorized person is denied entry and, even more important, that an outsider is falsely given access are required as potential customers would like to know them. Such tasks fall into the domain of statistical learning.

Therefore, in this part of the book, we focus on tasks (1), (2), and (4) and, additionally, due to space, to only **supervised learning**. Here, the data consist of two parts: an input and an output, e.g., in the face recognition example, we have images as inputs together with the information if the respective person is authorized or not. The goal is to develop a system that in the future allows to infer the unknown output from an input newly provided. In training an appropriate algorithm, we can use the performance on the given data with already observed outputs to find a good solution. In contrast, in unsupervised learning, only the inputs are given, and the task consists in discovering hidden structures, e.g., a partition of the data into groups consisting of similar objects each. Such data-analytic problems are also important, but as the majority of industrial applications belongs to the area of supervised learning, we only consider this part of machine learning in this book.

In the following, in presenting the models and methods, we do not strive for rigorous mathematical accuracy. In referring to mathematical results that help in understanding a method, we often use loose formulations like "under appropriate conditions ..." to get the main message across without burying it under a lot of technical details, and we also sometimes do not mention exceptions that are irrelevant for practical applications. For the reader interested in the theory of statistical learning, we provide references. Here, we preferably refer to more detailed textbooks and to review articles for further reading instead to the original papers as they are the recommended next step for readers of this first introduction who want to know more about a particular topic. We do not attempt to cover the rich literature on statistical learning as such information can be found in the referred books and reviews.

Finally, let us mention that, except for the next section, we use artificial examples for illustrating the described ML procedures. In such a manner, we can highlight the essential points without the dirty details typical of real real-life experiments, and the low-dimensional inputs allow for an easy graphical representation of the results. As a software, we use basic MATLAB and its toolbox *Statistics and Machine Learning*. For readers who, for the first time, are interested in applying statistical learning procedures themselves, we would recommend as a starting point R or Python due to the large variety of ready-made routines instead. Good introductions to statistical learning going beyond the following concise exposition and containing also many computational exercises are [47] for R and [48] for Python.

1.1 Three Typical Examples

The following three examples are simplified versions of real applications from projects with industry. They illustrate the three main types of problems covered in this book: regression, classification, and forecasting.

The goal of *regression* in data analysis is in its most basic form the estimation of an unknown function from independent experiments. We consider an example from press hardening of sheet metal that, in a somewhat more refined version, is studied in [54], where also more details are given. The goal of this manufacturing process is the production of a formed metal part satisfying a minimum hardness condition. The blank metal sheet is heated in a chamber furnace with temperature T_f and then transported by an industrial robot into the cooled forming tool where it is finally formed and quenched. There are four process parameters under the control of the operator: T_f, the handling time t_h of the robot, the press force F_p during forming, and the quenching time t_q. We combine them to a four-dimensional input vector $x = (T_f, t_h, F_p, t_q)$.

To achieve the prescribed hardness, the operator has to find a suitable combination of the input components. For this purpose, we need to learn the function $y = m(x)$ describing the dependence of the hardness y on the components of x, i.e., on the process parameters. The available data consist of only $N = 60$ observed input-output pairs $X_j, Y_j, j = 1, \ldots, N$, as the cost of the experiments does not allow for a larger training set. Due to measurement errors and other minor influences, the data only satisfy $Y_j = m(X_j) + \varepsilon_j$ with small errors ε_j, which are 0 on average.

Numerical simulation of the heating, hardening, and cooling process would be possible in principle but also too expensive, and we would have to check the validity of the used model anyhow comparing the results with experimental data. Statistical learning provides a satisfactory solution, which allows to predict the resulting hardness for a whole range of input values in real time such that the final algorithm may be used during production. Scarcity of data is balanced by using the knowledge of press hardening experts in training the algorithm (compare Sects. 2.6 and 4.2 for more details).

The second large group of tasks, which may be solved with the help of statistical learning, consists of *classification* problems. In industrial applications, they typically appear in all kinds of quality control. Here, we consider examples from surface inspection. In the past, experienced workers were needed to visually judge the quality of, e.g., an autobody sheet or a veneer layer. In the last decade, they have been more and more replaced by optical systems together with particular image analysis software. Modern statistical learning methods like convolutional neural networks (compare Sect. 13.2) allow for further improvements of such systems.

In the simplest case, a classification algorithm uses an image of the surface as input and, based on this information, decides if the product is defect or not. There are, however, more complicated situations. The quality inspection of furniture panels, e.g., is not an easy task as wood is a natural material with a complicated surface allowing for a lot of variations due to the fiber structure of the material. There is not only one kind of defect to detect here, but we have to consider scratches and dents as well as color or texture variations, i.e., the algorithm has to have several different outputs corresponding to the various kinds of defects. An industrial application is described in the chapter *Machine Learning for Inline Surface Inspection Systems* in this book. Additionally, we do not only want to know if the product is good or bad, but we want to have a grading into various classes of quality for which we can demand different prices.

A related problem is the sorting of wood, intended for furniture production or parquet floors, with respect to texture and color such that it can be packed into batches of similar appearance. Statistical learning algorithm, nowadays, solve such problems well and fast such that they can be used for inline sorting.

In industrial applications, a typical problem involving *forecasting* is predictive maintenance. Here, the data frequently consist of streams of measurements from a production plant, i.e., the inputs are observations from a vector-valued time series. The goal is to predict imminent changes in the production environment such that a timely maintenance can by scheduled. One possibility is to develop an online algorithm for detecting a changepoint in the production parameters as an early indicator for consecutive malfunctions (compare Sect. 14.5).

In [88], another approach is considered, which allows to consecutively monitor the probability of failure. As an example, a stream of input data is observed from a two-fluid injection nozzle, consisting of the pressure of the liquid, the compressed air pressure, and volumetric flow rate of liquid, i.e., the inputs are given as observations from a three-dimensional time series. The goal is the reduction of unplanned shutdowns due to blocking of the nozzle by replacing maintenance on a regular schedule by adaptive maintenance whenever it is necessary. A satisfactory solution is achieved again by combining a standard statistical learning method, here neural networks, with the operators' experience, here quantified in terms of fuzzy logic.

1.2 Notions and Notation of Basic Statistics

The following text assumes as a prerequisite a very basic training in probability theory and statistics for engineers. Nevertheless, we briefly review a few essential notions and concepts, which are used repeatedly, and on this occasion introduce some notation.

A real-valued *random variable* X is considered as a model of a measurement resulting in a value that is not completely determined in advance but fluctuates between repeated experiments, e.g., due to observational error or other unknown influence. The *distribution* of X allows to calculate the probability $\mathrm{pr}(a < X < b)$ that the random variable assumes values between a and b. If the range of values

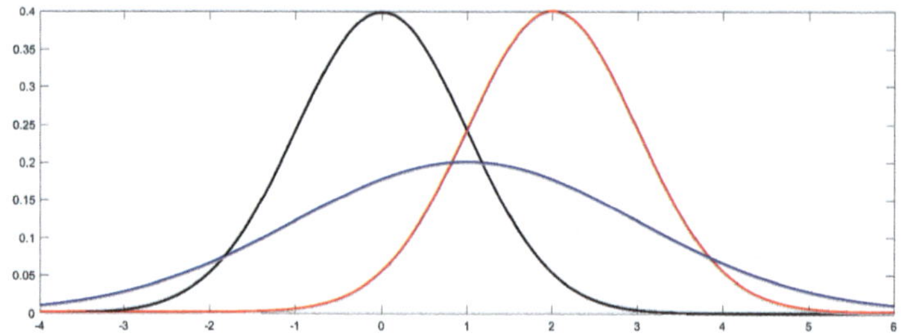

Fig. 1 Gaussian densities with $\mu = 0, \sigma^2 = 1$ (black), $\mu = 2, \sigma^2 = 1$ (red), and $\mu = 1, \sigma^2 = 4$ (blue)

of X is continuous, the distribution is characterized by the *probability density* $p(x)$ such that

$$\mathrm{pr}(a < X < b) = \int_a^b p(x)dx.$$

Among many distributions that are useful as models for experimental data, the *normal* or *Gaussian distribution* plays a special role. It depends on two real-valued parameters, the mean μ and the variance $\sigma^2 > 0$, and the probability density of X is given by

$$p(x) = \frac{1}{\sqrt{2\pi\sigma^2}} \exp(\frac{(x-\mu)^2}{2\sigma^2},$$

i.e., the Gaussian bell-shaped curve. Figure 1 shows examples for three different parameter sets $\mu = 0, \sigma^2 = 1$ (black), the so-called *standard normal* density, $\mu = 2, \sigma^2 = 1$ (red), and $\mu = 1, \sigma^2 = 4$ (blue). Gaussian random variables are frequently good (though not completely exact) models for many kinds of data.

In general, the *mean* or *expectation* $\mathrm{E}\,X = \int xp(x)dx$ of a real random variable X describes where its values will be located on the average. More generally, the mean of a function g of the data is given by

$$\mathrm{E}\,g(X) = \int g(x)p(x)dx.$$

The *variance* $\mathrm{var}\,X = \mathrm{E}\,(X - \mathrm{E}\,X)^2$ is a mathematically convenient measure of the amount of random variability of the values of X from repeated experiments around the mean. $\sigma(X) = \sqrt{\mathrm{var}\,X}$, the so-called *standard deviation*, is, however, to be preferred for the purpose of interpretation as it has the same scale as the data X themselves.

The mean is usually unknown and often a prime quantity of interest, which we would like to learn or estimate from data. In the most basic setup, we assume that we have a sample $X_1, \dots, X_N$ of independent realizations of the random variable X, typically as results from a sequence of *identical* experiments under the same conditions. Independence of X_i and X_j here is understood in the strong sense, i.e., there is no way how the outcome of one experiment may influence another one. Mathematically, we may define independence by means of factorization properties of probabilities. If these assumptions are fulfilled, the data $X_1, \dots, X_N$ are called *independent, identically distributed*. Then, we may estimate the expectation $\mu = \mathrm{E}\, X$ by the *sample mean*

$$\hat{\mu} = \frac{1}{N} \sum_{j=1}^{N} X_j \to \mu \quad \text{for } N \to \infty,$$

which converges to the true unknown mean for increasing sample size N. This is the simplest version of the **law of large numbers**, which is the core of the justification for most statistic learning procedures. In our basic setting, we see this from an interpretation of the sample mean as a **least-squares estimate**. Minimizing the squared empirical loss

$$\frac{1}{N} \sum_{j=1}^{N} (X_j - \mu)^2 = \min_{-\infty<\mu<\infty} !$$

results in $\hat{\mu}$ as solution of this optimization problem.

Of course, the average of functions $g(X_1), \dots, g(X_N)$ of the data correspondingly converges to $\mathrm{E}\, g(X)$; in particular, we get the *sample variance*

$$\widehat{\sigma}^2 = \frac{1}{N} \sum_{j=1}^{N} (X_j - \hat{\mu})^2 \to \sigma^2 \quad \text{for } N \to \infty,$$

as a reasonable estimate of the variance var X.

The great importance of the law of large numbers for statistical learning is due to the fact that the assumptions can be considerably weakened such that they cover a lot of realistic situations. We still have convergence of sample means to the true mean if $X_1, \dots, X_N$ are dependent, though not too strongly. The data also are allowed to have different distributions, satisfying some restrictions, as long as they all have the same mean $\mathrm{E}\, X_j = \mu, j = 1, \dots, N$. However, these deviations from the basic assumptions have an influence on the kind of inference discussed in the next paragraph, in particular on the size of confidence intervals. Moreover, there are stronger versions like *uniform laws of large numbers* that provide uniform convergence of the sample means $\frac{1}{N} \sum_{j=1}^{N} g(X_j)$ to the means $\mathrm{E}\, g(X_1)$ over whole classes of functions $g(x)$, which is an essential key to showing that, e.g., neural networks work.

In engineering, we often are not satisfied with only getting the estimated value $\hat{\mu}$ of the quantity of interest μ, but we also want to have information about the precision of this estimate. Typically, this is provided in terms of a *confidence interval* $[\hat{\mu}_\ell, \hat{\mu}_u]$, which includes the true unknown μ with prescribed large probability, i.e.,

$$\mathrm{pr}(\hat{\mu}_\ell < \mu < \hat{\mu}_u) \geq \gamma$$

for some level of confidence $\gamma < 1$, but close to 1. The lower and upper bounds $\hat{\mu}_\ell, \hat{\mu}_u$ are functions of the data like the estimate $\hat{\mu}$ itself. For example, exploiting the fact that under the ideal conditions the sample mean is roughly normally distributed for large enough N, we have $[\hat{\mu} - \frac{2}{\sqrt{N}}\widehat{\sigma}, \hat{\mu} + \frac{2}{\sqrt{N}}\widehat{\sigma}]$ as an approximate 95 %-confidence interval ($\gamma = 0.95$).

In many applications, the data are not simply real valued but consist of vectors and, hence, are modeled as random vectors X_j with values in d-dimensional space. Then, the means and variances μ_k, σ_k^2 of the coordinates $X_{jk}, k = 1, \ldots, d$, again describe their average location and variability. For a complete description of the basic properties of a random vector, we, however, need also some simple measure for the amount of interaction between coordinates X_{jk} and $X_{j\ell}$. Here, we can rely on the notions of covariance or correlation.

If U, V are two real-valued random variables, their *covariance* is defined as

$$\mathrm{cov}\,(U, V) = \mathrm{E}\,(U - \mathrm{E}\,U)(V - \mathrm{E}\,V),$$

where obviously $\mathrm{cov}\,(U, U) = \mathrm{var}\,U$. The *correlation* is the scale-invariant version of the covariance

$$\mathrm{corr}\,(U, V) = \mathrm{cov}\,(U, V)/\sqrt{\mathrm{var}\,U \mathrm{var}\,V}$$

which only assumes values between -1 and $+1$. If U, V are independent, then their covariance vanishes, and $\mathrm{corr}\,(U, V) = 0$, i.e., they are *uncorrelated.* On the other hand, there are examples where $V = f(U)$ with some suitable nonlinear function f, i.e., V is completely determined by U, but the correlation vanishes nevertheless.

If, on the other hand, $V = aU + b$, i.e., we have a perfect linear dependence, then $\mathrm{corr}\,(U, V) = \pm 1$ with the sign depending on the sign of the coefficient a. In general, if we try to approximate $V - \mathrm{E}\,V$ as good as possible linearly by $a(U - \mathrm{E}\,U)$, then the coefficient a is proportional to $\mathrm{corr}\,(U, V)$ with a factor depending only on $\mathrm{var}\,U$ and $\mathrm{var}\,V$.

Summarizing, the covariance or, better, the correlation measures how strongly V depends **linearly** on U. However, it does not take into account nonlinear dependencies, with one notable exception: if the two random variables involved are jointly Gaussian, then uncorrelatedness implies independence in every respect.

Now, we assume that we have repeated independent observations $(U_j, V_j), j = 1, \ldots, N$, of the two random variables. As for the variance, the law of large numbers immediately provides simple estimates, the *sample covariance* and the *sample correlation* given by

$$\hat{c}_{UV} = \frac{1}{N}\sum_{j=1}^{N}(U_j - \hat{\mu}_U)(V_j - \hat{\mu}_V) \to \operatorname{cov}(U, V),$$

$$\hat{r}_{UV} = \frac{\hat{c}_{UV}}{\sqrt{\widehat{\sigma}_U^2 \widehat{\sigma}_V^2}} \to \operatorname{corr}(U, V),$$

for $N \to \infty$, where $\hat{\mu}_U, \hat{\mu}_V$ and $\widehat{\sigma}_U^2, \widehat{\sigma}_V^2$ denote the sample mean, respectively, the sample variance of the U_j and of the V_j, $j = 1, \ldots, N$.

There are several possibilities to also quantify general (non-linear) dependencies, e.g., mutual information (compare the chapter *Machine Learning for Predictive Maintenance in Production Environments* in this book), but they require information about the whole joint distribution of the random variables U, V.

The role of the normal distribution for real-valued random variables is assumed by the *multivariate normal distribution* for d-dimensional random vectors. It is characterized by the probability density

$$p(x) = \frac{1}{(2\pi)^{d/2} \det \Sigma} \exp\left(-\frac{1}{2}(x - \mu)\Sigma^{-1}(x - \mu)^T\right), \tag{1}$$

where we have assumed that the random X and its possible values x are written as row vectors. The parameters are the *mean vector* $\mu = (\mu_1, \ldots, \mu_d)$, where $\mu_k = \mathrm{E}\, X_k$, and the $d \times d$ *covariance matrix* Σ with entries $\Sigma_{k\ell} = \operatorname{cov}(X_k, X_\ell)$, $k, \ell = 1, \ldots, d$. In particular, all coordinates X_k of a Gaussian random vector X are normally distributed real random variables with mean μ_k and variance Σ_{kk}).

2 Statistical Learning: Basic Concepts and Pitfalls

In this section, we introduce the terminology and the general approach of statistical learning. We show that the three problems discussed in the introduction can be subsumed under the general task of estimating an unknown function from data. Hence, the basic learning methods used for regression, classification, and forecasting are quite similar. We start with regression and, then, discuss some major problems of statistical learning, which practitioners have to be aware of. Finally, we illustrate how classification and forecasting are closely related to regression such that the same kind of problems have to be solved with rather similar approaches.

2.1 Regression

As in the first example of Sect. 1.1, we consider a quantity of interest y, which we assume to be a real number. We know that the value of y is mainly determined up to measurement errors and other small disturbances by d explanatory variables $x_1, \ldots, x_d$, which we combine to a vector $x = (x_1, \ldots, x_d)$. We assume that all x_ℓ have numerical values, i.e., qualitative variables have already been transformed appropriately (compare Sect. 12.1). We are looking for a regression function m that describes the dependence of y on x well such that $y \approx m(x)$.

To learn the regression function m, we use a sample of data

$$(X_j, Y_j),\ j = 1, \ldots, N,$$

from an experiment that is the so-called **training set**. The values X_j of the vector of explanatory variables are called **input**, **predictor**, or **feature vector**, and the corresponding Y_j are the **outputs** or **responses**.

We assume that the data can be modeled by the regression relationship

$$Y_j = m(X_j) + \varepsilon_j, \quad j = 1, \ldots, N, \tag{2}$$

where the **errors** or **residuals** ε_j are independent and identically distributed with mean 0, and ε_j does not depend on the respective value X_j of the feature vector. The **regression function** m is unknown, and it may be quite general.

Our goal is to use the model and the data to predict the output of a future experiment given the feature vector X_{N+1}. As under our modeling assumptions, we do not have any information about the value of ε_j, except that it is 0 on average, and it is plausible to predict Y_{N+1} by $m(X_{N+1})$. More precisely, if we consider functions $f(X_{N+1})$ as possible predictors of the corresponding output Y_{N+1}, we want the **prediction error** $Y_{N+1} - f(X_{N+1})$ to be small. As this quantity is random, we cannot expect that a fixed prediction function $f(x)$ performs well under any circumstances, but we want it on the average to be good. This requirement is made concrete by looking at a popular and mathematically convenient measure of performance, the **mean-squared prediction error**

$$\text{mspe}(f) = \text{E}\left(Y_{N+1} - f(X_{N+1})\right)^2.$$

In our model (2), the best prediction function, minimizing mspe(f), is given by the regression function m. However, the latter is unknown such that the prediction of Y_{N+1} by $m(X_{N+1})$ is not feasible.

To get a practically useful prediction, we have to estimate or **learn** the function $m(x)$ from the training set. If $\widehat{m}(x)$ denotes such an estimate, then our final approximation of the output Y_{N+1} of the new experiment is

$$\widehat{Y}_{N+1} = \widehat{m}(X_{N+1}).$$

The toolbox of statistical learning contains many different methods like classical series expansions, regression trees, neural networks, local smoothers, etc. which we will discuss later on in detail. Here, we only consider quite a simple example that already allows to highlight some important features and problems shared by all kinds of learning methods.

In contrast to classical parametric statistics, we do not specify a simple form of the regression function $m(x)$, which is known up to a few parameters to be estimated from data. The function $m(x)$ may be arbitrary up to some very general assumptions concerning, e.g., its smoothness. Nevertheless, many regression approaches in statistical learning are based on parametric function classes, but with a twist.

For this first exposition, we consider input dimension $d = 1$, i.e., the inputs $X_j, j = 1, \ldots, N$, are just real numbers. We look at an experiment where the X_j are prescribed and chosen to be equidistantly spread over some interval, say [0,4]. We estimate $m(x)$ by polynomial regression, i.e., by functions of the form

$$m_p(x; b(p)) = \sum_{k=0}^{p} b_k(p)x^k, \quad p \geq 1. \tag{3}$$

Here, the degree p of the polynomial is **not fixed** as in a classical parametric regression model, but it ranges principally over all $p \geq 1$. In practice, we consider only values of p up to some upper-bound $p_{\max}$ depending on the size N of the training set. Moreover, we do not assume that any of the polynomial functions m_p exactly reproduces the unknown regression function m, but it only provides a good enough approximation for large enough p.

For every p, the function $m_p(x; b(p))$ is known up to the $p + 1$ coefficients $b_k(p)$, which we combine to the parameter vector $b(p) = (b_0(p), \ldots, b_p(p))$. To get a good fit to the data, we may, e.g., estimate the parameters by the familiar least-squares estimates $\hat{b}_k(p)$, which minimize the residual sum-of-squares (RSS)

$$\mathrm{RSS}(b(p)) = \sum_{j=1}^{N} \big(Y_j - \sum_{k=0}^{p} b_k(p)X_j^k\big)^2 = \min! \tag{4}$$

with respect to the parameters $b_0(p), \ldots, b_p(p)$. By choosing p appropriately, we expect to get a good approximation

$$m(x) \approx \widehat{m}(x) = m_p(x; \hat{b}(p)). \tag{5}$$

This reasoning is based on a **universal approximation** property of polynomials. Referring, e.g., to the well-known Taylor expansion of infinitely often differentiable functions $m(x)$, we know that such functions can be arbitrarily well approximated by polynomials of large enough degree. The quality of the approximation (5) depends of course on the choice of p. All statistical learning approaches to function

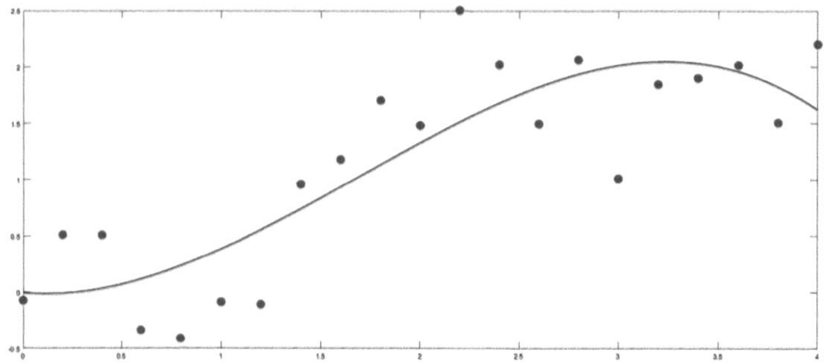

Fig. 2 Data and cubic polynomial fit

estimation depend on one or more of such **hyperparameters**, which have to be chosen in tuning the method to the data.

There sometimes are rules of thumb how to select, e.g., p depending on the sample size N under certain circumstances. If there are no such reliable rules based on past experience with similar data, then the hyperparameters may be selected adaptively from the data that come at a certain computational cost. We discuss such methods later on.

Figure 2 shows the data and the least-squares estimate $m_p(x; \hat{b}(p))$ using $p = 3$. In view of the relatively large random errors, the fit of the function estimate to the data looks quite good.

Remark For exposition, we have considered a pretty simple situation where the output is just real number and where the residuals are particularly simple. These assumptions may be considerably relaxed in many respects without changing the principle approach to estimating the regressions function.

For example, in many applications, the variance of the residuals ε_j is not stable as the precision of measurements varies over the range of output values, and it may even depend on the input, i.e., on the whole setup of the specific experiment. An ordinary least-squares approach often still leads to a reasonable estimate of $m(x)$, though sometimes a slightly more complicated weighted least-squares method leads to considerable improvements. Nonhomogeneity of the residual variances, however, has to be taken into account in earnest if we are interested in further inference like confidence bounds for the estimates.

The usage of least squares is tied to residuals that are approximately Gaussian. In engineering, more heavy-tailed distributions are also occurring in some applications, e.g., in reliability studies where Weibull distributions are popular for the lifetime of equipment. In other situations, the training set may contain outliers, i.e., observations where the measurement errors are much larger than usual due to unknown influences on the particular experiments. For such data, one would replace common residual sum of squares like (4) by other more robust training functions.

Finally, if the output Y_j of interest has more than one component, then $m(x)$ also assumes values in the corresponding, say, I-dimensional space, and we could just replace the square in, e.g., (4) by the squared norm. Alternatively, it often suffices to solve our regression problem component-wise, i.e., to learn real-valued regression functions $m_i(x), i =, 1, \dots, I$, separately from each component-wise training set $Y_{j,i}, X_j, j = 1, \dots, N$, and to use them as components of the I-variate function $m(x)$.

2.2 *Generalization, Stability, and Extrapolation*

Estimates like $m_p(x; \hat{b}(p))$ of the regression function depend on two things. First, the data analyst selects a learning method, in our example the class of polynomials as approximating functions together with least squares as the parameter estimation method and a choice of the hyperparameter p, resp., a method for data-adaptive selection of the latter. Second, the data determine the final result of estimation, in the example by means of the parameter estimates $\hat{b}_k(p)$. The latter aspect guarantees that the function estimate $\widehat{m}(x)$ will provide a good approximation to $m(x)$ for feature values $x = X_1, \dots, X_N$, which are part of the training sample. However, this is not really the goal of regression. We want to use $\widehat{m}(x)$ as an approximation of $m(x)$ also **out-of-sample**, i.e., at new feature values x, which were not part of the experiment generating the training data. If a learning method is up to this task, then we say that it **generalizes well**.

Usually, if the unknown function $m(x)$ is not too irregular and the learning method is decent and applied correctly, generalization will hold reasonably well for values of x in an area where we have enough data. In the example, we have made experiments for feature values X_j covering the interval [0,4], and we expect a good generalization for other x in this range.

The capability of generalization is related to the concept of **stability** of a statistical learning approach (compare [85]). A stable method should lead to similar results, if the training set is perturbed without significantly changing the structure of the data. For example, if we replace the training set with different independent, identically distributed data, then, e.g., the regression function estimate $\widehat{m}(x)$ should not change too much. The same requirement holds if we calculate $\widehat{m}(x)$ from similar pseudodata, which are generated from the original training set by resampling methods like crossvalidation (Sect. 3.4) or bootstrap (Sect. 9.1) or independently from a distribution that is close to that of the original input-output pairs. In the latter sense, stability generalizes the concept of *robustness* of a statistical procedure (for a brief review of the latter, compare [40]).

As a stable method leads to function estimates $\widehat{m}(x)$ or forecasts $\widehat{Y}_{N+1} = \widehat{m}(X_{N+1})$, which do not change significantly if the training set is replaced by similar data, it is less prone to overfitting. Hence, we expect it to also generalize well. From the viewpoint of industrial development and production, there is also another reason why stability of a statistical learning approach is a desirable property. If we make small changes to a machine or a production line, e.g., adding or replacing a few components, then a stable data-based method developed and optimized for the original configuration frequently will usually be still good enough such that we do not immediately have to start the whole training process from scratch.

Another related big problem of statistical learning is **extrapolation**. In our example, we use polynomials for approximating the regression function, but we are aware that $m(x)$ may be a function of quite a different type, e.g., an exponential function. The polynomial model therefore is misspecified. In the interval [0,4], where we have data, it nevertheless may provide a good approximation to $m(x)$, but it is always quite dangerous to use $\widehat{m}(x)$ for feature values x that are not in or at least close to the area covered by the feature values $X_1, \ldots, X_N$ of the training set. There may be situations where some additional knowledge, e.g., from well-studied similar experiments or from some law of physics, provides some confidence in such extrapolations based on statistical learning, but usually they have to be treated with extreme caution.

In our example of Fig. 1, the fitted polynomial $m_3(x; \hat{b}(3))$ approximates the true underlying function $m(x)$ reasonably well for $0 \leq x \leq 4$. On the right-hand side, we know from the limited shape options for cubic polynomials that $m_3(x; \hat{b}(3))$ decreases rather rapidly for $x > 4$. On the other hand, the true $m(x)$ is an increasing function for $x > 4$ with a linear rate of growth. Hence, the discrepancy between this function and its polynomial approximation becomes large fast if we increase x beyond 4. In such a case, extrapolation will result in gross errors.

If one is really interested in using a function estimate $\widehat{m}(x)$ for x far away from the original training set, then the right way to do it is to extend the sample by taking a few additional measurements with feature values around x and, then, update $\widehat{m}(x)$ based on the extended training set.

2.3 *Overfit, Underfit, and the Bias-Variance Dilemma*

The desired good generalization of a statistical learning procedure is directly related with the issues of overfit and underfit, which are closely connected with the bias-variance dilemma. To avoid wrong conclusions from a data analysis, it is important to understand these inherent features of statistical learning.

We again consider the general regression model

$$Y_j = m(X_j) + \varepsilon_j, \ \ j = 1, \ldots, N,$$

where the residuals $\varepsilon_1, \ldots, \varepsilon_N$ are independent and identically distributed with mean 0, finite variance σ_ε^2, and independent of $X_1, \ldots, X_N$.

We want to estimate $m(x)$ from the training data to get good predictors for new data, i.e., for a future experiment with given input $X_{N+1} = x$, we want to predict the output Y_{N+1} well or estimate the value $m(x)$ of the underlying regression function well. Both goals are essentially identical, as the mean-squared error (mse) of estimating $m(x)$ by $\widehat{m}(x)$ and the mean-squared prediction error (mspe) of approximating Y_{N+1} by $\widehat{Y}_{N+1} = \widehat{m}(X_{N+1}) = \widehat{m}(x)$ differ only by the constant

σ_ε^2. The latter is the variance of the residuals, about which we know nothing except that they are 0 on average. This represents an unavoidable source of error. Finding a good estimator $\widehat{m}(x)$ and a good predictor $\widehat{Y}_{N+1}$ therefore is the same data-analytic problem.

Unfortunately, in solving that problem, we have to cope with two conflicting goals as

$$\text{mspe}\,\widehat{Y}_{N+1} = \text{E}\left(Y_{N+1} - \widehat{Y}_{N+1}\right)^2 = \text{E}\left(m(x) + \varepsilon_{N+1} - \widehat{m}(x)\right)^2 \tag{6}$$

$$= \text{E}\left(m(x) - \widehat{m}(x)\right)^2 + \text{E}\ \varepsilon_{N+1}^2 = \text{mse}\,\widehat{m}(x) + \sigma_\varepsilon^2$$

$$= \text{var}\,\widehat{m}(x) + \text{bias}^2\widehat{m}(x) + \sigma_\varepsilon^2, \tag{7}$$

using $\text{E}\ \varepsilon_{N+1} = 0$ and the independence of ε_{N+1} from training set and, hence, from $\widehat{m}(x)$. The error measures mspe and mse, which we want to become small by choosing an appropriate estimate, depend on two components, the variance of the estimate and the squared bias where

$$\text{var}\,\widehat{m}(x) = \text{E}\left(\widehat{m}(x) - \text{E}\,\widehat{m}(x)\right)^2, \qquad \text{bias}\,\widehat{m}(x) = \text{E}\,\widehat{m}(x) - m(x).$$

The variance represents the size of the random component of the estimation error $\widehat{m}(x) - m(x)$, which is due to the stochastic nature of the training set. The bias measures the systematic error, which is caused by the misspecification of the model used for our statistical learning procedure.

In the polynomial regression example, the bias depends on the quality of approximation of $m(x)$ by polynomials of degree p. If the $m(x)$ happens to be a polynomial of unknown degree p_0 itself, then the bias of our estimate $\widehat{m}(x) = m_p(x, \hat{b}(p))$ will vanish for all $p \geq p_0$. In general, even if the polynomial model is misspecified for all degrees p, the bias at least becomes smaller with increasing p. A larger p provides more freedom for the function estimate $\widehat{m}(x)$ to get close to the true function $m(x)$ on average.

On the other hand, if p increases, we have to estimate more and more parameters $b_k(p), k = 0, \ldots, p$, which lets the variability of the data-dependent function estimate $m_p(x, \hat{b}(p))$ grow, i.e., the variance of $\widehat{m}(x)$ increases with p. As a side effect, $\widehat{m}(x)$ becomes rougher, showing more local fluctuations.

The bias-variance dilemma describes these conflicting goals. On the one hand, we would like to have p large to reduce the systematic error, whereas on the other hand, p should be small to keep the size of the random error in check. A good estimate is one that achieves a balance between both error components and an overall small estimation error.

All methods of statistical learning depend on some hyper- or tuning parameters like p controlling the complexity of the function estimate. Their choice is crucial for achieving a balance between squared bias and variance and hence a good estimate. They are also important for finding a compromise between overfit and underfit.

Overfit is the tendency of an estimate, e.g., $\widehat{m}(x)$, to adapt too well to the random particularities of the training set. By minimizing a residual sum of squares like (4), the estimated function tries to reduce the distance $|Y_j - \widehat{m}(X_j)|$ between output and prediction. If $\widehat{m}(x)$ is allowed to be quite complex, it uses the ample degrees of freedom not only to get close to the target function, i.e., to the $m(X_j)$, $j = 1, \ldots, N$, but partly tries to mimic the random residuals ε_j too. This leads to quite a rough function estimates, and in the extreme, if $\widehat{m}(X_j)$ uses more parameters than the sample size N, this leads to interpolation of the outputs: $\widehat{m}(X_j) = Y_j$. Of course, there is a price to be paid for this perfect or almost-perfect fit to the training data: such estimates do not generalize well. If an overfitted prediction function is applied to new data, the results usually will be quite bad. In terms of the error decomposition, overfitted estimates will have a small bias and a large variance.

Underfit, on the other hand, occurs if the estimate $\widehat{m}(x)$ is too smooth, resp., underparametrized. In this case, it is not flexible enough to adapt to the form of the true prediction function $m(x)$. A rather extreme case would be a polynomial estimate with $p = 1$, where the fitted straight line has no chance to adapt to curvatures of the true $m(x)$. The systematic error, reflected in the bias, is dominating here, and the variance is relatively small. Such estimates also do not generalize well usually, as the true underlying $m(x)$, which is the optimal prediction function, is not approximated well.

Example We consider in detail the polynomial regression estimate (5) for a sample of size $N = 21$. The larger the hyperparameter p, the rougher (more complex) the functions may be, and the larger becomes the contribution of variance in relationship to squared bias to the mean squared error. Figure 3 show the same data as in Fig. 2 together with estimates $\widehat{m}_p(x)$ for increasing degree $p = 2, 4, 6$ and 14 where the last one ($p = 14$) is already almost an interpolation:

Figure 4 shows the function estimates $\widehat{m}_p(x)$ for various values of p together with the true function $m(x)$ in red. For $p \leq 3$, the shape of the true function with a peak at $x = 2$ is not recovered which is a typical case of underfit. For $p = 4, 5$, the estimates look visually reasonable, though they show a wrong local minimum around $x = 0.5$; the latter is a data artifact that cannot be avoided for that particular

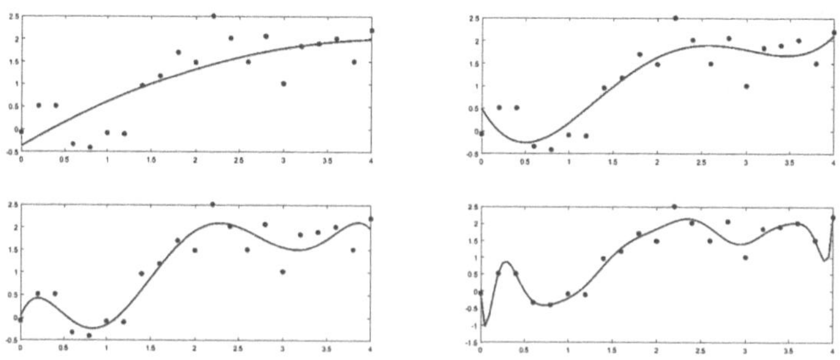

Fig. 3 Data and polynomial fits for increasing order $p = 2, 4, 6, 14$

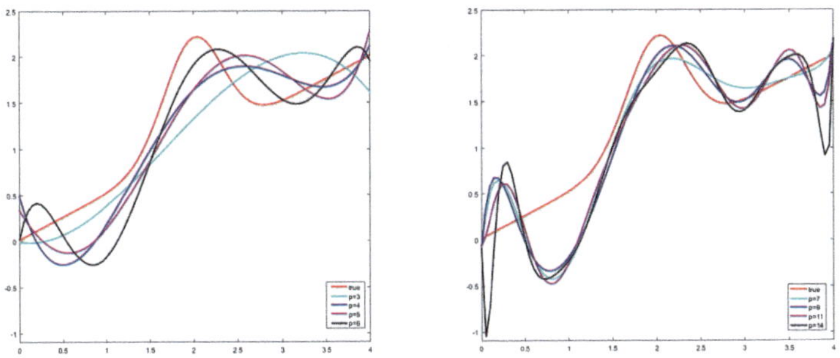

Fig. 4 True regression function m and polynomial fits of increasing order p

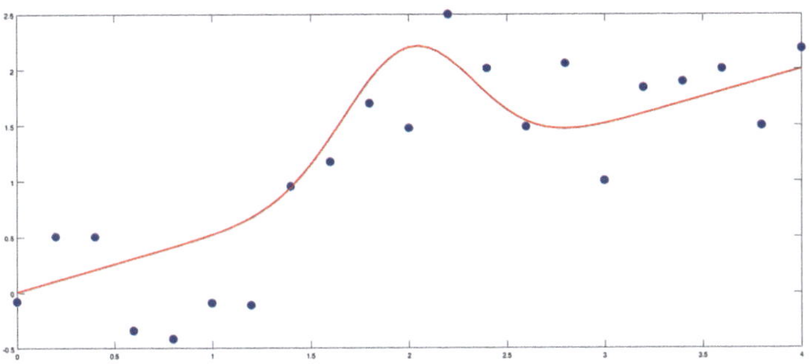

Fig. 5 Data and true regression function m

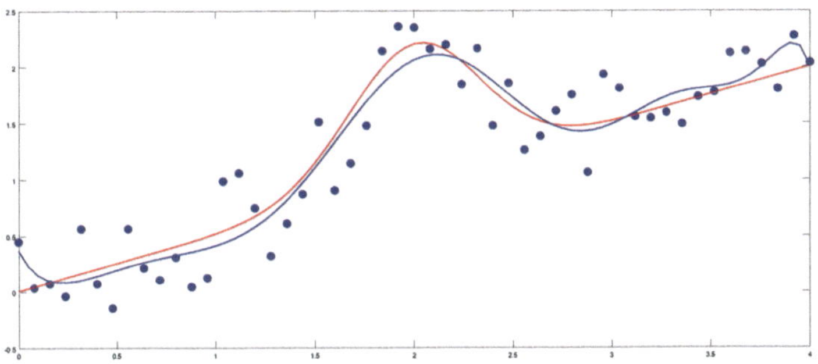

Fig. 6 Sample of size $N = 51$, true regression function m and polynomial fit of degree $p = 11$

sample. For $p > 5$, more data artifacts show up due to overfitting, and the estimated functions become rougher.

For comparison, Fig. 5 shows the true function together with the data. Here, the reason for the wrong minimum around 0.5 is apparent: four residuals, roughly for $0.6 \leq x \leq 1.2$, by pure chance all are negative and pretty large. This may happen in spite of $\mathrm{E}\, \varepsilon_j = 0$.

To highlight the effects of overfit and underfit, the sample size $N = 21$ in the example has been chosen pretty small. Figure 6 shows the same data-generating mechanism, but a larger sample size. For the larger N, larger p becomes reasonable, here $p = 11$, and the overall result looks much better. Of course, we always expect that we can estimate better with more data that allow for more complex learning models.

As a second example, we look at another dataset with again equidistant fixed feature values X_j. We now consider a quite different approach to statistical learning, which, instead of using approximation by a function class, relies on **local smoothing**. Again, the goal is to remove the effect of the random residuals and extract the regression relationship out of the training set. Here, the function value $m(x)$ is estimated by the average of those outputs Y_j for which the corresponding feature values X_j lie in a small interval $[x-h, x+h]$ around x:

$$\widehat{m}(x;h) = \frac{1}{2hN} \sum_{j;|X_j - x| \le h} Y_j = \frac{1}{2hN} \sum_{j=1}^{N} Y_j 1_{[-h,+h]}(X_j - x), \tag{8}$$

where $1_A(u)$ denotes the indicator function of the set A, i.e., it assumes the values 1 for x lying in A and 0 else. The intuition behind this estimate is derived from the decomposition based on (2)

$$\widehat{m}(x;h) = \frac{1}{2hN} \sum_{j;|X_j - x| \le h} m(X_j) + \frac{1}{2hN} \sum_{j;|X_j - x| \le h} \varepsilon_j .$$

As the mean of the residuals is 0, the second term as an average of ε_j will be close to 0 by the law of large numbers, provided that the average is taken over enough terms, which requires h to be not too small. On the other hand, we usually expect the regression function to be reasonably smooth, i.e., it does not change much over small intervals. Therefore, if h is not too large, we have $m(X_j) \approx m(x)$ if $|X_j - x| \le h$, and, therefore, the summands of the first term will all be approximately identical. As the scaling of the sum by $2hN$ is chosen appropriately, we have

$$\widehat{m}(x;h) \approx m(x), \qquad \text{if } h \text{ not too small and not too large.}$$

Again, this statistical learning algorithm produces an estimate of the regression function m, which depends on a hyperparameter, here $h > 0$. To get an idea of its influence, we look at Figs. 7 and 8, which show the $N = 51$ data, the true but unknown regression function $m(x)$, and the local averaging estimate $\widehat{m}(x;h)$ for $h = 0.02, 0.05$, resp., $h = 0.10, 0.25$.

Figure 7 clearly shows on the left-hand side an overfit where $\widehat{m}(x,h)$ is almost as rough as the data themselves. The smoothness of $\widehat{m}(x;h)$ increases with h until finally we get an underfitted situation on the right-hand side of Fig. 8, where $\widehat{m}(x;h)$ is no longer able to capture the maximum and minimum of the true function $m(x)$ at $x = 0.25$ resp. $x = 0.75$ well. For $h = 0.1$, the estimate looks quite reasonable, and a good balance between bias and variance is achieved. It is a bit wiggly, but an easy

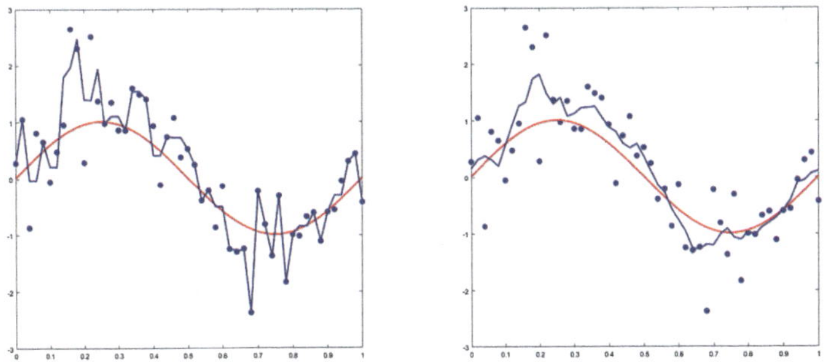

Fig. 7 Data, true regression function m (red), and local averages $\widehat{m}$ (blue) for $h = 0.02, 0.05$

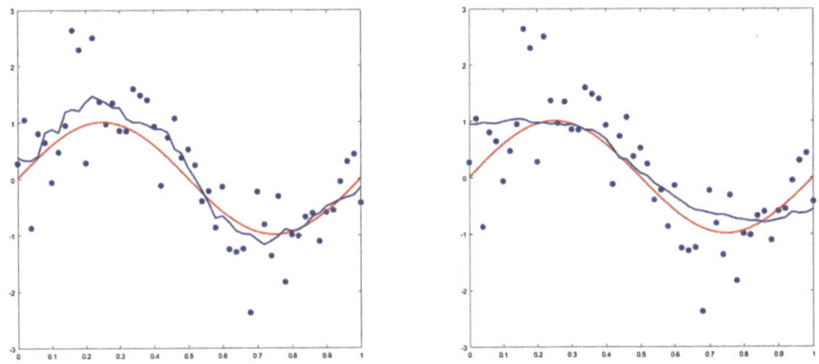

Fig. 8 Data, true regression function m (red), and local averages $\widehat{m}$ (blue) for $h = 0.10, 0.25$

remedy would be to consider weighted averages instead of simple averages, which we will discuss later on in Sect. 10 in more detail.

2.4 Classification

Applying statistical learning to a classification problem, we again start from a training set of experimental data

$$(X_j, Y_j),\ j = 1, \dots, N.$$

As in regression, the feature vectors X_j usually are d-dimensional. The known feature values $X_{jk}, k = 1, \dots, d$, of the jth experimental object, are used for inferring to which of a finite number M of classes this object belongs. For the training set, we also know the correct classification given by the output Y_j. The values of Y_j may just be the labels of the classes, but for the sake of convenience, we transform them to numerical values, e.g., the class numbers $1, \dots, M$.

A classification rule is characterized by a mapping $c(x)$ from the input space $\mathbb{R}^d$ to the class numbers. If a new object has feature values X_{N+1}, then we classify it as belonging to class number $\widehat{Y}_{N+1} = c(X_{N+1})$. Statistical learning looks, based on the training data, for a reasonable classification rule in the sense that $\widehat{Y}_{N+1} = Y_{N+1}$ with high probability for a large range of input values where Y_{N+1} is the true unknown class of the new object. The mean-squared prediction error as a performance measure of regression is here replaced by the **misclassification probability**

$$\mathrm{pr}(Y_{N+1} \neq \widehat{Y}_{N+1}) = \mathrm{pr}(Y_{N+1} \neq c(X_{N+1})).$$

As in regression, we know a theoretically best classification rule achieving the minimum of this performance measure, the so-called **Bayes classifier** $c_B(x)$. Let

$$p_\ell(x) = \mathrm{pr}(Y_{N+1} = \ell | X_{N+1} = x), \quad \ell = 1, \dots, M,$$

be the (conditional) probability that object no. $N + 1$ belongs to class ℓ if we know that the corresponding feature vector is given as x. Then, the Bayes classifier $c_B(x)$ is that class for which $p_\ell(x)$ is maximal w.r.t. ℓ, i.e.,

$$c_B(x) = \ell_{\max} \qquad \text{where} \quad p_{\ell_{\max}}(x) = \max_{1 \le \ell \le M} p_\ell(x). \tag{9}$$

This classification rule is quite intuitive as it assigns the object to that class, which is most likely given the value of the feature vector. However, it is not directly applicable in practice, as we do not know the conditional class probabilities $p_\ell(x)$. Statistical learning enables us to estimate those functions without any restrictive assumptions about their form. If $\hat{p}_\ell(x)$ denote such function estimates from the training set, then, mimicking the Bayes classifier, we immediately get also a reasonable classification rule

$$\hat{c}(x) = \ell_{\max} \qquad \text{where} \quad \hat{p}_{\ell_{\max}}(x) = \max_{1 \le \ell \le M} \hat{p}_\ell(x).$$

As a simple example, we consider a classification algorithm that is similar to the linear regression example of Sect. 2.3. We look at a binary classification problem with only $M = 2$ classes, which, for convenience, we label as 0 and 1 instead of 1 and 2. Let

$$p_\ell(x) = \mathrm{pr}\{Y_{N+1} = 1 | X_{N+1} = x\}, \quad \ell = 0, 1.$$

As probabilities sum to 1, $p_0(x) = 1 - p_1(x)$ is determined by $p_1(x)$, and we only have to estimate the latter function.

Assuming a purely linear form of $p_1(x)$ seems impractical as values outside of the range [0,1] of probabilities are possible. Therefore, we first apply a monotone transformation and, only then, assume linearity w.r.t. the features. For some given monotone **link function** $g : [0, 1] \to \mathbb{R}$, we consider

$$g(p_1(x)) = b_0 + \sum_{k=1}^{d} b_k x_k, \qquad \text{resp.} \qquad p_1(x) = g^{-1}\big(b_0 + \sum_{k=1}^{d} b_k x_k\big). \tag{10}$$

This is a special case of the large class of **generalized linear models** (GLIM), here for 0–1-response variables. If we choose, in particular,

$$g(p) = \frac{p}{1-p}, \qquad g^{-1}(u) = \frac{e^u}{1+e^u} = \frac{1}{1+e^{-u}},$$

then we get the **logistic regression** model. Essentially, this represents the most simple form of a neural network for classification with only one hidden layer and logistic activation function g^{-1} (compare Sect. 8).

As in regression, we **estimate** the unknown parameters b_k of the generalized linear model from the training set $X_j, Y_j, j = 1, \ldots, N$, by, here nonlinear, least squares. The estimates $\hat{b}_0, \ldots, \hat{b}_d$ are determined by minimizing the residual sum of squares

$$\text{RSS}(b) = \sum_{j=1}^{N} \big(Y_j - g^{-1}(b_0 + \sum_{k=1}^{d} b_k X_{jk})\big)^2 = \min!$$

w.r.t. $b_0, \ldots, b_d$. The resulting estimate of the probability function $p_1(x)$ is

$$\hat{p}_1(x) = g^{-1}\big(\hat{b}_0 + \sum_{k=1}^{d} \hat{b}_k x_k\big), \qquad \hat{p}_0(x) = 1 - \hat{p}_1(x),$$

and the corresponding classification rule derived from the training set is

$$\hat{c}(x) = 1, \text{ if } \hat{p}_1(x) > \hat{p}_0(x), \qquad \hat{c}(x) = 0, \text{ if } \hat{p}_1(x) < \hat{p}_0(x).$$

On the **decision boundary**, i.e., the set of all feature values x with $\hat{p}_1(x) = \hat{p}_0(x)$ or, equivalently $\hat{p}_1(x) = 0.5$, both classes are equally likely, and we cannot make a clear-cut decision. If all components of the feature vector have a continuous range of values, however, that will occur in practice only with probability 0.

The classification rule based on the generalized linear approach (10) sometimes works quite well, but in many applications, it is too simple to match the complex relationship between feature values and class assignment. In terms of Sect. 2.3, it suffers from a severe case of underfit. This consideration was a major reason for developing neural networks as an important tool in statistical learning. For the simple case of feedforward networks with only one hidden layer, H versions of the functions on the right-hand side of (10) with different choices of parameters b_k

are combined to get a more flexible and complex description of the function $p_1(x)$. The number H of building blocks or neurons acts as hyperparameter, which has to be chosen to get a balance between overfit and underfit (compare Sect. 8 for further details).

2.5 Forecasting

Many data in engineering are repeated measurements of the state of a plant or of a physical or chemical process over time usually but not necessarily made at equidistant points in time. The resulting sequence of data $Z_{t_1}, \ldots, Z_{t_T}$ is called a **time series** or signal. The single observations Z_{t_i} may be real or integer valued or data vectors.

Often, a major goal of data analysis is **forecasting** future values $Z_{t_T+n}, n > 0$, from the available past data up to time t_T. It would also be possible to additionally use data from other so-called exogenous time series $Z^*_{t_1}, \ldots, Z^*_{t_T}$ as inputs of the forecasting algorithm that would make no principal difference in the process of statistical learning.

To keep this exposition simple, we assume that the Z_t are real numbers with a continuous range of values; that they are observed at equidistant points $t = 1, \ldots, T$, in time; and that we only want to forecast the next observation Z_{T+1} by, say, $\widehat{Z}_{T+1}$ using no other information but past observations $Z_t, t \leq T$.

We also pretend that the data are part of a **stochastic process** in discrete time, i.e., of a sequence $\ldots, Z_{-1}, Z_0, Z_1, \ldots, Z_T, \ldots$ running from the infinite past into the infinite future. We make the crucial assumption that this process is **stationary** over time, which means that the data-generating process and, hence, the underlying physical process, does not change over time. More precisely, we require that for all $m \geq 1$, the distribution of the random vector $(Z_{1+t}, \ldots, Z_{m+t})$ of m consecutive observations from the signal is the same independently of t. This assumption is not always true in practice, but frequently, it can at least be achieved by appropriate preprocessing of the data before starting the statistical learning machinery.

Stationary time series can be well approximated by so-called (nonlinear) autoregressions for large enough order p, i.e., by models of the form

$$Z_t = m(Z_{t-1}, \ldots, Z_{t-p}) + \varepsilon_t, \tag{11}$$

where the residuals or **innovations** ε_t have mean 0 and are independent of observations $Z_s, s < t$, before time t. Z_t is decomposed into the first autoregressive term, which is completely determined by the past before time t, and the innovation ε_t describing new impulses to the process at time t independent of the past. For an autoregressive process (11), the optimal forecast of Z_t given the past data $Z_s, s < t$ obviously is

$$\widehat{Z}_t = m(Z_{t-1}, \ldots, Z_{t-p}),$$

as the past provides no information at all about ε_t, and we only know about it that is 0 on average. Of course, this is no feasible forecasting procedure as the autoregression function is not known.

The similarity to the regression model (2) is rather obvious, if we set $d = p$, $N = T - p$, and

$$Y_t = Z_t, \ X_t = (Z_{t-1}, \dots, Z_{t-p}), \ t = p+1, \dots, T.$$

Therefore, the same kind of statistical learning procedures that we use for regression are also used for forecasting. We estimate m from the training set $(X_t, Y_t), t = p+1, \dots, T$, i.e., from the time series data $Z_1, \dots, Z_T$, and then, we use the estimate $\widehat{m}$ for forecasting Z_{T+1} by

$$\widehat{Z}_{T+1} = \widehat{m}(Z_T, \dots, Z_{T-p+1}).$$

Though, on the first glance, forecasting does not look different from regression, as the same kind of algorithms for calculating an estimate $\widehat{m}$ from the data are used, we have to be aware of two major differences that have an impact on the validity of some procedures and on advanced inference, e.g., if we do not only want to calculate a forecast $\widehat{Z}_{T+1}$ but also want to have information about its precision. On the one hand, we have to deal with the autoregressive order p as tuning parameter additional to the hyperparameters of the function estimate $\widehat{m}$, as usually p is not fixed but has to be chosen appropriately on the basis of the data. On the other hand, the data pairs (Y_t, X_t) of the training sample are no longer independent, as, e.g., the feature vector X_t consists of $Y_{t-1}, \dots, Y_{t-p}$, and, similarly, the residual ε_t is independent of past $Z_s, s < t$, but not of future $Z_s, s > t$, as the latter depend on Z_t.

These differences arising from the dependency of time series data can lead to severe problems if regression algorithms developed for training sets of independent data are blindly applied to forecasting. Some common methods, e.g., straightforward crossvalidation (compare Sect. 3.4) for selecting hyperparameters in regression, may fail for time series, and precision measures of $\widehat{Y}_{T+1}$ as an approximation of Y_{T+1} like estimates of the mean-squared prediction error or confidence intervals for Y_{T+1} strongly depend on the degree of dependence (compare [43] for an example).

For illustration, we consider a class of linear models as in regression. If, for $p \geq 1$, a real-valued time series satisfies

$$Z_t = m_p(Z_{t-1}, \dots, Z_{t-p}; a(p)) = \sum_{k=1}^{p} a_k(p) Z_{t-k} + \varepsilon_t \tag{12}$$

with independent and identically distributed residuals or innovations ε_t having mean 0, then, the stochastic process is called a **linear autoregression** of **order** p or AR(p)-process. If we want to forecast Z_{T+1} from past data $Z_t, t \leq T$, then we fit such models to the data in the training set $Z_1, \dots, Z_T$ by least squares, i.e., we replace the unknown autoregressive coefficients $a_k(p)$ by estimates $\hat{a}_k(p)$, solving the minimization problem

$$\mathrm{RSS}(a(p)) = \sum_{t=p+1}^{T} \left(Z_t - \sum_{k=1}^{p} a_k(p) Z_{t-k}\right)^2 = \min!$$

Then we have to select the hyperparameter p appropriately to get a balance between the bias and the variance component of the forecasting error, say by $\hat{p}$, and we get the following one-step ahead predictor of Z_{T+1}:

$$\widehat{Z}_{T+1} = \sum_{k=1}^{\hat{p}} \hat{a}_k(\hat{p}) Z_{t-k}.$$

Model (12) requires the data Z_t to have mean 0, which usually is not the case in practice. In that case, we first estimate the mean μ, e.g., by the sample mean

$$\hat{\mu} = \frac{1}{T} \sum_{t=1}^{T} Z_t$$

and, then, apply the autoregressive algorithm to the centered observations $Z_t^c = Z_t - \hat{\mu}$ to get a forecast $\widehat{Z}_{T+1}^c$ of $Z_{T+1} - \mu$. Then, we have immediately $\widehat{Z}_{T+1}^c + \hat{\mu}$ as a prediction of Z_{T+1}.

As an example, we consider the artificial data set of Fig. 9 mimicking the behavior of monthly data like, e.g., energy consumption in production observed over a period of 12 years, i.e., $T = 144$. We estimate the mean by the sample mean, fit AR(p)-processes of increasing order to the centered data, and select the hyperparameter as $\hat{p} = 18$ by minimizing the Bayes information criterion BIC_p as a function of p (compare Sect. 3.5 for details). Figure 11 shows on the left-hand side the one-month-ahead forecasts $\widehat{Z}_s, s = 145, \ldots, s = 156$ (as blue squares), together with the true observations for the 12 months following the training period. The performance looks quite good.

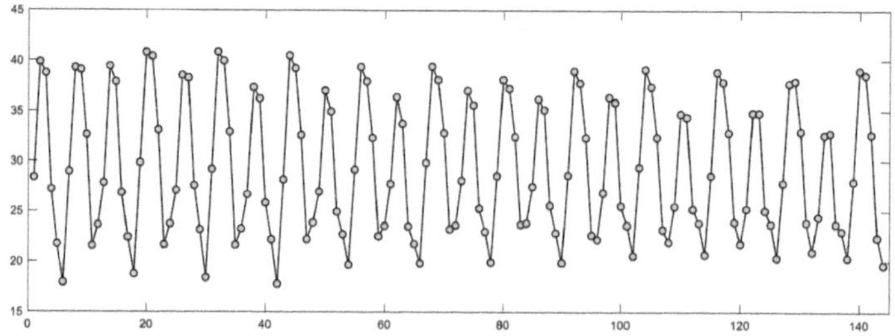

Fig. 9 Monthly time series data Z_t over a period of 12 years

Preprocessing the Data

In the last example, we applied a generic forecasting algorithm to the raw data and happened to get rather good results. We should worry a bit about potential overfit, as the number of parameters of the final forecasting function is rather large compared to the sample size: 18 autoregressive coefficients and additionally the variance of the innovations, for which we also get an estimate out of this kind of procedure, needed for further inference like quantifying the average forecasting error. Data predicted in Fig. 11 were not used in calculating the estimates, but they are far from independent of the training set due to the rather strong dependence of the Z_t over short and medium ranges of time. So, forecasting in the near future after the training period may profit from overfit, but it is not guaranteed to show a good performance in the more remote future resp. if applied to another independent time series of identical structure.

In general, however, it is not guaranteed that learning from the raw data will produce the desired performance in regression and classification and even more in forecasting, due to the more complicated data structure of time series. Often, the data have to preprocessed, i.e., transformed in an appropriate manner, before it is fed into a machine learning algorithm. This kind of data engineering has a long history in statistics, and its careful application is often crucial for success and requires a lot of experience. We discuss preprocessing of data later on in more detail (compare Sects. 12 and 14.4). Here, we only have a further look at our time series example.

Forecasting time series based on general autoregressive models (11) requires stationarity of the data. Looking at Fig. 9, however, we see a strong seasonal pattern with two peaks in every year. Such a behavior may be caused by a stationary process that is close to nonstationarity, but it is more plausible to assume that we have an annual pattern where the stochastic structure of observations changes over the course of the year. For such quasi-periodic data, a classical preprocessing method is **deseasonalization** or seasonal adjustment. As we have monthly data, it is plausible that the data-generating mechanism has a period of 12 years if at all. We preprocess the raw data by seasonal differencing, i.e., we transform them to

$$D_t = Z_t - Z_{t-12}, \ t = 13, \ldots, 144,$$

which also removes automatically the nonvanishing mean. Figure 10 shows the transformed data that no longer show any prominent deviation from stationarity.

Then, we apply autoregressive forecasting to the transformed data where now a much smaller hyperparameter $\hat{p} = 2$ is selected, and the forecast of D_{T+1} is given as

$$\widehat{D}_{T+1} = \hat{a}_1(2)D_T + \hat{a}_2(2)D_{T-1}.$$

We immediately get $\widehat{Z}_{T+1} = Z_{T-11} + \widehat{D}_{T+1}$ as forecast of $Z_{T+1} = Z_{T-11} + D_{T+1}$. Figure 11 shows on the right-hand side these forecasts for the validation period $t = 145, \ldots, 156$ (red circles) together with the forecasts from the AR(18)-model (blue squares) and the true data. The performance looks similar with a slight advantage

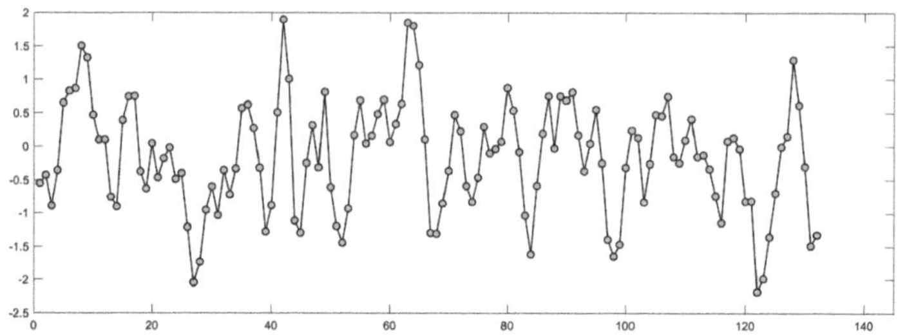

Fig. 10 Deseasonalized data D_t

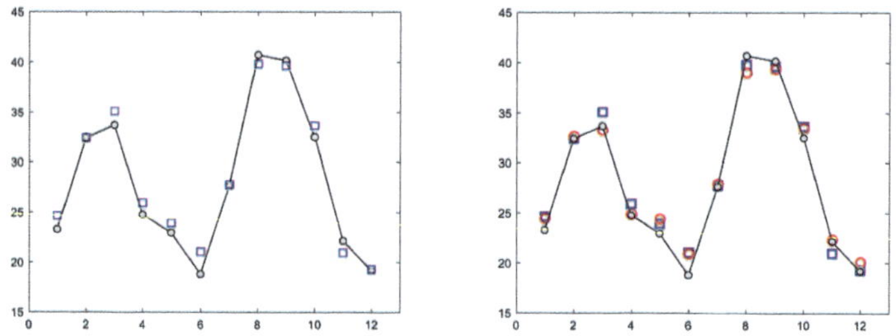

Fig. 11 Validation of AR(18)-forecasts. Validation of deseasonalization together with AR(2)-forecasts

for the simpler model using preprocessing first. This visual impression is correct, as the average squared prediction error over the validation period, i.e., the mean of $(Z_t - \widehat{Z}_t)^2, t = 145, \ldots, 156$, is 1.12 for the brute force AR(18)-forecast and only 1.03 for the forecast combining deseasonalization and an AR(2)-model. Moreover, the latter depends only on two coefficient estimates, i.e., the model is much simpler and definitely not prone to overfitting.

2.6 *Black, White, and Gray: Engineering Expertise and Statistical Learning*

Many successful applications of machine learning in regression, classification, or forecasting are not using specific information about the data to be analyzed. Often, this is due to a lack of corresponding knowledge. In image recognition, e.g., there is to our knowledge no detailed scientific theory how the human eye and brain distinguish between a duck and a chicken. Nevertheless, machine learning algorithms are able to solve this classification problem based on images. They can do it as they have access to extremely large training sets to learn an appropriate classification function. However, such classification rules achieve their results in a quite different manner than humans, as, e.g., they can be led astray by particular minor modifications of images that have no influence at all on human recognition.

Machine learning methods that are complex, but generic and not tailor-made for the particular application, find solutions, but the final tool is a **black box**. We observe the input and the output of the system, but we only have a vague idea, how both are related beyond a purely formal mathematical representation and why and when the results are satisfactory. In industrial applications, such black box approaches frequently suffer from a major problem, the lack of massive data for training, which are needed to train a generic system that is complex enough to provide good solutions. In an engineering context, available training samples are usually small or medium sized due to the costs of respective experiments.

A contrast to black box solutions are systems based on a **white-box** approach. This is the classical situation, where, e.g., in regression, the relationship between input and output is modeled by a relatively simple function, known up to a fixed number of parameters. This is function is derived from physical or chemical principles and engineering expertise. However, a white-box approach that works well in the framework of idealized laboratory experiments often does not scale well if applied to a related large-scale industrial process, which is more complex and where a lot more variables are involved, not all under the control of the experimenter. There may even be situations where in principle a very large white-box model for a complex system would be possible, but where there is no hope to get enough data to identify all of its parameters and where we are forced to work with simpler, but identifiable approximate models.

Gray-box methods, also called *physics-informed machine learning*, try to combine the advantages of both approaches. Using the knowledge about data-generating processes and the underlying scientific theory, provided by experienced engineers, together with the flexibility of machine learning, which is able to adapt to data features that we do not fully understand, such gray-box approaches often are able to train good algorithms even from smaller data samples.

There are various ways on how to combine engineering expertise and statistical learning in a gray-box approach. In, e.g., regression, we could start with a well-tried white-box model from engineering, i.e., with a fixed function $m^o(x; c)$, which is known up to a low-dimensional parameter vector $c = (c_1, \ldots, c_q)$. This model is fitted to the training data, e.g., by replacing c by a least-squares estimate $\hat{c}$. If the real-world situation is more complex than the physical model from which the relationship $m^o(x; c)$ has been derived, then prediction errors $Y_{T+1} - m^o(X_{T+1}, \hat{c})$ for out-of-sample data will be too large on average. To improve the regression model further, we could look at the residuals from the white-box approach and consider a corresponding general regression model (2) for them, i.e.,

$$R_j = Y_j - m^o(X_j, c), \qquad R_j = m(X_j) + \varepsilon_j, \qquad j = 1, \ldots, N.$$

We replace the unknown R_j by the approximations $\widehat{R}_j = Y_j - m^o(X_j, \hat{c})$ and use the training set $(X_j, \widehat{R}_j)$, $j = 1, \ldots, N$, for learning the unknown function $m(x)$ by some black-box method. We end up with

$$\widehat{Y}_{T+1} = m^o(X_{T+1}, \hat{c}) + \widehat{m}(X_{T+1})$$

as predictor of Y_{T+1} consisting of one part from the known underlying physical principles and another part taking care of the unexplainable features of a complex production system.

Beyond gray-box models, there are many other possibilities to incorporate the experience of engineers into a statistical learning system for achieving a satisfactory performance in an industrial context. Such information does not always come in the form of simple white-box models but may be of a more qualitative nature. Engineers know perhaps that the output is a non-negative function or that it depends in a monotone or convex manner on certain components of the feature vector. Such constraints can be incorporated into learning algorithms but require special tools from numerical optimization to keep the computational load in check. An example is the regression problem described in Sect. 1.1, which motivated the development of a new efficient algorithm for fitting monotone functions to data in [54] resp. for more general shape constraints in [57]. The forecasting problem from the previous section is also an example of the use of information about the data-generating process in finding a good and reliable procedure. There, we used the information that the data likely show a seasonal pattern that engineers working regularly with such systems will know.

The most important step toward satisfactory machine learning methods in an industrial context is, however, a close cooperation between engineers and data scientists in the selection of features, which has to be done at the very beginning of a project. Engineers know from their past experience which variables influence the output and which of them are easy to measure such that a large enough and informative training set can be generated. Only then can we expect that statistical learning will find a good regression, classification, or forecasting algorithm without requiring an excess of data. Most of the case studies in the second part of this book use such expert knowledge extensively to achieve a satisfactory performance of the machine learning procedures.

2.6.1 Synthetic Training Data from White-Box Models

Gray-box approaches to statistical learning constitute one possibility to combine white-box models from classical engineering with generic machine learning methods. Here, we briefly sketch another important use of engineering expertise in the form of white-box models.

As already pointed out, many applications of machine learning to problems in industry suffer from a severe lack of training data. If, e.g., we want to optimize the output of a production line with respect to several input parameters under our control, which we combine to the feature vector x, then we have to learn the function $m(x)$ describing the input-output relationship and find the location of its maximum. Often, performing enough experiments at various values $X_1, \ldots, X_N$ of x are expensive or right out impossible in the real world.

In such cases, sometimes, large physical or chemical models are known, which approximate the industrial process reasonably well. For example, they are given in the form of an extremely large system of differential algebraic equations. Theoretically, we could try to learn $m(x)$ by solving this system numerically, but, in practice, for already a single input value x, this could take days or even weeks of computation even on high-performance computers. Finding the maximum of $m(x)$ would, however, require a lot of those numerical calculations, which again is often impractical.

A possibility for overcoming this dilemma is the combination of the large white-box model with a generic statistical learning approach. In a first step, the model is used to calculate the outputs Y_j for a still computationally manageable number of input values X_j, $j = 1, \ldots, N$. As the white-box model is only an approximation of the real process, we can interpret X_j, Y_j as input and output of a general regression relationship (2), where the errors ε_j here measure numerical errors as well as deviations between model and reality. In the next step, we use, e.g., a neural network (compare Sects. 8 and 13) to learn the input-output relationship. This results in an estimate $\widehat{m}(x)$ of the function $m(x)$, which, in the example, would be the function described by a neural network. In the final step, $\widehat{m}(x)$ would be maximized w.r.t. x using suitable optimization algorithms. The latter is computationally feasible, as computing the network function $\widehat{m}(x)$ can be done much faster than numerical calculation of the model-based approximation of $m(x)$.

This approach of generating synthetic training data from white-box models widens the range of applications of statistical learning in industry considerably. Examples of practical applications can be found in several chapters of the second part of this book.

3 Extensions of Linear Regression

In this chapter, we consider some ideas from statistics, which have become major ingredients of now popular machine learning methods. Here, we focus on regression problems and start with the classical linear model. If, as in Sect. 2.1, we allow for several models of various complexity to be fitted to the training set, then several extensions of linear regression help in finding good sparse representations of the data, which rely only on a reasonable number of significant parameters.

3.1 Linear Models in Regression

Again, we consider a training set (X_j, Y_j), $j = 1, \ldots, N$, where the output Y_j is real valued and the p-dimensional input vector X_j has real-valued components. The Y_j are still the variables of interest, but the X_{jk}, $k = 1, \ldots, p$, are not necessarily the original measurements of explanatory variables from the jth experiment, say $Z_j = (Z_{j1}, \ldots, Z_{jq})$. More generally, we may use transformations

$$X_{jk} = g_k(Z_j),\ j = 1, \ldots, N,$$

with arbitrary, but known functions g_k, as inputs for the general linear regression model

$$Y_j = b_0 + \sum_{k=1}^{p} b_k X_{jk} + \varepsilon_j,\ j = 1, \ldots, N, \tag{13}$$

where $\varepsilon_1, \ldots, \varepsilon_N$ are independent identically distributed zero-mean residuals.

The model is linear in the parameters $b_0, \ldots, b_p$, but not necessarily in the explanatory variables. For example, we can write the polynomial regression model of Sect. 2.1

$$Y_j = m_p(Z_j; b) + \varepsilon_j = b_0 + \sum_{k=1}^{p} b_k Z_j^k + \varepsilon_j$$

with one-dimensional feature values Z_j in the form (13) by setting $X_{jk} = Z_j^k, k = 1, \ldots, p$.

Linear models have the big advantage that minimizing the residual sum of squares

$$\mathrm{RSS}(b) = \sum_{j=1}^{N} \big(Y_j - b_0 - \sum_{k=1}^{p} b_k X_{jk}\big)^2 = \min! \tag{14}$$

can be solved analytically instead of only numerically, resulting in an explicit least-squares estimate $\hat{b}$ of the parameter vector $b = (b_0, \ldots, b_p)$.

For calculating $\hat{b}$, let $\mathbf{X}$ denote the so-called $N \times (p+1)$ **design matrix**:

$$\mathbf{X} = (X_{jk})_{j=1,\ldots,N,k=0,\ldots,p} \qquad \text{with } X_{j0} = 1,\ j = 1, \ldots, N.$$

Its row no. j contains all the feature values of the jth input from the training set. Its columns contain all values of the kth input component throughout the training set.

Setting $Y = (Y_1, \ldots, Y_N)^T$, the residual sum of squares can be written as

$$\mathrm{RSS}(b) = ||Y - \mathbf{X}b||^2 = \min!.$$

The solution satisfies the so-called normal equations

$$\mathbf{X}^T Y = \mathbf{X}^T \mathbf{X} \hat{b}.$$

If $\mathbf{X}^T \mathbf{X}$ is nonsingular, we get the least-squares estimate of b as

$$\hat{b} = (\mathbf{X}^T \mathbf{X})^{-1} \mathbf{X}^T Y. \tag{15}$$

General linear models with suitable inputs derived from the original measurements and combined with least squares have been and are still successfully applied in many situations, in particular if the number p of input values is not too large.

During the recent decades, however, there are more and more practical applications, where we are confronted with a lot of input variables $X_{jk}, k = 1, \ldots, p$, which are potentially useful as predictors of the output Y_j and, on the other hand, the data are limited, i.e., N is not so large compared to p. Using all predictors may lead to a considerably overfitted estimate

$$\widehat{m}(x) = \hat{b}_0 + \sum_{k=1}^{p} \hat{b}_k x_k, \tag{16}$$

of the prediction function analogously to the discussion in Sect. 2.3. Additional problems are caused by the usual considerable dependencies between the many input variables. The $X_{jk}, k = 1, \ldots, p$ likely show high correlations among themselves. This is conceptually difficult as we are looking at several variables that more or less describe the same information about likely values of Y_j.

Consider, e.g., a situation where $X_{jk} \approx cX_{j\ell}, j = 1, \ldots, N$, for some $1 \leq k \neq \ell$. For some Δ, set $\tilde{b}_i = \hat{b}_i, i \neq k, \ell$, and $\tilde{b}_k = \hat{b}_k + \Delta, \tilde{b}_\ell = \hat{b}_\ell - \Delta/c$. The predictors using both sets of coefficients are almost equal

$$\sum_{i=0}^{p} \hat{b}_i X_{ji} \approx \sum_{i=0}^{p} \tilde{b}_i X_{ji}, \quad \text{such that } \operatorname{RSS}(\hat{b}) \approx \operatorname{RSS}(\tilde{b})$$

for arbitrary large $|\Delta|$. Both input variables $X_{jk}, X_{j\ell}$ almost contain the same information about Y_j, and only one of them is needed. The corresponding model parameters b_k, b_ℓ are almost not identifiable. This redundancy also causes numerical instability in calculating the least-squares estimates. As a consequence, $\widehat{m}(x)$ will show high variability. On the other hand, for large p, the bias is usually small as the amount of information exhibited in the large number of input variables allows $\widehat{m}(x)$ to capture the main structure of the optimal prediction function $m(x)$.

As an example typical for industrial production, where some output of the process depends on the relative amount of material components used, we consider a small classical data set of size $N = 13$. Here, the total heat Y_j evolved by Portland cement after hardening over a period of 180 days depends on the mineral components of the particular mixture. The inputs $X_{j1}, \ldots, X_{j4}$ are the four main cement klinker components, e.g., tricalcium-aluminate, in percent. The data are known under the name of *Hald's cement data* and have been discussed in [38]. Since then, they have repeatedly been used as benchmark example in regression research. The original source of the data is [82].

To investigate relationships between the feature components $X_{j1}, \ldots, X_{j4}$, we estimate their correlations as measures of (linear) dependency (compare Sect. 1.2). Table 1 shows the results in the form of a sample correlation matrix.

Table 1 Sample correlation matrix of the input components in Hald's cement data

	x_1	x_2	x_3	x_4
x_1	1.00	0.23	−0.82	−0.26
x_2	0.23	1.00	−0.14	−0.97
x_3	−0.82	−0.14	1.00	0.03
x_4	−0.26	−0.97	0.03	1.00

We see that x_1 and x_3, resp., x_2 and x_4, are strongly correlated with a negative sign. In the second case, the absolute value is close to the maximum 1, i.e., X_{j4} is almost identical to $-cX_{j2}$ for some $c > 0$, and we definitely do not need both feature components for forecasting the outputs Y_j linearly.

A seeming solution for the problems described would be the reduction of p, i.e., the elimination of some input variables from the model. However, usually, one has no information which input variables are less important. Engineering knowledge about the context of the data does not help necessarily as the experts already have communicated that all input variables may perhaps be relevant for the output of a complex system. Therefore, in classical regression analysis and much more in machine learning from limited data, we are looking for data-adaptive methods to enforce **sparsity** of the prediction function estimate, i.e., here for the linear model $\hat{b}_k = 0$ for most k, which means that the kth input variable is eliminated from the predictor.

Enforcing sparsity can be formulated as **selecting a subset** of input variables for which $\hat{b}_k \neq 0$ is allowed. Ideally, we would investigate **all subsets** of predictor variables:

$$X_{jk_1}, \dots, X_{jk_q}, \; 1 \le k_1 < \dots < k_q \le p, \; q = 0, \dots, p,$$

and fit linear regression models to Y_j and those reduced input vectors. The prediction performance has to be evaluated on an independent **test set** or **validation set** of data, say $\tilde{Y}_j, \tilde{X}_j, j = 1, \dots, M$, which allows to approximate the mean-squared prediction error by

$$\frac{1}{M} \sum_{j=1}^{M} \big(\tilde{Y}_j - \hat{b}_0 - \sum_{i=1}^{q} \hat{b}_{k_i} \tilde{X}_{jk_i}\big)^2.$$

Note that the $\hat{b}_{k_i}$ depend on the whole subset of indices. Then, one chooses the subset of input variables, which minimizes this empirical mean squared prediction error on the validation set.

Note that the empirical mean squared prediction error on the training set itself is not a good performance measure as, due to overfitting, larger models (subsets) will automatically be better than smaller ones. If, for not so large samples, all data have to be used for parameter estimation and we cannot afford to put some of them aside for validation, then a solution is provided by methods like crossvalidation discussed later on.

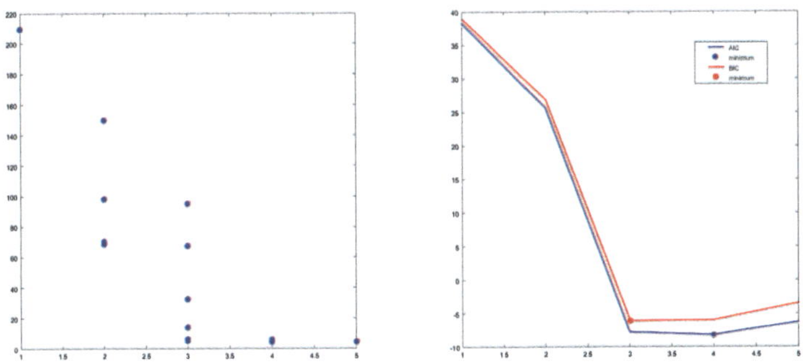

Fig. 12 Minimal RSS-values for all models fitted to Hald's cement data (left); AIC (blue) and BIC (red) curves for Hald's cement data (right)

Table 2 Minimal RSS-values for each p based on Hald's cement data

min(RSS)	p	Optimal input(s)
67.99	2	x_4
4.45	3	x_1, x_2
3.69	4	x_1, x_2, x_4 or, almost as good, x_1, x_2, x_3
3.68	5	$x_1, \ldots, x_4$ (full model)

As there are 2^p subsets of $\{1, \ldots, p\}$, the computational effort of **all subsets regression** increases rapidly with p, but the *leaps and bounds algorithm* of [32] provides a numerically feasible solution up to at least $p = 30$. For much larger p, however, all subset regression becomes numerically intractable even in the linear case.

After fitting all subset models to data, we still have to select a suitable parsimonious one, which finds a balance between overfit and underfit. The left-hand side of Fig. 12 shows for Hald's cement data the RSS-values of all fitted subset models ordered w.r.t the number of parameters, where $p = 5$ corresponds to the full model using all four feature components $X_{j1}, \ldots, X_{j4}$. The minimal achievable RSS-value for each given number of parameters p is of course decreasing with p. The more features we use for fitting the data, the smaller the in-sample error becomes due to the increasing tendency to overfit. From a certain point onward, it does not make sense to add more input components as the gain in terms of reduced RSS-values will become small. For Hald's cement data, the minimum RSS-values are shown in Table 2.

Obviously, increasing the number of parameters p from 2 to 3, i.e., the number of used feature components from 1 to 2, leads to a considerably better performance. If it is worthwhile to add one additional parameter and input component is not an easy decision. Using the full model, however, is not a reasonable choice at all, as the reduction in RSS is minimal compared to simpler models. In this example, that can be easily explained by the fact that $X_{j1}, \ldots, X_{j4}$ correspond to the main cement components and, as percentages, add up to almost 100 % in each experiment.

Therefore, we almost have a deterministic linear relationship, making one of the involved feature components superfluous for the data analysis.

The RSS-values are not really suitable for comparing models of differing complexity, as they systematically decrease with increasing number of parameters. The decision, if adding more feature components is reasonable, requires some experience. Therefore, more automatic model selection procedures have been developed. The right-hand side of Fig. 12 shows the AIC_p- and BIC_p-values of the best models, shown in Table 2, as a function of the number of parameters $p = 2, \ldots, 5$. We can use these curves by selecting that value of p where they assume their minimum. For Hald's cement data, this would be $p = 3$ for BIC and $p = 4$ for AIC. The ratio behind these model selection methods and the differences between the two variants are explained in Sect. 3.5.

To avoid the computational effort of considering and fitting a large number of submodels to the data and simultaneously take precautions against overfit, **thresholding** of the coefficients has been proposed in the past, in particular in the context of function estimates based on series expansions like polynomial regression (compare Sect. 6.1). Here, we start from the maximal model (16), using all input variables. Then, we only keep the inputs for which the coefficients are above a threshold $\tau > 0$, which is the hyperparameter of this learning method, i.e., the final function estimate is of the form

$$\widehat{m}(x; \tau) = \hat{b}_0 + \sum_{k; |\hat{b}_k| > \tau} \hat{b}_k x_k .$$

In contrast to estimates based on series expansions, where such an approach has been successfully applied, for general linear models, it shares one disadvantage with all subsets regression. Both are in a certain sense discontinuous as an input variable is either completely in or out of the fitted model, which leads to a higher variability in model selection. Sometimes, adding one additional observation to the training set may lead to a completely different selection of predictor variables.

As a solution to those problems, one can **relax** the **sparsity** requirement to $\hat{b}_k \approx 0$ for most k instead. There are various approaches leading to such predictors, one of them being **dimension reduction by shrinkage**, which uses a constraint on the total size of the parameter vector.

Before going into the details in the next section, we remark that an important aspect of data adaptive selection of predictor variables is standardization. It makes the methods scale invariant, such that all candidates as input variables are treated fairly independently of the ranges of numerical values, which the original data have. In the following, we assume that we first **rescale** the original input variables X^o_{jk} such that they all are of comparable size, e.g., by setting for $j = 1, \ldots, N$,

$$X_{jk} = \frac{X^o_{jk}}{\max_{1 \le i \le N} |X^o_{jk}|} \in [-1, +1], \; k = 1, \ldots, p. \tag{17}$$

3.2 Ridge Regression

The classical shrinkage method known as ridge regression uses a quadratic penalty on the size of the parameter vector. The estimate $\hat{b}^r$ is a **penalized least-squares estimate** solving the minimization problem

$$\mathrm{RSS}_\lambda(b) = \frac{1}{N}\sum_{j=1}^{N}\Big(Y_j - b_0 - \sum_{k=1}^{p} b_k X_{jk}\Big)^2 + \lambda \sum_{k=1}^{p} b_k^2 = \min! \tag{18}$$

The penalty term does not contain the component b_0, which covers the constant in the regression relationship and does not correspond to a particular predictor variable. The same kind of penalty is used, e.g., for the common **weight decay** approach in neural network regression [67, Section 5.3].

As a hyperparameter, the coefficient $\lambda \geq 0$ of the penalty controls the balance between the goal to fit the data well, i.e., to keep the first term, the residual sum-of-squares, small, and the goal of having a not too complex function and to avoid overfit, which is represented by a not too large penalty term. For $\lambda = 0$, we get the usual least-squares estimate.

An equivalent formulation (via the Lagrange multiplier technique) of the penalized least-squares approach is in terms of the constrained minimization problem

$$\frac{1}{N}\sum_{j=1}^{N}\Big(Y_j - b_0 - \sum_{k=1}^{p} b_k X_{jk}\Big)^2 = \min! \quad \text{under the constraint} \quad \sum_{k=1}^{p} b_k^2 \leq s.$$

If the hyperparameter s is small (resp. λ is large), most parameter estimates $|\hat{b}_k|$ solving this minimization problem will be small. Only a few X_{jk} contribute to the prediction of Y_j in an essential manner, i.e., a relaxed sparsity requirement is fulfilled. Due to using a quadratic penalty term, the penalized least-squares estimator can also be calculated explicitly.

As in calculating the least-squares estimate (compare Sect. 3.1), we write the penalized residual sum of squares as

$$\mathrm{RSS}_{\mathrm{pen}}(b) = \frac{1}{N}||Y - \mathbf{X}b||^2 + \lambda||b^{-0}||^2 = \min! \qquad \text{with } b^{-0} = (b_1, \ldots, b_p)^T.$$

For ease of notation, we first center inputs and outputs around 0. Let the original values be Y_j^o, X_{jk}^o and subtract the sample means

$$Y_j = Y_j^o - \frac{1}{N}\sum_{i=1}^{N} Y_i^o, \quad X_{jk} = X_{jk}^o - \frac{1}{N}\sum_{i=1}^{N} X_{ik}^o, \; k = 0, \ldots p.$$

We get, in particular, $X_{j0} = 0$ such that now we look at a linear regression without intercept, i.e., $b_0 = 0$:

$$Y_j = \sum_{k=1}^{p} b_k X_{jk} + \varepsilon_j,\ j = 1, \ldots, N.$$

The penalized least-squares estimate here solves

$$\text{RSS}_\lambda(b) = ||Y - \mathbf{X}b||^2 + \lambda||b||^2 = \min!$$

The solution and the corresponding function estimate are

$$\hat{b}^r = (\mathbf{X}^T\mathbf{X} + \lambda\mathbf{I}_p)^{-1}\mathbf{X}^T Y, \qquad \widehat{m}^r(x) = \hat{b}_0^r + \sum_{k=1}^{p} \hat{b}_k^r x_k \tag{19}$$

where $\mathbf{I}_p$ denotes the $p \times p$ unit matrix. From this explicit formula similar to (15), we see that ridge regression just regularizes the matrix $\mathbf{X}^T\mathbf{X}$ for a more stable calculation of the inverse.

As an example, we consider the training set already used in Figs. 2, 3, 4, and 5 and fit a polynomial of degree $p = 11$ to the data, using ridge regression with a well-chosen $\lambda = 12$. Figure 13 shows the resulting function estimate $\widehat{m}^r(x)$ (blue) together with the true function $m(x)$ (red) and the data. Recall that the unconstrained least-squares estimate $\widehat{m}(x)$ shown as the magenta curve on the left-hand side of Fig. 4 resulted in a considerable overfit. The penalized least-squares approach of ridge regression results in a considerable improvement as can be seen from the right-hand side of Fig. 13 where the estimation errors $\widehat{m}^r(x) - m(x)$ (blue) and $\widehat{m}(x) - m(x)$ (black) are plotted. Looking at the average squared errors

$$\frac{1}{N}\sum_{j=1}^{N}(\widehat{m}^r(X_j) - m(X_j))^2 = 0.060 \quad \text{and} \quad \frac{1}{N}\sum_{j=1}^{N}(\widehat{m}(X_j) - m(X_j))^2 = 0.141,$$

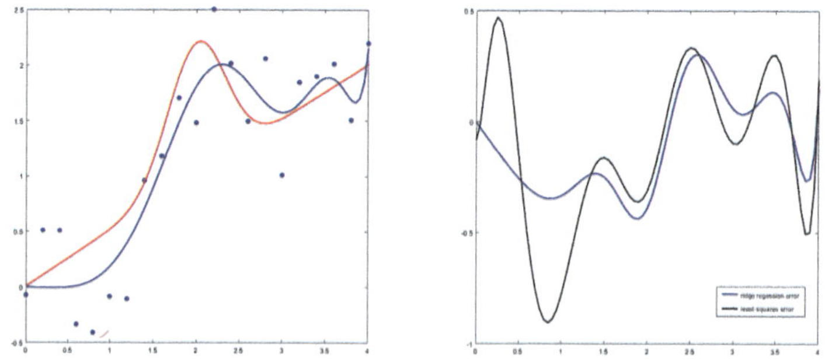

Fig. 13 Data, ridge regression estimate and true $m(x)$ (left) and estimation errors (right)

we also get that the ridge regression estimate is a better estimate. It is even better than the polynomial estimate of degree $p = 4$, which with 0.106 achieves the smallest average squared error among all unconstrained polynomial least-squares estimates.

3.3 LASSO and the Elastic Net

The **LASSO** (*least absolute shrinkage and selection operator*) is a popular estimate of prediction coefficients in machine learning. The motivation and formulation is the same as for ridge regression. The only difference is the penalty term: The size of the parameter vector is not measured by the square of its Euclidean (or L^2) norm but by its L^1 norm, i.e., the LASSO estimate $\hat{b}^\ell$ is defined as the solution of the penalized least-squares problem

$$\mathrm{RSS}_\lambda^\ell(b) = \frac{1}{N}\sum_{j=1}^{N}\big(Y_j - b_0 - \sum_{k=1}^{p} b_k X_{jk}\big)^2 + \lambda \sum_{k=1}^{p} |b_k| = \min!$$

or, equivalently, as the solution of

$$\frac{1}{N}\sum_{j=1}^{N}\big(Y_j - b_0 - \sum_{k=1}^{p} b_k X_{jk}\big)^2 = \min! \quad \text{under the constraint} \quad \sum_{k=1}^{p} |b_k| \leq t \tag{20}$$

where $\lambda > 0$ resp. $t > 0$ act as hyperparameters controlling the smoothness of the final function estimate

$$\widehat{m}_t(x) = \sum_{k=0}^{p} \hat{b}_{t,k} x_k.$$

Due to the nondifferentiability of $|.|$ at 0, there is no explicit solution to this problem, and $\hat{b}^\ell$ has to be calculated numerically, for which quadratic programming algorithms are used.

Note that for $t > t_0 = |\hat{b}_1| + \ldots + |\hat{b}_p|$, LASSO estimate $\hat{b}^\ell$ and common least-squares estimate $\hat{b}$ coincide. For $t \approx t_0/2$, the least-squares parameter estimates are shrunk by 50% on the average. For small t, $\hat{b}_k^\ell = 0$ exactly for many k, i.e., the LASSO enforces sparsity in the strict sense, which is one reason for its popularity in case of high-dimensional input vectors X_j (compare Section 3.4 of [44] for more details).

Due to its tendency to select sparser models, the LASSO performs better than ridge regression if only a small number of features have a strong impact on the output. If, on the other hand, many input variables are relevant for Y_j, ridge regression usually results in better predictors. As in applications, one usually does not know in advance which of both cases is present; it is advisable to use both techniques and compare the resulting predictors by some validation technique, e.g., crossvalidation. Alternatively, one can use the **elastic net**. It combines ridge regression and the LASSO into one general penalized least-squares estimate. It uses two hyperparameters λ_1, λ_2, and the estimate of b is given as the solution of

$$\mathrm{RSS}_\lambda^{en}(b) = \frac{1}{N}\sum_{j=1}^{N}\big(Y_j - b_0 - \sum_{k=1}^{p} b_k X_{jk}\big)^2 + \lambda_1 \sum_{k=1}^{p}|b_k| + \lambda_2 \sum_{k=1}^{p} b_k^2 = \min!$$

3.4 Validation and Crossvalidation

We have repeatedly mentioned that machine learning crucially depends on the choice of hyperparameters. If the values of those method-specific parameters are specified well, then it usually is of secondary importance which particular learning method is applied for estimating, e.g., a regression function. The selection of good hyperparameter values requires a lot of experience with similar data or a lot of trial-and-error experimentation. Therefore, automatic selection methods, which adapt the hyperparameter to the actual data, have been investigated quite early in the context of data-adaptive statistical model selection (compare, e.g., [1]).

In this section, we discuss the popular crossvalidation approach for getting good values for hyperparameters from the data. As an illustration, we consider general regression models (2), where a major goal is forecasting new data Y_{N+1} from the corresponding input X_{N+1}. As in Sect. 2.1, we calculate a prediction function estimate $\widehat{m}(x)$ from the training set $\mathscr{T} = (Y_1, X_1, \ldots, Y_N, X_N)$ and use the forecast $\widehat{Y}_{N+1} = \widehat{m}(X_{N+1})$. We are interested in good predictors, which, again, we interpret as those resulting in a small mean-squared prediction error

$$\mathrm{mspe}(\widehat{m}) = \mathrm{e}\{\big(Y_{N+1} - \widehat{m}(X_{N+1})\big)^2 | \mathscr{T}\}, \qquad \text{where } \mathscr{T} = (Y_1, X_1, \ldots, Y_N, X_N) \tag{21}$$

Note that we treat the prediction function $\widehat{m}(x)$ as given, i.e., we have to condition on the data used for calculating it, i.e., on $\mathscr{T}$. The mspe only involves the expectation with respect to the randomness in Y_{N+1}, X_{N+1}. If it is small, then the forecasting error $Y_{N+1} - \widehat{Y}_{N+1}$ will be small on average.

$\mathrm{mspe}(\widehat{m})$ depends on the hyperparameters of the learning method, e.g., in polynomial regression (3) on the degree p of the polynomial, or on the weight λ in the loss function of ridge regression (18). For selecting good values for these unknown hyperparameters of the estimation procedure, we have to estimate mspe.

A first idea would be to look at the minimal empirical squared forecasting error on the training set

$$\frac{1}{N}\text{RSS}(\widehat{m}) = \frac{1}{N}\sum_{j=1}^{N}\left(Y_j - \widehat{m}(X_j)\right)^2.$$

However, this does not approximate mspe($\widehat{m}$) well even for large N, but it remains systematically smaller. The reason is again an overfitting effect, as $\widehat{m}(X_j)$, which is calculated from $\mathcal{T}$ tends to be closer to Y_j than the true optimal predictor $m(X_j)$ if the data Y_j, X_j are part of the training set.

To avoid this effect, we may use a separate **validation set** $\mathcal{V}$ consisting of data $\tilde{Y}_j, \tilde{X}_j, j = 1, \ldots, M$, which are independent of the training set $\mathcal{T}$. To approximate the mspe, we use the residual sum of squares calculated from $\mathcal{V}$:

$$\widehat{\text{mspe}}(\widehat{m}) = \frac{1}{M}\sum_{j=1}^{M}\left(\tilde{Y}_j - \widehat{m}(\tilde{X}_j)\right)^2.$$

This is a reliable estimate of mspe($\widehat{m}$), i.e., it converges to mspe($\widehat{m}$) for $M \to \infty$ by the law of large numbers.

As an example for an application to hyperparameter selection by **validation**, let $\widehat{m}_\lambda^r$ be the ridge estimate of m from (19) using hyperparameter λ and calculated from the training set of size N. Then, we get a data-adaptive choice $\widehat{\lambda}$ of λ by solving the minimization problem

$$\widehat{\text{mspe}}(\widehat{m}_\lambda^r) = \min_{\lambda>0}!$$

Validation does not use the available data in an efficient manner, as part of it has to be set aside as the validation set and may not be used for improving the estimate $\widehat{m}(x)$. As a rule of thumb, 80% of all the data are forming the training set, and the remaining 20% are reserved for validation. For big data, which are available in many general machine learning situations, this is not really a problem, but for small and medium-sized datasets, which are more common in engineering problems, we can do better by the crossvalidation technique (for the price of higher computational expense, which becomes an issue for really large datasets, but for them, simple validation usually is good enough).

The original idea of crossvalidation was splitting the sample into two parts. First, one of them acts as training set to get the estimate $\widehat{m}(x)$ and the other one as validation set for calculating $\widehat{\text{mspe}}(\widehat{m})$ as a performance measure. Then, the two roles are interchanged such that one gets two estimates for $m(x)$ resp. the mean-squared prediction error, which are averaged (with weighting corresponding to the relation of the two subsample sizes). In this manner, all the data are used for estimation as well as for validation, and the respective independence between training set and validation set is still retained.

Due to improved computational facilities in the 1970s, it became popular to repeat the estimation and validation steps many times with various splits of the data into two subsamples. We consider the common **leave-one-out technique** first. For a sample of size N, we consider for each $i = 1, \dots N$,

$$(Y_j, X_j),\ j \neq i, \text{ as training set of size } N-1, \quad (Y_i, X_i) \text{ as validation set of size } 1.$$

Let $\widehat{m}^{(-i)}$ denote the prediction function estimate using the ith training set (which does not contain (Y_i, X_i)). The corresponding estimate of the mean squared prediction error is simply

$$\widehat{\text{mspe}}(\widehat{m}^{(-i)}) = \big(Y_i - \widehat{m}^{(-i)}(X_i)\big)^2.$$

Finally, we average all those error estimates to get the crossvalidatory approximation of $\text{mspe}(\widehat{m})$:

$$\widehat{\text{mspe}}_{\text{CV}}(\widehat{m}) = \frac{1}{N}\sum_{i=1}^{N}\big(Y_i - \widehat{m}^{(-i)}(X_i)\big)^2.$$

Considering again the ridge regression estimate $\widehat{m}^r_\lambda$, we get a crossvalidation choice $\widehat{\lambda}_{\text{CV}}$ of the hyperparameter λ by solving

$$\widehat{\text{mspe}}_{\text{CV}}(\widehat{m}^r_\lambda) = \text{CV}(\lambda) = \min_{\lambda > 0}!$$

This hyperparameter selection method is not only based on intuition, but it can be justified by mathematical theory. For $N \to \infty$, we have on the one hand for the training sample error $\frac{1}{N}\text{RSS}(\widehat{m}^r_\lambda) \ll \text{mspe}(\widehat{m}^r_\lambda)$ and on the other hand for all $\lambda > 0$

$$\frac{\text{CV}(\lambda)}{\text{mspe}(\widehat{m}^r_\lambda)} \to 1$$

or in an even stronger sense of approximation

$$\frac{\text{mspe}(\widehat{m}^r_{\widehat{\lambda}_{\text{CV}}})}{\min_{\lambda \in C_N} \text{mspe}(\widehat{m}^r_\lambda)} \to 1,$$

for compact subsets C_N, which grow to $(0, \infty)$ for $N \to \infty$. In this sense, crossvalidation provides an asymptotically optimal way to select the hyperparameter λ from the data (compare [33] for such a result, using generalized crossvalidation). The asymptotic optimality of this hyperparameter selection method in ridge regression even holds, if the models become more and more complex with increasing N, i.e., in the case of a so-called dimension asymptotics where $p = p(N) \to \infty$ for $N \to \infty$.

Patil et al. [63] presents such a result even in the case where p grows with the same rate as N and may be even larger, i.e., for $p/N \to c, 0 < c < \infty$.

Crossvalidation is not the only method who has this kind of asymptotic optimality. For example, hyperparameter selection based on Akaike's Information Criterion (AIC) is asymptotically equivalent to crossvalidation such that one can use whatever is more convenient (compare [77])

As an example, we consider the same data as in Fig. 13. We want to fit a polynomial of degree $p = 7$ using ridge regression. The performance of the estimate $\widehat{m}(x;\lambda) = \widehat{m}^r_\lambda(x)$ depends on the weight of the penalty term in the regularized least-squares approach. To automatically select a reasonable value of this hyperparameter, we apply leave-one-out crossvalidation, i.e., we minimize

$$\mathrm{CV}(\lambda) = \frac{1}{N}\sum_{i=1}^{N}\left(Y_i - \widehat{m}^{(-i)}(X_i;\lambda)\right)^2$$

w.r.t. λ, where $\widehat{m}^{(-i)}$ denotes the ridge regression estimate based on the training set with the ith observation left out. The function $\mathrm{CV}(\lambda)$ together with the location of the minimum at $\widehat{\lambda}_{\mathrm{CV}} = 0.175$ is shown as the blue curve in Fig. 14. For comparison, we also show as the red curve the average squared estimation error

$$\mathrm{ase}(\lambda) = \frac{1}{N}\sum_{j=1}^{N}\left(m(X_j) - \widehat{m}(X_j;\lambda)\right)^2,$$

which achieves its minimum at $\lambda_{\mathrm{ase}} = 0.3$. The average squared error, which we can only calculate for artificial data with known true function $m(x)$, is a good approximation to $\mathrm{mspe}(\widehat{m})$ such that λ_{ase} is close to the optimal hyperparameter value.

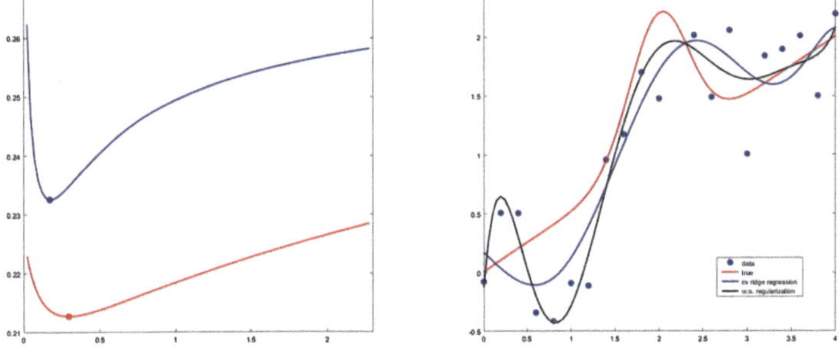

Fig. 14 Crossvalidation function $\mathrm{CV}(\lambda)$ (blue) and average squared error $\mathrm{ase}(\lambda)$ (red) on the left-hand side; data, true function $m(x)$ (red), ridge regression estimate $\widehat{m}(x;\widehat{\lambda}_{\mathrm{CV}})$ (blue), and simple least-squares estimate $\widehat{m}(x;0)$ (black) on the right-hand side

On the right-hand side of Fig. 14, the function estimate $\widehat{m}(x; \widehat{\lambda}_{CV})$ (blue) is shown together with the true function $m(x)$ (red) and, for comparison, the plain least-squares polynomial fit $\widehat{m}(x; 0)$ of degree $p = 7$ (black), which corresponds to $\lambda = 0$.

Note that the error function ase is rather flat around its minimum, which is rather typical and makes the exact choice of hyperparameter less critical as long as it lies in a neighborhood of the optimum. In our example, the crossvalidatory choice is almost as good as the optimal weight as $\text{ase}(\widehat{\lambda}_{CV}) = 0.84 \approx 0.83 = \text{ase}(\lambda_{\text{ase}})$. For comparison, without regularization, i.e., with $\lambda = 0$, the error $\text{ase}(0) = 1.24$ is about 50% larger.

The original polynomial regression estimate (black curve in Fig. 14) shows a considerable overfit, in particular for the smaller values of x, where it tries to get close to data that, by chance, deviate rather strongly from $m(x)$. Using ridge regression together with crossvalidation, we get a smoother function estimate, which also has a notably smaller estimation error.

To save computations, in particular for larger samples, leave-one-out is replaced by **leave-L-out**. Here, the sample is split into K roughly equal subsamples of size $\approx L$ each, i.e., $LK \approx N$. One subsample after the other serves as validation set, whereas the rest of the data serves as training set. For the example of the LASSO estimate, let $\widehat{m}_t^{-\kappa}$ denote the regression function estimate calculated from the data without the κth subsample, $\kappa = 1, \ldots, K$. The corresponding mspe-estimates based on the excluded subsamples as validation sets are averaged, which results in the crossvalidation criterion function for the leave-L-out approach

$$\text{CV}_L(t) = \frac{1}{N} \sum_{j=1}^{N} \left(Y_j - \widehat{m}^{(-\kappa(j))}(X_j)\right)^2,$$

where $\kappa(j)$ denotes the number of the subsample containing observation no. j. As the $\widehat{m}_t^{-\kappa}$ are based on samples of size $N - L$, the estimation error is for larger values of L a bit too large than for a sample of size N, and $\text{CV}_L(t)$ overestimates the prediction error a bit. However, the location of the minimum w.r.t. t, which we need for hyperparameter selection, usually does not depend much on the size of L.

Regarding data-adaptive choice of hyperparameters, we like to finish with a word of **warning** about an issue that is common to all such selection procedures. We focus on ridge regression with hyperparameter λ and validation as the method for choosing λ, but the problems are the same for other learning methods and selection procedures.

The validation set $\mathscr{V}$ is independent of the training set $\mathscr{T}$, and, hence, we may use $\widehat{\text{mspe}}(\widehat{m}_\lambda^r)$ for comparing the performance of various weights λ and selecting the best one $\widehat{\lambda}$. In presenting their final result, sometimes, researchers use $\widehat{\text{mspe}}(\widehat{m}_{\widehat{\lambda}}^r)$ to illustrate how well the method performed in the data analysis. This is not legitimate! The reason is kind of a **secondary overfit** effect. The selected hyperparameter $\widehat{\lambda}$ has been calculated from the validation set $\mathscr{V}$ and, therefore, depends on the estimate

$\widehat{\text{mspe}}$, which has been evaluated on the same data. $\widehat{\lambda}$ is, by definition, a particularly good choice for the validation data, but it may perform less well for other data.

For a fair evaluation of the quality of a statistical learning method with data-adaptive hyperparameter, we essentially need three independent datasets $\mathcal{T}$, $\mathcal{V}$ and an additional independent **test set** $\mathcal{P}$ of input-output pairs $X_j^*, Y_j^*, j = 1, \ldots, K$. $\mathcal{T}$ is used for estimating the model parameters ($b_k, k = 1, \ldots, p$, in the framework of this chapter), $\mathcal{V}$ is used for choosing the hyperparameter (λ for ridge regression), and $\mathcal{P}$ is used for calculating a performance measure of the method using the data-adaptive hyperparameter ($\widehat{\lambda}$ for ridge regression) by the average prediction error on the test set

$$\widehat{\text{mspe}}^*(\widehat{m}) = \frac{1}{K}\sum_{j=1}^{K}\left(Y_j^* - \widehat{m}(X_j^*)\right)^2.$$

For ridge regression, the estimate $\widehat{m}^r_{\widehat{\lambda}}$ with data-dependent $\widehat{\lambda}$ has to be applied to data which have not been used at all in learning the estimate **including** its hyperparameter. Only then can we judge reliably how well the regression estimate will perform in the intended future applications.

3.5 Information Criteria

The residual sum of squares (RSS) measures how well a regression function fits the training data (compare Sect. 2.1). We already have discussed that it decreases if we add more and more free parameters to the model, leading finally to a severe overfit. Based on Akaike's seminal work [1, 62], this can be avoided by adding a complexity penalty that does not address the total size of the parameters as in ridge regression but directly the model structure and, in particular, the number of parameters. As an illustration, we consider polynomial regression as in Sect. 2.1 with one-dimensional inputs X_j. Here, the residual sum of squares is

$$\text{RSS}_p(b) = \sum_{j=1}^{N}\left(Y_j - b_0 - \sum_{k=1}^{p} b_k X_j^k\right)^2$$

Let $\hat{b}(p)$ denote the least-squares estimate of the model parameter vector for given polynomial degree p. As we have more and more parameters, $\text{RSS}_p(\hat{b}(p))$ decreases with p. If we add twice the number of parameters as a penalty related to the complexity of the regression model, then, dropping some additive constants independent of p from the expression, we get Akaike's information criterion (**AIC**) for this simple situation

$$\text{AIC}_p = N \log \widehat{\sigma}_\varepsilon^2(p) + 2p, \qquad \text{with } \widehat{\sigma}_\varepsilon^2(p) = \frac{1}{N}\text{RSS}_p(\hat{b}(p)),$$

where the latter is an estimate of the residual variance σ_ε^2 (more precisely, the conditional maximum likelihood estimate given the X_j assuming that the residuals ε_j are normally distributed).

For increasing p, the first term of the information criterion AIC_p decreases, and the second one increases. Choosing $\hat{p}_{\text{AIC}}$ as a minimizer of the criterion function results in a data-adaptive choice of the hyperparameter p providing a good balance between bias and variance and, hence, a good compromise between overfit and underfit.

Another popular information criterion is the Bayes information criterion (**BIC**), also due to Akaike [2]

$$\text{BIC}_p = N \log \widehat{\sigma}_\varepsilon^2(p) + p \log N,$$

which uses a complexity penalty depending on the size of the training set. The minimizer $\hat{p}_{\text{BIC}}$ is smaller than the hyperparameter chosen by AIC as soon as $N \geq 8$.

Figure 15 shows AIC_p and BIC_p for the sample of size $N = 21$ shown on the left-hand side of Fig. 13. Both criteria happen to select the same degree $\hat{p} = 7$, which is a clear decision for BIC, whereas AIC assumes an only marginally larger value at $p = 10$. In this artificial example, we know the true function $m(x)$, which is not a polynomial of finite degree, such that we may calculate the average squared estimation error as in Sect. 3.2, getting 0.124, whereas the best choice of degree, exploiting knowledge of the true regression function $m(x)$, would be $p = 4$.

Another example for the application of AIC and BIC is discussed in the context of Hald's cement data (compare Fig. 12). The result illustrates the fact that BIC often chooses smaller models than AIC and is preferable if one is mainly interested in using not too many input variables, e.g., to reduce computational complexity of the final forecasting algorithm.

There are a lot of other hyperparameter selection methods that have been shown to be either almost equivalent to AIC or to BIC in many different situations for large enough N. This even holds for crossvalidation though that is not immediately

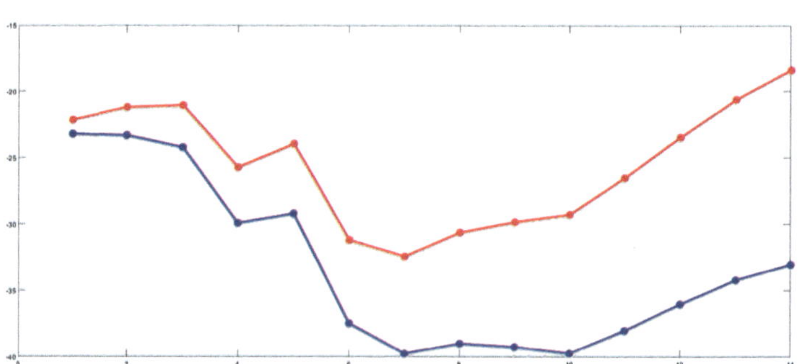

Fig. 15 AIC_p (blue) and AIC_p (red) for polynomial regression using the data shown in Fig. 13

obvious [72, 77]. However, for a given training set, there may be considerable differences in performance.

Generally, BIC is preferable if we look for a parsimonious regression function that uses as few as possible parameters but still captures the data structure well. AIC is on the other hand better, if we are accepting a slight overfit and mainly are interested in good forecasts of future values. There are theoretical results that underpin these recommendations [41, 73].

There are lots of different versions of AIC and BIC that try, e.g., to improve the small sample performance or that are based on different kinds of intuition but show a similar behavior. In particular, information criteria may be used for choosing hyperparameters of statistical learning algorithms. For example, [61] generalized the AIC to neural networks and proposed a neural information criterion (**NIC**).

4 Bayesian Learning for Regression

Bayesian approaches in statistics are often advertised by their ability to incorporate prior knowledge about the data-generating mechanism. Principally, this is correct, but in practice, such prior information frequently is not available in a form, which may be immediately used. We present the main ideas of Bayesian learning and briefly discuss how the Bayesian viewpoint may be used to incorporate the expert knowledge of engineers into a machine learning procedure.

The main advantage of Bayesian learning is however the easy availability of predictive distributions derived from the training set. In regression, we have in the previous chapter only discussed forecasts like $\widehat{Y}_{N+1} = \widehat{m}(X_{N+1})$ for new data Y_{N+1} given the input X_{N+1}. In engineering, it is frequently of interest to know how precise those forecasts are. The reliability of $\widehat{Y}_{N+1}$ as an approximation of the yet unobserved Y_{N+1} can be measured by an estimate of the mean-squared prediction error like those considered in the previous section. Even more informative would be a **confidence interval** $[\widehat{Y}^{\ell}_{N+1}, \widehat{Y}^{r}_{N+1}]$ derived from the training set, which contains the forecasts $\widehat{Y}_{N+1}$ and which is known to cover also the true value Y_{N+1} with a prescribed high probability like 95%.

Bayesian learning allows for calculating from the training set an estimate of the whole distribution of Y_{N+1} given the input X_{N+1}, the so-called **predictive distribution** from which we can derive various measures for the precision of forecasts $\widehat{Y}_{N+1}$.

4.1 Bayesian Linear Regression and Predictive Distributions

We illustrate the Bayesian learning approach first with the linear basis function regression model, i.e., we approximate the true regression function $m(x)$ by functions of the form

$$m_q(x; b) = \sum_{k=0}^{q} b_k g_k(x), \tag{22}$$

where the known $g_0, g_1, \ldots$ form a basis of the function space of interest. We choose q large such that the difference between the true regression function $m(x)$ and $m_q(x; b)$ for suitable chosen parameter values b_k is negligible. We do not have to worry about overfit, as the Bayesian approach has some inbuilt penalty similar or in some cases even identical to ridge regression. Therefore, we assume that q is large enough such that $m(x)$ can be approximate well, i.e., $m(x) \approx m(x; b^o)$ for a suitable b^o.

For the moment, let us assume that the variance σ_ε^2 of the residuals ε_j is known. We pretend that ε_j is Gaussian, i.e., normally distributed. Then, given the input X_j, $Y_j = m(X_j) + \varepsilon_j$ is Gaussian too with mean $m(X_j)$ and variance σ_ε^2. Replacing the regression function $m(x)$ by its approximation (22), we get the approximate density

$$p_j(y_j|b) = \frac{1}{\sqrt{2\pi\sigma_\varepsilon^2}} \exp\Big\{ -\frac{1}{2\sigma_\varepsilon^2}\big(y_j - m_q(X_j; b)\big)^2 \Big\}, \tag{23}$$

and the corresponding output vector $Y = (Y_1, \ldots, Y_N)$ has the approximate density $p(y|b) = \prod_{j=1}^{N} p_j(y_j|b)$ with $y = (y_1, \ldots, y_N)$, which is the conditional likelihood function given the input values X_j, $j = 1, \ldots, N$.

The parameter vector $b = (b_0, \ldots, b_q)$ is not known, and we want to learn it from the training set. However, we might have some idea about likely and less likely values of b, e.g., from experience with similar data or from engineering theory. In a Bayesian approach, we quantify this knowledge in terms of a **prior** probability **distribution** over the possible range of parameter values. In a certain sense, b is treated as random. This device has the purpose to measure the uncertainty about the values of this unknown quantity.

In the regression model with regression function (22) and Gaussian residuals, it is convenient to choose the prior distribution as Gaussian too, i.e., the **prior density** $p_0(b)$ is a multivariate normal density (1) on $(q+1)$-dimensional space with, say, mean vector β_0 and covariance matrix $\mathbf{S}_0$. Under this distribution, $\mathrm{e}b_k = \beta_k$, and $\mathrm{cov}\,(b_k, b_\ell) = S_{0,k\ell}$.

The training set provides new information about the parameter. As the input variables X_j, $j = 1, \ldots, N$, are treated as given, it is mainly provided by the output vector $Y = (Y_1, \ldots, Y_N)$. In view of this new information, the prior density no longer describes our state of knowledge about the parameter, but our assessment of likely and unlikely values of b has to be updated. This is quantified by the **posterior distribution** $p(b|Y)$ of b given the training data. From the Bayes rule, the posterior density is proportional to the product of the likelihood function and the prior density with a factor not depending on b.

As $p(y|b)$ and $p(b)$ both are Gaussian, the posterior density of b is Gaussian too (compare, e.g., (3.49) of [5]), with updated mean vector β_N and covariance matrix $\mathbf{S}_N$ where

$$\mathbf{S}_N^{-1} = \mathbf{S}_0^{-1} + \frac{1}{\sigma_\varepsilon^2}\mathbf{G}^T\mathbf{G}, \qquad \beta_N = \mathbf{S}_N\big(\mathbf{S}_0^{-1}\beta_0 + \frac{1}{\sigma_\varepsilon^2}\mathbf{G}^T Y\big). \tag{24}$$

Here, $\mathbf{G}$ denotes the $N \times (p+1)$ design matrix of our linear regression model (compare Sect. 3.1) with entries $\mathbf{G}_{jk} = g_k(X_j),\ j = 1, \dots, N,\ k = 0, \dots, p$.

The posterior density $p(b|Y)$ describes our knowledge about the unknown parameter b after we have taken into account the information in the training set. b is more likely to be found where the density is large. If we look for the value that maximizes the posterior density, we immediately get the so-called **maximum a posteriori** (MAP) **estimate** $\hat{b}_{\text{MAP}}$ of b, i.e.,

$$p(\hat{b}_{\text{MAP}}|Y) = \max_b p(b|Y).$$

As a Gaussian density is maximal at its mean, we have in our specific situation $\hat{b}_{\text{MAP}} = \beta_N$. The corresponding Bayes estimate of the regression function $m(x)$ is given as

$$\widehat{m}_q^B(x) = m_q(x; \beta_N) = \sum_{k=0}^{q} \beta_{N,k} g_k(x).$$

On a first glance, Bayesian estimation depends on having enough prior information about b to be translated into a prior density. This offers an opportunity to incorporate engineering expertise. For example, it may be known that b has to lie in a certain range B of values as otherwise the physical processes underlying our experiment would become unstable. Then, a prior density should reflect this information and be concentrated in this range. In our example, β_0 would be chosen in the center of B and $\mathbf{S}_0$ concentrated enough such that values outside B are extremely unlikely. Known dependencies between feature components X_{jk} and $X_{j\ell}$ lead to dependencies between b_k, b_ℓ and translate into larger values for their prior covariance $\mathbf{S}_{0,k\ell}$.

If no such information is available, it is always possible to fall back onto non-informative or uninformed priors, e.g., in our case, a Gaussian distribution with $\beta_0 = 0, \mathbf{S}_0 = \frac{1}{\alpha}\mathbf{I}_{q+1}$ where again $\mathbf{I}$ denotes a unit matrix. This often works surprisingly well. For this prior, we have from (24)

$$\mathbf{S}_N^{-1} = \alpha\mathbf{I}_{q+1} + \frac{1}{\sigma_\varepsilon^2}\mathbf{G}^T\mathbf{G}, \qquad \beta_N = \frac{1}{\sigma_\varepsilon^2}\mathbf{S}_N\mathbf{G}^T Y.$$

Comparing this with (19) and considering that, here, the design matrix is **G**, the MAP estimate β_N turns out to be identical to the ridge estimate $\hat{b}^r$ with hyperparameter $\lambda = \alpha\sigma_\varepsilon^2$.

Other priors complicate the matter as they no longer have the conjugate property. However, choosing as prior density

$$p(b) = c(\alpha, \gamma) \exp\Big\{ -\frac{\alpha}{2} \sum_{k=0}^{q} |b_k|^\gamma \Big\} \tag{25}$$

we get as MAP estimate again a penalized least-squares estimate (18), but with regularization term

$$\frac{\lambda}{2} \sum_{k=0}^{q} |b_k|^\gamma .$$

This confirms our statement that Bayesian linear regression contains an inbuilt safeguard against overfit such that we may choose q rather large to avoid a strong bias effect (compare Sect. 2.3). Note that the MAP estimate based on prior density (25) includes not only ridge regression as special case $\gamma = 2$ but also the LASSO for $\gamma = 1$ (compare Section 6.2 of [48].

Bayesian linear regression provides a different angle for justifying estimates like ridge regression or the LASSO. Its main advantage is, however, an easy access to quantifying the reliability of estimates and forecasts. The key is the predictive distribution for Y_{N+1} given $X_{N+1} = x$. As we have assumed normally distributed residuals, the density of Y_{N+1} given $X_{N+1} = x$ is Gaussian, too, with mean $m(x)$ and variance σ_ε^2. As in (23), we approximate it by

$$p_{N+1}(y_{N+1}|b) = \frac{1}{2\pi\sigma_\varepsilon^2} \exp\Big\{ -\frac{1}{2\sigma_\varepsilon^2} \big(y_{N+1} - m_q(x; b)\big)^2 \Big\}.$$

b is not known exactly, but the posterior distribution provides information about its likely values. Integrating $p_{N+1}(y_{N+1}|b)$ w.r.t. to this density, we get the **predictive distribution** of Y_{N+1} given $X_{N+1} = x$ and the training data

$$p_{N+1}(y_{N+1}|Y) = \int \cdots \int p_{N+1}(y_{N+1}|b) p(b|Y) db_0 \ldots db_q .$$

As both densities in the integrand are Gaussian with mean $m(x; b)$ and variance σ_ε^2 resp. mean vector β_N and covariance matrix $\mathbf{S}_N$, the resulting $p_{N+1}(y_{N+1}|Y)$ is a normal density, too, with mean and variance

$$m_{N+1}(x) = \widehat{m}_q^B(x) = \sum_{k=0}^{q} g_k(x)\beta_{N,k} = g(x)\beta_N, \quad \sigma_{N+1}^2(x) = \sigma_\varepsilon^2 + g(x)\mathbf{S}_N g^T(x),$$

where $g(x) = (g_0(x), \ldots, g_q(x))$ denotes the row vector of basis function values evaluated at x.

The density of the predictive distribution immediately provides the forecast $\widehat{Y}_{N+1} = m_{N+1}(x) = \widehat{m}_q^B(x)$ as the most likely value. It is just the approximation of $m(x)$ by $m_q(x; b)$ with b replaced by the parameter estimate $\beta_N = \hat{b}_{\mathrm{MAP}}$. The standard deviation $\sigma_{N+1}(x)$ of the predictive distribution provides a measure of precision of $\widehat{Y}_{N+1}$ as a forecast of Y_{N+1} given $X_{N+1} = x$. If from additional experiments we get more information, i.e., if the size N of the training set increases, the reliability of the forecast increases, as $\sigma_{n+1}^2(x) \leq \sigma_n^2(x)$ for all $n \geq N + 1$. For $N \to \infty$, we have $\sigma_{N+1}(x) \to \sigma_\varepsilon$, which is the unavoidable error due to the randomness and independence of the residual ε_{N+1}.

The predictive distribution also provides **confidence intervals** for Y_{N+1} as usual for Gaussian distributions. For example, a 95%-confidence interval is given by

$$m_{N+1}(x) - 1.96\sigma_{N+1}(x) \leq Y_{N+1} \leq m_{N+1}(x) - 1.96\sigma_{N+1}(x) \tag{26}$$

with probability 0.95.

As an example, we consider a similar training set of size $N = 51$ as in Fig. 6, but with inputs X_j, which are not lying on an equidistant grid but are uniformly distributed random variables on the interval [0,4]. We fit again a polynomial of degree $q = 11$ to the data such that $g(x) = (1, x, x^2, \ldots, x^q)$. Then, we apply the Bayes estimation procedure with uninformed prior distribution, i.e., a Gaussian distribution with $\beta_0 = 0, \mathbf{S}_0 = \frac{1}{\alpha}\mathbf{I}_{q+1}$. The hyperparameter is chosen as $\alpha = 20$, which results in a satisfactory estimate.

The left-hand side of Fig. 16 shows the training set, the true regression function $m(x)$ (red), the standard polynomial estimate $\widehat{m}_q(x)$ of degree $q = 11$ (black), and the corresponding Bayes estimate $\widehat{m}_q^B(x)$ (blue). The Bayes estimate is smoother and less prone to overfit compared to the plain least-squares estimate, illustrating the automatic regularization, which is part of the Bayesian method.

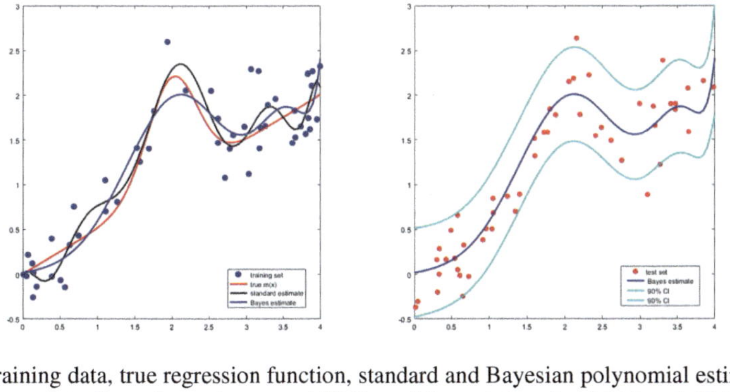

Fig. 16 Training data, true regression function, standard and Bayesian polynomial estimate (left-hand side), test data, Bayesian estimate, and 90 % confidence intervals (right-hand side)

The right-hand side of Fig. 16 shows $\widehat{m}_q^B(x)$ (blue) and the 90 % predictive intervals (light blue), corresponding to (26) with 1.645 replacing 1.96, together with an independent test set $(\tilde{X}_j, \tilde{Y}_j)$, $j = 1, \ldots, N$, of the same size $N = 51$ as the training set and generated in the same manner. The predictive intervals are constructed such that they should contain the true output $\tilde{Y}_j$ in 90 % of the cases. In the example, we have five data from the test set outside of the confidence intervals (where two are almost on the lower boundary), i.e., 46 out of 51, hence roughly 90%, of the data are covered by the confidence intervals, which confirms our statement about the reliability of forecasts.

Up to now, we have assumed that the residual variance σ_ε^2 is known, which only happens to be the case in rare applications, e.g., if the errors are pure measurement errors, and σ_ε^2 is given in the documentation of the measuring device. It is, however, possible to replace the residual variance by rough and ready estimates, in particular for large sample sizes N, with only a small effect on the performance of Bayesian regression. More refined methods are also available if necessary (compare [5], Section 3.3).

4.2 Gaussian Process Regression (GPR)

Gaussian process regression is an advanced learning method based on Bayesian principles. The main idea of this approach is the interpretation of the function, which describes the relationship between input and output in a regression model, as a random variable, in particular a Gaussian one. We denote this function by $F(x)$ such that inputs X_j and outputs Y_j, $j = 1, \ldots, N$ of the training set satisfy

$$Y_j = F(X_j) + \varepsilon_j, \quad j = 1, \ldots, N.$$

The noise variables $\varepsilon_1, \ldots, \varepsilon_N$ as usual are independent and identically distributed random variables with mean 0 and variance σ_ε^2, but, here, we assume that they are Gaussian too. The inputs X_j are d-dimensional random vectors, and the outputs Y_j are real valued.

To explain the concept of a Gaussian process, we first have a look at Gaussian random vectors. Let Z denote a N-dimensional random vector that has mean vector μ and $N \times N$ covariance matrix Σ with entries

$$\Sigma_{k\ell} = \text{cov}\,(Z_k, Z_\ell) = \text{e}(Z_k - \mu_k)(Z_\ell - \mu_\ell)$$

Z is called a Gaussian random vector if it has a multivariate normal density $p(z)$ as in (1). Each component Z_k of Z is a univariate normal random variable with mean μ_k and variance Σ_{kk}, and each subvector of Z is again a Gaussian random vector of corresponding lower dimension with mean and covariance matrix being a subvector of μ, resp., a submatrix of Σ.

We extend the concept of a Gaussian random vector to random functions, as $F(x)$ is a function on d-dimensional space. It is called a **Gaussian process** with mean function $m(x)$ and *covariance kernel* $C(\tilde{x}, x)$, if for all d-dimensional $x_1, \ldots, x_N$ and all $N \geq 1$, the random vector $\big(F(x_1), \ldots, F(x_N)\big)$ is a Gaussian random vector with mean vector $\mu = \big(m(x_1), \ldots, m(x_N)\big)$ and covariance matrix Σ having entries $\Sigma_{k\ell} = C(x_k, x_\ell)$. As an abbreviation, we write $F \sim \mathrm{GP}(m, C)$.

In GPR, we do not estimate the function $F(x)$, but we rather estimate the whole distribution, which, e.g., allows to calculate point estimates and confidence intervals for particular function values $F(x)$. The methodology is Bayesian, i.e., we start from a prior Gaussian process distribution $\mathrm{GP}(m_0, C_0)$ and, then, learn the posterior distribution $\mathrm{GP}(\widehat{m}, \widehat{C})$ from the training set, where $\widehat{m}, \widehat{C}$ are estimates of mean function m and covariance kernel C.

There are two ways to do statistical inference based on GPR as implemented, e.g., in the toolbox of [78]. In the fully Gaussian setting, there are analytical formulas for calculating $\widehat{m}, \widehat{C}$ (compare chapter 2 of [65] for details). For ease of notation, we only consider the case of one-dimensional inputs X_j, where $\mu(x)$ is a function on the real line and $C(\tilde{x}, x)$ is a function on the plane. As a prior, we choose mean $m_0(x) = 0$ for all x and a squared exponential covariance kernel

$$C_0(\tilde{x}, x) = \sigma^2 \exp\Big(-\frac{1}{2\lambda^2}(\tilde{x} - x)^2\Big) + \sigma_\varepsilon^2 \delta(\tilde{x}, x),$$

where σ^2 and λ act as hyperparameters for tuning the function estimate and $\delta(\tilde{x}, x) = 1$ for $\tilde{x} = x$, and $= 0$, else. As discussed at the end of the previous section, σ_ε^2 may be known in some cases but usually has to be inferred from the training data or considered as an additional hyperparameter, which has to be chosen by the data analyst.

The mean function and covariance kernel of the posterior distribution are given by

$$\widehat{m}(x) = Y\big(C_0^{(N)} + \sigma_\varepsilon^2 \mathbf{I}_N\big)^{-1} c_N^T \qquad \widehat{C}(\tilde{x}, x) = C_0(\tilde{x}, x) - \tilde{c}_N\big(C_0^{(N)} + \sigma_\varepsilon^2 \mathbf{I}_N\big)^{-1} c_N^T,$$

where $Y = (Y_1, \ldots, Y_N)$ is the vector of all outputs in the training set, $C_0^{(N)} = \big(C_0(X_i, X_j)\big)_{i,j=1,\ldots,N}$ denotes the covariance matrix of $F(X_1), \ldots, F(X_N)$ under the assumption of the prior distribution, $\mathbf{I}_N$ is as usual the $N \times N$ unit matrix, and

$$c_N = \big(C_0(x, X_1), \ldots, C_0(x, X_N)\big), \quad \tilde{c}_N = \big(C_0(\tilde{x}, X_1), \ldots, C_0(\tilde{x}, X_N)\big)$$

denote the vector of covariances between $F(x)$, resp., $F(\tilde{x})$ and $F(X_1), \ldots, F(X_N)$, again under the prior distribution.

Hence, the posterior distribution $\mathrm{GP}(\widehat{m}, \widehat{C})$ is explicitly known, given the training set and the prior distribution, and may be used to calculate estimates or confidence intervals. For example, the best predictor for the response Y_{N+1} at a new prescribed input X_{N+1} is just $\widehat{Y}_{N+1} = \widehat{m}(X_{N+1})$. Moreover, we know that, given the training data, $F(X_{N+1})$ and Y_{N+1} have a normal posterior distribution with the same mean $\widehat{m}(X_{N+1})$ and variance $\sigma^2_{\widehat{C}} = \widehat{C}(X_{N+1}, X_{N+1})$, resp., $\sigma^2_{\widehat{C}} + \sigma^2_\varepsilon$, such that we can calculate confidence intervals in the same manner as in (26).

Of course, the performance of this Bayesian learning method depends on the choice of prior and even more on the choice of the hyperparameters. λ, e.g., determines the regularity of $\widehat{m}(x)$ where larger values of λ result in smoother function estimates. In [65], several common choices of prior as well as data-adaptive methods for choosing the hyperparameters including σ^2_ε are discussed.

In applications of GPR, inference about nonlinear functions of the data are frequently of interest. Here, analytic calculations often turn out to be intractable, but the second method for using Gaussian process-based learning, implemented in [78], will still be applicable. It relies on simulating paths of Gaussian processes from the posterior distribution and approximates quantities of interest by Monte Carlo simulation. Here, the predictor for Y_{N+1} would, e.g., be

$$\widehat{Y}_{N+1} = \frac{1}{M}\sum_{k=1}^{M} \widehat{F}_k(X_{N+1}),$$

where $\widehat{F}_k(x)$ are independent Gaussian processes generated from the distribution $\mathrm{GP}(\widehat{m}, \widehat{C})$.

As an illustration, we discuss a method where certain engineering expertise can be incorporated in statistical learning. Often, it is known that the relationship between input and output of some technical system has to fulfill certain shape constraints. For example, in the framework of the general regression model (2), the regression function $m(x)$ is known to be positive, convex, or monotone or cannot exceed certain bounds. Here, we consider the case where the output Y_j is, up to the usually small observational error ε_j, an increasing continuous function of some components of the input X_j. An example is the press hardening of sheet metal described in Sect. 1.1 where the hardness is known to be an increasing sigmoid function of T_f, an increasing concave function of t_q, and a decreasing, convex function of t_h (compare [54]).

We start with the simple case of one-dimensional inputs and additionally assume that the X_j are equidistant in a fixed interval, say [0,1]. The described method and its theory are, however, applicable to nonequidistant or random X_j as well as to d-dimensional, $d > 1$, inputs. For many regression methods, there are variants called *isotonic* or *monotone regression* that take monotonicity constraints into account. For Gaussian process regression, such an approach has been investigated in [56].

Knowing that the random function $F(x)$ is increasing in x, its realizations are lying in the subset $\mathscr{M} = \mathscr{M}[0, 1]$ of increasing continuous functions on the interval [0,1]. Using standard GPR, the estimate of $F(x)$ would be $\widehat{m}(x)$, which usually is not increasing. A first idea would be to replace the function $\widehat{m}$ by its projection

$P\widehat{m}$ onto $\mathcal{M}$, where for functions f on [0,1], the function $Pf(x)$ is defined as the solution of the minimization problem

$$\int_0^1 (g(x) - f(x))^2 dx = \min_{g \in \mathcal{M}}!$$

If f is increasing, then $Pf = f$. If f is decreasing, then Pf is constant. If f is continuous, then Pf is continuous too. Moreover, the projection has a uniform contraction property for continuous functions on [0,1], which is useful for estimation:

$$\sup_{x \in [0,1]} |Pf(x) - Pg(x)| \leq \sup_{x \in [0,1]} |f(x) - g(x)|. \tag{27}$$

The *pooled adjacent violators algorithm* (PAVA), well-known from classical parametric isotonic regression, allows the fast numerical calculation of a good approximation of the projection $P\widehat{m}(x)$, where it only considers the values $\widehat{m}(X_j)$, $j = 1, \ldots, N$. Implementations of PAVA are available in common software packages, e.g., in R or MATLAB.

Now, **isotonic GPR** proceeds in the following manner:

Step 1: For given x and training set, generate M independent Gaussian paths $\widehat{F}_k(x), x \in [0, 1], k = 1, \ldots, M$, from the posterior distribution GP($\widehat{m}, \widehat{C}$).

Step 2: Project each $\widehat{F}_k(x)$ onto the set of monotone functions using the PAVA. Note that in this step, it is not necessary to use the original inputs as arguments for the $\widehat{F}_k(x)$, but any convenient set of values, e.g., equidistant ones even if the original input values were not. Use the monotone functions $P\widehat{F}_k(x), k = 1, \ldots, M$, for making inference based on Monte Carlo simulation.

For example, an estimate $\widehat{m}_P(X_{N+1})$ of the unknown monotone function $F(x)$ of interest at a new input X_{N+1} or, equivalently, a prediction of $Y_{N+1} = F(X_{N+1}) + \varepsilon_{N+1}$ is given by

$$\widehat{Y}_{N+1} = \widehat{m}_P(X_{N+1}) = \frac{1}{M} \sum_{k=1}^{M} P\widehat{F}_k(X_{N+1}).$$

Note that $\widehat{m}_P(x) \neq P\widehat{m}(x)$, as $\mathcal{M}$, is not a linear space. A 95% confidence interval for Y_{N+1} would be $[\hat{q}_\ell, \hat{q}_u]$ where $\hat{q}_\ell, \hat{q}_u$ are the 2.5%, resp., 97.5% sample quantiles of the sample $P\widehat{F}_1(X_{N+1}), \ldots, P\widehat{F}_M(X_{N+1})$.

This ad hoc approach to dealing with monotonicity constraints is justified by theory. From [56], it leads to approximations of the distribution of $F(x)$, which are almost optimal asymptotically, i.e., for $N \to \infty$. In practice, for large N, there will be no other estimate of the distribution of $F(x)$, which is noticeably better.

In addition to taking into account expert knowledge about monotonicity, isotonic GPR also leads to more precise estimates. The reason is the contraction property (27), which results in less variability of the $P\widehat{F}_k(x)$ than of the original $\widehat{F}_k(x)$.

In the general situation with d-dimensional inputs, expert knowledge may postulate monotonicity of $F(x)$ w.r.t. only **one** input coordinate, say the first one. Then the above procedure remains largely unchanged. In step 1, Gaussian processes on d-dimensional space are generated as in unconstrained GPR, and in step 2, the PAVA is only applied w.r.t. to the first coordinate for all fixed values of the other coordinates separately. To get enough function values, it is usually advisable to consider them on an equidistant grid for the application of the PAVA.

In the general case, where we have monotonicity w.r.t. all or at least more then one input coordinate, we have to apply an iterative version of the coordinate-wise PAVA. First, we apply it to the first coordinate to get function values that are monotone in this coordinate. Then we apply to this function the PAVA for the second coordinate and so on. Several passes over the coordinates are usually necessary. Lin and Dunson [56] give a detailed version of this algorithm for dimension $d = 2$; they report that it converged to a monotone solution in at most 20 iterations in the examples they have considered.

5 Linear Methods for Classification

As in the previous chapter for regression, we consider here classical linear methods for classification that form the basis of popular statistical learning methods, e.g., of support vector machines. The latter, in particular, depends on *linear discriminant analysis*, and we review here some concepts needed later on, in particular in Sect. 11. For a first reading, this brief chapter may be skipped.

We concentrate on classification based on independent, identically distributed training data (X_j, Y_j), $j = 1, \dots, N$, with d-dimensional feature vectors X_j and class labels $Y_j \in \{1, \dots, M\}$. The class probabilities given the predictor X_j are

$$p_\ell(x) = \text{pr}\{Y_j = \ell | X_j = x\}, \; \ell = 1, \dots, M.$$

Linear classifiers are optimal if there is some monotone transformation making the $p_\ell(x)$ linear in x. In that case, a classification rule is characterized by a partition of the feature space $\mathbb{R}^d$ into M subsets, where the decision boundaries, i.e., the boundaries of the subsets, are *hyperplanes*, i.e., the d-dimensional analog of planes in space and straight lines in the plane.

With the logistic regression model (compare Sect. 2.4), we already have discussed such a situation for the case $M = 2$. We consider the logistic model where

$$p_1(x) = \frac{\exp(b_0 + \langle b, x \rangle)}{1 + \exp(b_0 + \langle b, x \rangle)}, \quad p_2(x) = 1 - p_1(x) \qquad \text{with } \langle b, x \rangle = \sum_{i=1}^{d} b_i x_i,$$

where $b = (b_1, \ldots, b_d)$ and $\langle b, x\rangle$ is the scalar product between b and x. Considering the *logit transformation* $g(p) = \log(p/(1-p))$, we get that the **log odds**

$$\log \frac{p_1(x)}{p_2(x)} = b_0 + \langle b, x\rangle$$

are (affine) linear in x. The optimal decision boundary between the two subsets of the feature space $\mathbb{R}^d$, which correspond to the two classes, is the hyperplane $\{x; b_0 + \langle b, x\rangle = 0\}$. Here we tacitly assume that we use the *Bayes classifier* $c_B(x)$ (compare Sect. 2.4), i.e., we classify an object with feature vector x as belonging to class 1 if $p_1(x) > p_2(x)$.

As a linear regression model, e.g., polynomial regression, may contain input components that are nonlinear functions of the original features, we can also sometimes use linear classificators for genuine nonlinear situations by enlarging the feature space. If, e.g., we have two features X_{j1}, X_{j2} and we suspect that a good classifier also depends on quadratic functions, then we can augment the original features by the additional input components $X_{j1}^2, X_{j2}^2, X_{j1} \cdot X_{j2}$. Starting from the original two-dimensional feature space, we end with a five-dimensional one. If we find a good linear classifier in that space, we can transform it back to $\mathbb{R}^2$, and the hyperplane separating the two class subsets in $\mathbb{R}^5$ is mapped onto a nonlinear decision boundary in $\mathbb{R}^2$.

Consider, e.g., a situation where the two classes differ with respect to the size of the feature vector. We have $Y_j = +1$ if $||X_j|| > 1$ and $Y_j = -1$ if $||X_j|| < 1$, i.e., the optimal decision boundary is the boundary of the unit circle. Then, usually there will be no separating hyperplane, i.e., a straight line for which all objects in the training set belonging to class +1 lie on one side and all objects of class -1 on the other. If we augment the feature vectors by $X_{j1}^2, X_{j2}^2, X_{j1} \cdot X_{j2}$, then in five-dimensional space, both classes will be linearly separable, as with $Z_j = (X_{j1}, X_{j2}, X_{j1}^2, X_{j2}^2, X_{j1}X_{j2})$, we have $Y_j = +1$ if $\langle b, Z_j\rangle = ||X_j||^2 > 1$ for the weight vector $b = (0, 0, 1, 1, 0)$.

Figure 17 shows an example of such a dataset of total size $N = 50$ where 33 observations belong to class $+1$ (blue) and the rest to class -1 (red). On the left-hand side, only the two-dimensional data are shown, and it is obvious that there is no straight line separating the two classes. On the right-hand side, the decision boundary $\{x; ||x|| = 1\}$ (magenta) is added to the plot.

Linear classification in the original feature space is rarely optimal in practice (but sometimes good enough depending on the purpose), but mapping the X_j (non-linearly) into a higher-dimensional space may lead to an equivalent classification problem where linear classification is good or even optimal.

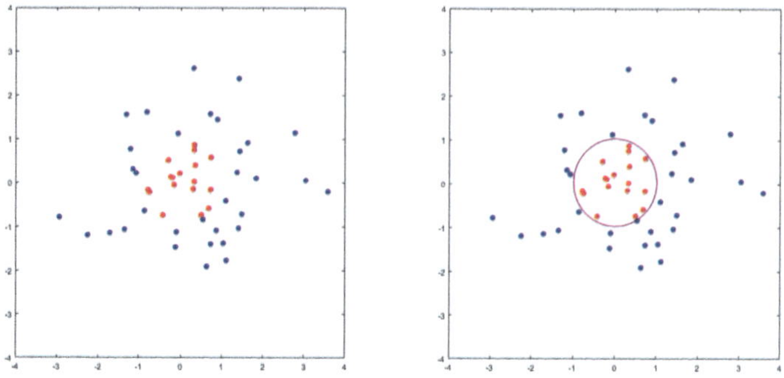

Fig. 17 Two-dimensional input data from two classes (blue resp. red) with, additionally, the circular decision boundary added on the right-hand side

5.1 Linear Discriminant Analysis

We consider the overall probability that an object belongs to class ℓ without considering the feature vector:

$$\pi_\ell = \text{pr}(Y_j = \ell),\ \ell = 1, \dots, M, \qquad \sum_{\ell=1}^{M} \pi_\ell = 1.$$

Assuming that the distribution of X_j is discrete, i.e., the feature vector can assume only a finite or (countably infinite) number of values, we get from Bayes' formula the relationship to the class probabilities given the feature value:

$$p_\ell(x) = \text{pr}\{Y_j = \ell | X_j = x\} = \frac{\text{pr}\{X_j = x | Y_j = \ell\}\pi_\ell}{\sum_{k=1}^{M} \text{pr}\{X_j = x | Y_j = k\}\pi_k}, \qquad \ell = 1, \dots, M.$$

For continuously distributed, e.g., Gaussian, feature vectors, a corresponding result for the probability densities can be shown to hold too:

$$p_\ell(x) = \frac{f_\ell(x)\pi_\ell}{\sum_{k=1}^{M} f_k(x)\pi_k} \tag{28}$$

where $f_\ell(x)$ is the conditional probability density of X_j given that the object belongs to class ℓ, i.e., $Y_j = \ell$.

We now assume that in each class ℓ, the feature vectors X_j are Gaussian with mean vectors μ_ℓ and $d \times d$ covariance matrices Σ_ℓ. From (1), for row vectors x, μ_ℓ, the density of X_j is for objects in class ℓ

$$f_\ell(x) = \frac{1}{(2\pi)^{d/2} \det \Sigma_\ell} \exp\Big(-\frac{1}{2}(x - \mu_\ell)\Sigma_\ell^{-1}(x - \mu_\ell)^T\Big). \tag{29}$$

In classical *linear discriminant analysis* (LDA), it is assumed that the classes differ only in terms of their means, and the random deviations from the respective means follow the same pattern in all classes, i.e., $\Sigma_\ell = \Sigma, \ell = 1, \ldots, M$. We consider the log odds for classes k, ℓ for given feature value x and get from (28)

$$\begin{aligned}\log \frac{p_k(x)}{p_\ell(x)} &= \log \frac{f_k(x)}{f_\ell(x)} + \log \frac{\pi_k}{\pi_\ell} \\ &= \log \frac{\pi_k}{\pi_\ell} - \frac{1}{2}(\mu_k + \mu_\ell)\Sigma^{-1}(\mu_k - \mu_\ell)^T + x\Sigma^{-1}(\mu_k - \mu_\ell)^T.\end{aligned}$$

Note that this expression is (affine) linear in x and all decision boundaries are hyperplanes.

Due to the ratio behind the Bayes classifier, we would prefer class k over class ℓ if $p_k(x) > p_\ell(x)$, i.e.,

$$\log \frac{p_k(x)}{p_\ell(x)} = \delta_k(x) - \delta_\ell(x) > 0.$$

where the linear **discriminant functions** $\delta_\ell(x)$ are given by

$$\delta_\ell(x) = x\Sigma^{-1}\mu_\ell^T - \frac{1}{2}\mu_\ell\Sigma^{-1}\mu_\ell^T + \log \pi_\ell, \;\; \ell = 1, \ldots, M,$$

and the decision boundaries, separating the classes, are given by the hyperplanes $\{x; \delta_k(x) = \delta_\ell(x)\}, k \neq \ell$. Finally, we use the Bayes classifier $\widehat{Y}_{N+1} = c_B(X_{N+1})$ for a new observation, which in linear discriminant analysis assumes the form

$$\widehat{Y}_{N+1} = \ell \;\text{ if }\; \delta_\ell(X_{N+1}) = \max_{k=1,\ldots,M} \delta_k(X_{N+1}).$$

As the discriminant functions depend on unknown parameters of the distribution, we need estimates. The unconditional class probabilities are estimated by the relative frequencies of occurrence

$$\hat{\pi}_\ell = \frac{N_\ell}{N} \quad \text{with} \quad N_\ell = \text{ number of data from class } \ell \text{ in the training set,}$$

and the Gaussian parameters are, as usual, estimated by the sample means resp. the unbiased version of the sample covariance matrix. For the latter, the data from all classes are pooled, as we have assumed a class independent Σ:

$$\hat{\mu}_\ell = \frac{1}{N_\ell}\sum_{j;Y_j=\ell} X_j, \; \ell = 1, \ldots, M, \;\; \widehat{\Sigma} = \frac{1}{N-M}\sum_{\ell=1}^{M}\sum_{j;Y_j=\ell}(X_j - \hat{\mu}_\ell)^T(X_j - \hat{\mu}_\ell).$$

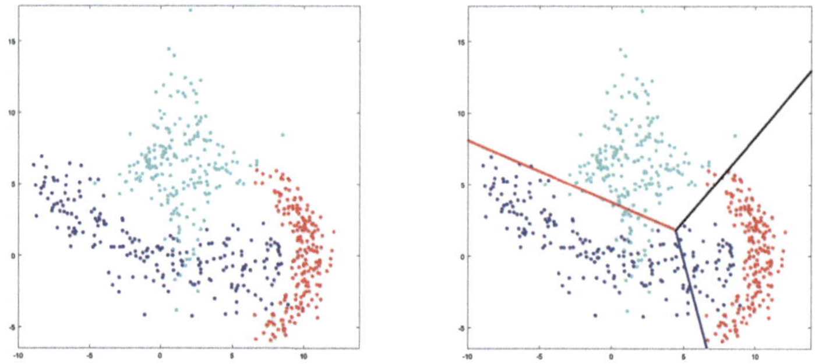

Fig. 18 Two-dimensional input data from three classes marked in different colors (left); data with decision boundaries from linear discriminant analysis (right)

As an example, we consider data from M = 3 classes (blue, red, and light blue), shown on the left-hand side of Fig. 18. The feature vectors X_j are two-dimensional. Applying linear discriminant analysis, we derive three hyperplanes, i.e., straight lines in dimension 2, which separate classes 1, 2 (blue) resp. 1, 3 (red) resp. 2, 3 (black) as well as possible. The result is shown on the right-hand side of Fig. 18 where we have only plotted those parts of the linear decision boundaries that are relevant for the final classification rule. The two-dimensional feature space is partitioned into $M = 3$ parts, where a new object with feature vector X_{N+1} from the upper left part is predicted as belonging to class $Y_{N+1} = 3$. Analogously, a feature vector in the lower left, resp. the right part, is classified as $Y_{N+1} = 1$, resp. $Y_{N+1} = 2$.

From looking at the training data, it is obvious that a linear classification rule is not ideal. The scatterplots corresponding to the classes 1 and 2 show an obvious curvature, and though the areas covered by them are almost disjoint, they cannot be separated by a straight line.

It is not always reasonable to assume equal covariance structure in all classes. In particular, this assumption is highly implausible for the data of Fig. 18. In such a case where the $\Sigma_\ell, \ell = 1, \ldots, M$, are arbitrary, the above approach leads to **quadratic discriminant analysis** (QDA), which uses quadratic instead of linear discriminant functions

$$\delta_\ell(x) = -\frac{1}{2} \log \det \Sigma_\ell - \frac{1}{2}(x - \mu_\ell)\Sigma_\ell^{-1}(x - \mu_\ell)^T + \log \pi_\ell, \ \ell = 1, \ldots, M.$$

Decision boundaries are no longer hyperplanes but given as solutions of quadratic equations $\delta_k(x) = \delta_\ell(x), k \neq \ell$. π_ℓ, μ_ℓ are estimated as in LDA. The different covariance matrices are estimated by their sample versions

$$\widehat{\Sigma}_\ell = \frac{1}{N_\ell - 1} \sum_{j;Y_j=\ell} (X_j - \hat{\mu}_\ell)^T (X_j - \hat{\mu}_\ell), \ \ell = 1, \ldots, M.$$

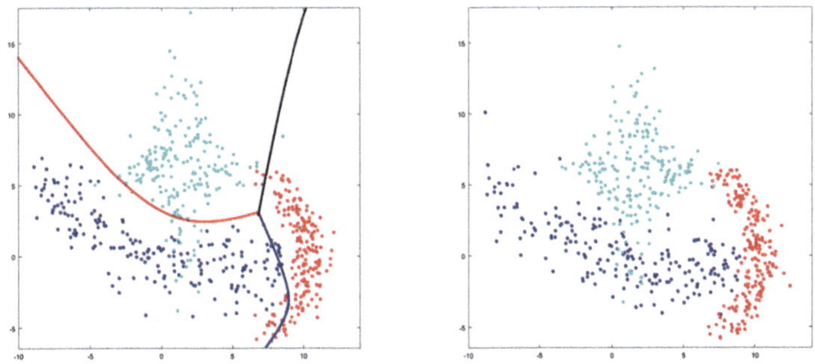

Fig. 19 Training data with decision boundaries from quadratic discriminant analysis (left); independent validation data (right)

Table 3 Misclassification rate in training and test sample of Figs. 18 and 19

	LDA	QDA
Training sample	13.8 %	7.7 %
Validation sample	14.3 %	7.7 %

Applying quadratic discriminant analysis to the data of Fig. 18, we get the curved decision boundaries shown on the left-hand side of Fig. 19, where again we only show those parts of the curves that are relevant for the resulting classification and that again partition the feature space into three regions. Visually, QDA performs better than LDA, as the curved decision boundaries can adapt better to the curvature of the data from classes 1 and 2.

Objectively, the performance of QDA is really considerably better for this dataset. Table 3 shows the percentages of misclassifications in the training set in the first line where the error rate is almost halved. Of course, this may be an effect of overfit, as QDA fits more parameters to the data (by estimating three instead of only one covariance matrices) than LDA. To check this, we applied the classification rule also to an independent validation set of the same size and composition and generated in the same manner as the training set. The second line of Table 3 shows the percentages of misclassifications for these data, and due to the independence and the law of large numbers, this is a valid approximation to the true probability for misclassification.

The error rate of LDA is a bit high but still reasonable in view of the obvious violation of the assumptions underlying the LDA methods. QDA with an error rate well below 10 % is quite acceptable but may still be not good enough for some applications. However, we cannot expect other methods to perform much better as long as we stick only to the two features X_{j1}, X_{j2}. Whereas class 2 (red) is rather well separated from the other two classes, there is a considerable overlap of feature vectors from the other two classes. To resolve this problem, we would have to take into account additional features, which, in practical applications, would mean to get additional data from a different source.

In practice, QDA and even LDA frequently perform very well even compared to more complex machine learning methods in spite of the fact that data, as in the example, are often rather not Gaussian (which is the basis of the classical discriminant analysis approach) and covariance matrices Σ_ℓ are not all equal. The reason for this observation is that sample sizes in many applications are not large enough to handle more complicated than linear or quadratic decision boundaries, and QDA/LDA give stable results, i.e., they (in particular LDA with its few parameters) result in rather small variability for the price of larger model misspecification (bias). For larger samples, the more advanced support vector machines, which retain some of the ideas of discriminant analysis, often perform better. In particular, they use separating hyperplanes, though not in the original feature space. Therefore, we have a closer look at those in Sect. 5.3.

5.2 Logistic Regression

We consider the logistic model, discussed briefly at the start of this chapter for $M = 2$, now for arbitrary class numbers with d-dimensional weight vectors $b_1, \ldots, b_{M-1}$:

$$p_\ell(x) = \frac{\exp(b_{0\ell} + \langle b_\ell, x\rangle)}{1 + \sum_{k=1}^{M-1} \exp(b_{0k} + \langle b_k, x\rangle)}, \quad \ell = 1, \ldots, M-1,$$

$$p_M(x) = 1 - \sum_{k=1}^{M-1} p_k(x) = \frac{1}{1 + \sum_{k=1}^{M-1} \exp(b_{0k} + \langle b_k, x\rangle)}.$$

Again, the log odds are linear in x

$$\log \frac{p_\ell(x)}{p_M(x)} = b_{0\ell} + \langle b_\ell, x\rangle, \; \ell = 1, \ldots, M-1. \tag{30}$$

We combine all parameters $b_{0\ell}, b_\ell, \ell = 1, \ldots, M-1$, into a parameter vector ϑ of dimension $(M-1)(d+1)$ and stress the dependence of the class probabilities on those parameters by writing $p_\ell(x) = p_\ell(x; \vartheta)$.

We estimate the model parameters in ϑ by a maximum likelihood method, i.e., we maximize the conditional log likelihood of the observed class labels $Y_1, \ldots, Y_N$ given the corresponding feature values $X_1, \ldots, X_N$, which, using the independence of the training data (X_j, Y_j), is

$$L(\vartheta) = \sum_{j=1}^{N} \log p_{Y_j}(X_j; \vartheta).$$

The resulting maximum likelihood estimate $\hat{\vartheta}$ has to be calculated numerically. An alternative to generic optimization algorithms is the *iteratively weighted least-squares* algorithm, which exploits the particular structure of the target function $L(\vartheta)$ to be maximized [12].

Formally, the models of linear discriminant analysis (LDA) and logistic regression look identical, as for both, the log odds (30) are linear in x. However, the parameters $b_{0\ell}, b_\ell$ are estimated differently resulting in different practical classifications rules. Logistic regression is more general than LDA as there are no restrictions about the distribution of the feature vectors X_j, whereas LDA assumes that they are a mixture of Gaussians (29), all with the same covariance. Hence, if a new feature vector X_{N+1} is observed without knowing the class, its probability density $f(x)$ is given by $f(x) = \sum_{\ell=1}^{M} \pi_\ell f_\ell(x)$ with $\Sigma_\ell = \Sigma, \ell = 1, \ldots, M$.

Frequently, LDA (even when misspecified, i.e., the assumptions about Gaussianity and/or equal covariances are not rigorously fulfilled) and logistic regression give similar results. The latter is, however, more robust in case of more severe violations of the LDA assumptions.

5.3 Separating Hyperplanes

Linear discriminant analysis and logistic regression result in linear decision rules with hyperplanes as decision boundaries. In this section, we discuss procedures for finding hyperplanes that, in the training set, separate observations belonging to different classes as well as possible.

A **hyperplane** or *affine set* in d-dimensional space is given by

$$H = \{x; b_0 + \langle b, x \rangle = 0\}$$

for some real number b_0 and d-dimensional vector b, which, due to the arbitrariness of scaling, we may assume to be a unit vector, i.e., $||b|| = 1$. Then,

(a) $\langle b, x_1 - x_2 \rangle = 0$ for all x_1, x_2 in H, i.e., b is orthogonal to the surface of H.
(b) For any d-dimensional vector x and arbitrary x_0 in H:

$$\langle b, x - x_0 \rangle \quad \text{is the signed distance of } x \text{ to } H. \tag{31}$$

To keep the exposition simple, we focus in the rest of the section on the special case with only $M = 2$ classes, which, for convenience, are labeled as $+1$ resp. -1. For a given training set (X_j, Y_j), $j = 1, \ldots, N$, a hyperplane $H = \{x; b_0 + \langle b, x \rangle = 0\}$ is called a **separating hyperplane**, if for x_0 in H

$$\langle b, X_j - x_0 \rangle > 0 \text{ for all } j \text{ with } Y_j = +1, \text{ and}$$
$$\langle b, X_j - x_0 \rangle < 0 \text{ for all } j \text{ with } Y_j = -1.$$

i.e., we have

$$Y_j = \operatorname{sgn}\langle b, X_j - x_0\rangle = \operatorname{sgn}(\langle b, X_j\rangle + b_0).$$

The existence of a hyperplane, which perfectly separates the objects in the training set belonging to one class from those belonging to the other class, is an ideal situation that is not guaranteed to happen. If, however, it exists, then there are many similar hyperplanes that also separate the two classes in the training set. In the following subsection, we present an algorithm that searches for a separating hyperplane. Then, in case where separating hyperplanes exist, we discuss how to characterize a unique optimal one.

5.3.1 The Perceptron Learning Algorithm

The perceptron of [68] was originally a mathematical model for a biological neuron, which assumes two different states depending on inputs from other cells (compare Sect. 8.1). This can be formulated as a classification task with $M = 2$ classes. The perceptron learning algorithm or *Widrow-Hoff algorithm* describes how the neuron learns a perfect classification from training data, provided there is one. The algorithm tries to find a separating hyperplane in input space by minimizing

$$D(b, b_0) = -\sum_{j\in\mathcal{M}} Y_j\big(\langle b, X_j\rangle + b_0\big),$$

where $\mathcal{M}$ denotes the indices of misclassified data, i.e., for which either $Y_j = +1, \langle b, X_j\rangle + b_0 < 0$ or $Y_j = -1, \langle b, X_j\rangle + b_0 > 0$. From (31) above, $D(b, b_0)$ is the total distance of misclassified points to the corresponding decision boundary $H(b, b_0) = \{x; b_0 + \langle b, x\rangle = 0\}$.

The gradient of D w.r.t. to b and the partial derivative w.r.t. b_0 are given as

$$\operatorname{grad}_b D(b, b_0) = -\sum_{j\in\mathcal{M}} Y_j X_j, \qquad \frac{\partial D(b, b_0)}{\partial b_0} = -\sum_{j\in\mathcal{M}} Y_j.$$

We could use these derivatives for solving the minimization problem by a common gradient descent algorithm where in each iteration we shift the current value of the parameters b, b_0 a bit into the direction where D becomes smaller. The Widrow-Hoff algorithm follows a similar philosophy but does not consider the training data simultaneously. Instead, it cycles through the training sample and iteratively updates the current value of the parameter by

$$b^{\text{new}} = b + \eta Y_j X_j 1_{\{j\in\mathcal{M}(b,b_0)\}}, \qquad b_0^{\text{new}} = b_0 + \eta Y_j 1_{\{j\in\mathcal{M}(b,b_0)\}}.$$

$\mathcal{M}(b, b_0)$ denotes the set of indices of those training data that are misclassified if we use the parameters b, b_0. $\eta > 0$ is the usual relaxation parameter of gradient descent methods. In each step, the perceptron learning algorithm does not change the current parameter value if it classifies the jth observation correctly. After cycling through all $j = 1, \ldots, N$, the algorithm continues from $j = 1$ again. It can be shown that the iterative procedure converges in a finite number of steps to the parameters of a separating hyperplane if one exists.

The Widrow-Hoff algorithm has a few drawbacks. If there is a separating hyperplane, the number of iterations until convergence is finite but may be very large. If, otherwise, in the training set, the two classes cannot be separated by a hyperplane, the algorithm does not converge and continues to cycle through the data over and over again, which may be hard to detect.

5.3.2 Optimal Separating Hyperplanes

Let $H = \{x;\, \langle b, x\rangle + b_0 = 0\}$ be a separating hyperplane with $||b|| = 1$. Following [14], we call it **optimal** if it maximizes the distance to the closest point in the training set.

This notion has two advantages: it characterizes one separating hyperplane in a unique manner, and the corresponding classification rule can be expected to generalize better, i.e., showing a better performance on an independent test data set.

Introducing the **margin** of width $2C$ of a hyperplane H:

$$\{z;\, z = x + tb \text{ for some } x \in H,\, |t| < C\},$$

the problem of finding the optimal separating hyperplane is equivalent to maximizing the margin, i.e., to solve:

$$\text{find } \max_{b_0, b, ||b||=1} C \quad \text{under the constraints} \quad Y_j(\langle b, X_j\rangle + b_0) \geq C,\, j = 1, \ldots, N. \tag{32}$$

Note that from (31), $Y_j(\langle b, X_j\rangle + b_0)$ is the distance between X_j and the hyperplane H. Hence, we are trying to find a separating hyperplane such that the margin of width C does not contain any object from the training set and such that width of this region is as large as possible.

To solve this maximization problem, it is advantageous to work with arbitrary, not necessarily unit vectors b (and rescaling b_0 accordingly), i.e., the constraints change to

$$Y_j(\langle b, X_j\rangle + b_0) \geq C||b||,\, j = 1, \ldots, N.$$

Then, b, b_0 are unique only up to a factor, which in retrospect we can choose, such that $||b|| = C^{-1}$, and then (32) becomes

$$\frac{1}{2}||b||^2 = \min_{b,b_0}! \qquad \text{under the constraints} \qquad Y_j(\langle b, X_j\rangle + b_0) \geq 1,\ j = 1, \ldots, N.$$

This is a convex optimization problem (with quadratic target function and linear constraints). Using Lagrange multipliers $\alpha_1, \ldots, \alpha_N \geq 0$, we can rewrite it as

$$L(b, b_0) = \frac{1}{2}||b||^2 - \sum_{j=1}^{N} \alpha_j \big(Y_j(\langle b, X_j\rangle + b_0) - 1\big) = \min_{b,b_0}! \tag{33}$$

with $\alpha_j = 0$ or $Y_j(\langle b, X_j\rangle + b_0) = 1$. Setting the partial derivatives of L w.r.t. $b_k, k = 0, \ldots, d$, to 0, we get

$$b = \sum_{j=1}^{N} \alpha_j Y_j X_j \qquad \text{and} \qquad 0 = \sum_{j=1}^{N} \alpha_j Y_j. \tag{34}$$

Plugging these relations into the optimization problem (33), we get its dual form

$$\sum_{j=1}^{N} \alpha_j - \frac{1}{2} \sum_{i,j=1}^{N} \alpha_i \alpha_j Y_i Y_j \langle X_i, X_j\rangle = \max_{\alpha_1,\ldots,\alpha_N \geq 0}!$$

$$\text{under the constraint} \qquad \sum_{j=1}^{N} \alpha_j Y_j = 0.$$

This is a convex optimization problem in standard form and can be solved numerically with standard software. In particular, if $\hat{b}, \hat{b}_0$ denote the solutions of (33), we have:

(a) If $\alpha_j > 0$, then $Y_j(\langle \hat{b}, X_j\rangle + \hat{b}_0) = 1$, i.e., X_j lies on the boundary of the margin around the separating hyperplane. Such X_j are called **support points** or **support vectors**.
(b) If $Y_j(\langle \hat{b}, X_j\rangle + \hat{b}_0) > 1$, then $\alpha_j = 0$. In this case, X_j lies outside of the margin as its distance to the hyperplane is larger than 1.

From (34), we see that $\hat{b}, \hat{b}_0$ only depends on the support points, which leads to a considerable reduction in computational load in relation to sample size N. However, one should not forget that, in finding the support points, all data have to be considered.

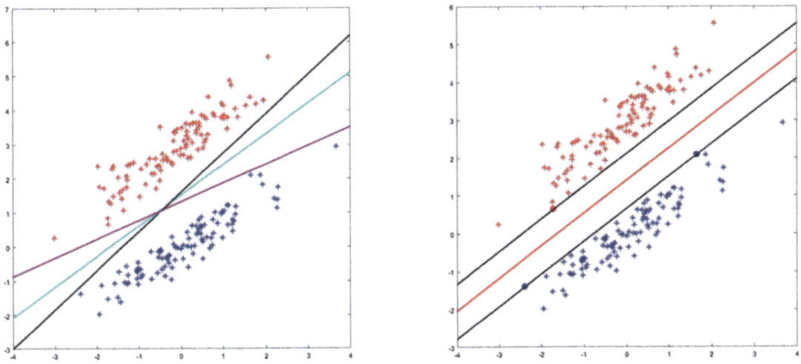

Fig. 20 Data from two classes and three separating hyperplanes (left); data, optimal separating hyperplane (red), margin boundaries (black), and three support vectors (marked as larger dots)

As the class indicators are +1 or −1, for the optimal separating hyperplane with parameters $\hat{b}, \hat{b}_0$, the classification rule for a new feature vector X_{N+1} has the simple representation

$$\widehat{Y}_{N+1} = \hat{c}(X_{N+1}) \qquad \text{with} \qquad \hat{c}(x) = \operatorname{sgn}(\langle \hat{b}, x \rangle + \hat{b}_0).$$

As an example, we consider two-dimensional feature vectors from $M = 2$ classes that can be separated by hyperplanes, i.e., lines. Figure 20 shows the data and three different separating hyperplanes on the left. The right-hand side shows the optimal separating hyperplane (red) together with the boundaries of the corresponding maximal margin (black). Only three out of $N = 200$ feature vectors are support points, lying on the margin boundaries and marked as larger dots. They completely determine $\hat{b}, \hat{b}_0$ and, hence, the classification rule.

6 Sieve Estimates: A General Approach to Machine Learning

In this chapter, we mainly consider the nonparametric regression model

$$Y_j = m(X_j) + \varepsilon_j, \ \ j = 1, \ldots, N,$$

with independent, identically distributed data (X_j, Y_j), $j = 1, \ldots, N$, and, for sake of simplicity, $\varepsilon_1, \ldots, \varepsilon_N$ independent, identically distributed residuals with mean 0 and variance $\sigma_\varepsilon^2 < \infty$, independent of $X_1, \ldots, X_N$. The outputs Y_j are real valued, whereas input vectors X_j are d-dimensional. The prediction function $m : \mathbb{R}^d \to \mathbb{R}$ is rather arbitrary; depending on the context, we only impose certain rather weak regularity assumptions.

To enforce a reasonable function estimate for $m(x)$, which achieves a balance between overfit and underfit, we do not apply regularization as in Sects. 3.2 or 3.3, but instead we restrict the estimate to be an element of a parametrized, not-too-complicated function class $\mathscr{F}_N$, where the complexity of the latter may increase with sample size N and is controlled by a hyperparameter. Such function estimates are called **sieve estimates** (compare [35]) and include neural network regression, wavelet regression, and many other popular approaches from machine learning.

As a first simple example, we have a look at orthogonal series expansions where the one based on Fourier transforms is still popular and frequently performs quite well in many engineering applications. Then, we present the general framework of sieve estimates to illustrate the common theme and properties of many machine learning algorithms like tree-based methods (compare Sect. 7) or neural networks (compare Sect. 8).

6.1 Orthogonal Series Expansions for Regression

For the sake of simplicity, we only consider one-dimensional inputs X_j in this section. Recall the simple example of **polynomial regression** (3), where we have fitted a polynomial of degree p to the data. This hyperparameter controls the number of parameters $b_k(p), k = 0, \ldots, p$ and hence the complexity of the function $m_p(x; b(p))$. The latter is a linear expansion in the powers $x^k, k = 0, \ldots, p$, of x with coefficients $b_k(p)$, which are calculated from the training set by a least-squares approach, i.e., by minimizing the residual sum-of-squares RSS. This results in a sieve estimate of $m(x)$, where the function class $\mathscr{F}_N$ is the class of all polynomials of degree p, where the latter may depend on N.

Usually, we get more numerically and statistically stable results from polynomial regression, if we replace the powers x^k by a system of orthogonal polynomials $g_k(x)$ of degree k where $g_k, g_\ell, k \neq \ell$, are orthogonal with respect to some scalar product in a space of functions, e.g., L^2 with the scalar product (35) introduced below. $\mathscr{F}_N$ remains unchanged by a switch of the generating system of functions but now is represented by functions

$$m_p(x; b(p)) = \sum_{k=0}^{p} b_k(p) g_k(x) \quad \text{with } p = p_N \text{ depending on } N.$$

The family of polynomial function classes $\mathscr{F}_N, N \geq 1$, has a *universal approximation property* if $p_N \to \infty$ for increasing sample size N, i.e., rather general functions $m(x)$ may be arbitrarily well approximated by functions from $\mathscr{F}_N$ for large enough N. This fact is implied by the well-known Taylor's theorem for infinitely often differentiable functions and by a theorem of Weierstrass for the much larger class of continuous functions $m(x)$.

Other sets of orthogonal functions may be used too for this kind of estimation procedure. We have a closer look at the widely used **Fourier series** expansions. For ease of presentation, we consider generally complex-valued functions $m(x)$ on the interval $[0, 2\pi]$, but any other finite interval may be used similarly. We assume that the functions under considerations are square-integrable, i.e.,

$$||m||^2 = \frac{1}{2\pi} \int_0^{2\pi} |m(x)|^2 dx < \infty.$$

We call the space of those functions $L^2[0, 2\pi]$, which we abbreviate in this chapter to L^2. For functions in L^2, we consider the scalar product

$$\langle m, f \rangle = \frac{1}{2\pi} \int_0^{2\pi} m(x)\overline{f(x)}\, dx, \tag{35}$$

where $\overline{z}$ denotes the complex conjugate of the complex number z. A well-known basis of L^2, which is orthogonal w.r.t. (35) and whose elements are scaled to $||e_n|| = 1$, is given by the trigonometric functions or complex exponentials

$$e_n(x) = e^{inx} = \cos(nx) + i \sin(nx), \qquad \text{for integer } n, -\infty < n < \infty.$$

Any $f \in L^2$ has a representation as a *Fourier series*

$$m(x) = \sum_{n=-\infty}^{\infty} c_n e^{inx} \tag{36}$$

with *Fourier coefficients*

$$c_n = \langle m, e_n \rangle = \frac{1}{2\pi} \int_0^{2\pi} m(x) e^{-inx} dx.$$

Note that for real-valued function $m(x)$, the Fourier coefficients satisfy $c_{-n} = \overline{c}_n$.

We consider the function class $\mathscr{F}_N$ of *trigonometric polynomials* of degree M

$$m_M(x; c(M)) = \sum_{k=-M}^{M} c_k(M) e^{ikx} \quad \text{with } M = M_N \text{ depending on } N.$$

These function classes again have a *universal approximation property*, if $M_N \to \infty$ for increasing N, as any function $m(x)$ in L^2 is the limit of

$$\sum_{n=-m}^{m} c_n e^{inx}$$

for increasing m. We consider the regression model with fixed equidistant input values $X_j = x_j$ on $[0, 2\pi]$, i.e.,

$$Y_j = m(x_j) + \varepsilon_j\,,\ \ x_j = \frac{2\pi j}{N}\,,\quad j = 1, \ldots, N.$$

with independent, identically distributed zero-mean residuals ε_j with finite variance σ_ε^2. We want to estimate $m(x),\ 0 \le x \le 2\pi$. Assuming that the regression function lies in L^2, we use its Fourier series expansion and note that the Fourier coefficients satisfy

$$c_k = \langle m, e_k\rangle = \frac{1}{2\pi}\int_0^{2\pi} m(x)e^{-ikx}dx \approx \frac{1}{N}\sum_{j=1}^{N} m(x_j)e^{-ikx_j}\,,$$

approximating the integral by a Riemann sum.

As E $Y_j = m(x_j)$, we could try to replace $m(x_j)$ by Y_j in the above formula to get an estimate of c_k. This leads to $\hat{c}_k = D_N(\omega_k)$, where the latter results from the **discrete Fourier transform** (DFT) of the output data

$$\hat{c}_k = D_N(\omega_k) = \frac{1}{N}\sum_{j=1}^{N} Y_j e^{-ij\omega_k}\,, \qquad \omega_k = \frac{2\pi k}{N}.$$

The $D_N(\omega_k)$ are complex-valued but satisfy certain symmetry relationships as $\hat{c}_{N-k} = \bar{\hat{c}}_k = \hat{c}_{-k}$ for $0 \le k \le \frac{N-1}{2}$. Therefore, they are completely determined by only N real numbers. The discrete Fourier transform is a one-to-one transformation of the data: $Y_1, \ldots, Y_N \ \longleftrightarrow\ \hat{c}_1, \ldots, \hat{c}_N$. Using the inverse discrete Fourier transform, we can reconstruct the data Y_j from the coefficients $\hat{c}_k$, i.e., we have

$$Y_j = \sum_{|k|\le (N-1)/2} \hat{c}_k e^{ikx_j}\,, \quad j = 1, \ldots, N. \tag{37}$$

For $M = M_N$, we consider the sieve estimate of the regression function

$$\widehat{m}_N(x) = \sum_{k=-M}^{M} \hat{c}_k e^{ikx}. \tag{38}$$

For, e.g., odd N and $M_N = (N-1)/2$, we have from the inverse discrete Fourier transform that $\widehat{m}_N(x_j) = Y_j$ for all j, i.e., the estimate provides a perfect fit to the noisy data and, hence, an undesirable massive overfit. If we, however, choose $M = M_N \to \infty$ such that $M^\gamma/N \to 0$ for some suitable $0 < \gamma < 1$, then under appropriate assumptions on $m(x), \varepsilon_j$, the Fourier regression estimate $\widehat{m}_N(x)$ will converge to $m(x)$. As usual, we have to somehow select a suitable value for the hyperparameter M_N of this sieve method.

The feasibility of the Fourier approach in function estimation has a nice interpretation if we look at the problem from the viewpoint of **signal analysis**. We interpret the output data Y_j as observations of a signal $m(x_j)$, observed at equidistant points in time, say $j = 1, \dots, N$, and contaminated by noise ε_j. Equation (37) is interpreted as a decomposition of the noisy signal into its harmonic components, which, for real-valued data, are essentially cosine functions with frequencies $2\pi k/N$, $0 \le k \le (N-1)/2$, using the particular form of the x_j. In (38), we truncate the sum (37) at M, which is much smaller than $(N-1)/2$, i.e., we remove the high-frequency components from the data Y_j.

Due to the additive structure of our model, a Fourier coefficient of the observed signal $Y_1, \dots, Y_N$ is the sum of the corresponding Fourier coefficients of the smooth signal and of the noise. The smooth signal, represented by the smooth function $m(x)$, consists mainly of low-frequency components. Its Fourier coefficients at higher frequencies $2\pi k/N$ are negligible. Due to the independence of the noise variables ε_j, their Fourier coefficients are not so large but equally spread out over the whole frequency spectrum. By deleting only the high-frequency components, we essentially remove a major part of the noise from the data, whereas the impact on the smooth signal is not worth mentioning. Hence, a Fourier regression may be interpreted as denoising the signal $Y_1, \dots, Y_N$ such that the resulting smoothed signal $\widehat{m}_M(x_1), \dots, \widehat{m}_M(x_N)$ is much closer to the function $m(x)$ of interest.

A slightly more general approach uses some weights $w_k \ge 0, k = 0, \dots, M$, which decrease with k. Setting $w_{-k} = w_k$, the resulting estimate based on a *windowed Fourier transform* is

$$\widehat{m}_N(x; w) = \sum_{k=-M}^{M} w_k \hat{c}_k e^{ikx}.$$

This approach avoids the sudden change between frequency components retained to 100 percent and those which are completely deleted.

In orthogonal series expansion, a further alternative to truncating the expansion with estimated coefficients at M is thresholding, which already has been mentioned at the end of Sect. 3.1. Here, components are not removed w.r.t. to their index k but with respect to the size of the coefficients, i.e., of $|\hat{c}_k|$ for Fourier regression. A threshold $\tau > 0$ is chosen as the hyperparameter of the method, and then we consider either **hard-thresholding**

$$\hat{c}_k^* = \hat{c}_k, \quad \text{if } |\hat{c}_k| > \tau, \quad \text{and } \hat{c}_k^* = 0, \text{ else.}$$

or **soft-thresholding**

$$\hat{c}_k^* = \hat{c}_k \cdot \frac{|\hat{c}_k| - \tau}{|\hat{c}_k|}, \quad \text{if } |\hat{c}_k| > \tau, \quad \text{and } \hat{c}_k^* = 0, \text{ else.}$$

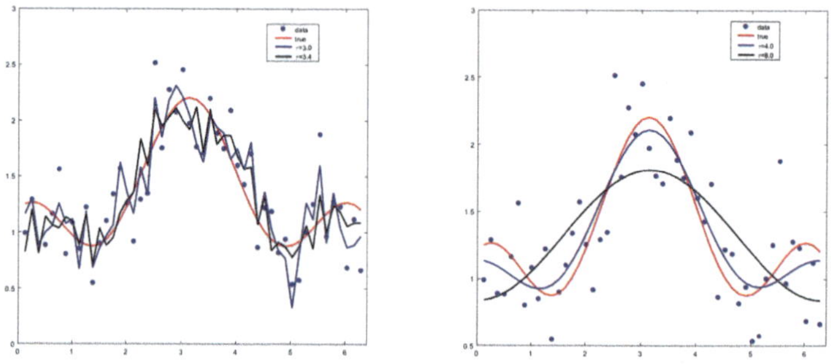

Fig. 21 Orthogonal series expansions with hard-thresholding and increasing τ

The resulting estimate is

$$\widehat{m}_N(x;\tau) = \sum_{|k| \le (N-1)/2} \hat{c}_k^* e^{ikx}.$$

If $\tau \to 0$ for increasing N with an appropriate rate, this estimate will also converge to $m(x)$. Usually, it will be much sparser, i.e., containing less parameters and frequency components, than a comparable estimate (38).

Figure 21 shows the result of estimating a function $m(x)$ (red) by a Fourier regression estimate $\widehat{m}_N(x;\tau)$ from $N = 50$ observations (blue) where the regularization is achieved by hard-thresholding. Choosing $\tau = 0$ leads to exact interpolation as expected. For $\tau = 3$ and $\tau = 3.4$ (blue resp. black on the left-hand side), we still have a considerable overfit. On the right-hand side, the effect of larger thresholds $\tau = 4$ (blue) and $\tau = 8$ (black) is shown where the former results in a rather good estimate, whereas the latter is underfitted and does not recover the basic shape of the underlying $m(x)$ well. Note that, e.g., between $\tau = 4$ and $\tau = 7$, the estimates $\widehat{m}_N(x;\tau)$ are identical as there are no $|\hat{c}_k|$ in this range.

Fourier regression works well for smooth functions, but if $m(x)$ has irregular local features like jumps or sharp peaks, then its Fourier series converges very slowly. We would need many terms for a decent approximation by a finite trigonometric polynomial, which, due to the large number of parameters, are prone to overfit in estimation from noisy data. This problem led to attempts of a "local" Fourier analysis where the basis functions are not only specified by frequency (k in the above example) but also by their location. Bases of **wavelets** have this property, and they provide a very flexible generalization to Fourier analysis. A typical wavelet basis is of the form $\{\phi_\ell(x), \psi_{k\ell}(x), k \ge 0, -\infty < \ell < \infty\}$, where k denotes frequency and ℓ location, i.e., the corresponding function is essentially concentrated in a neighborhood of the location, denoted by ℓ. There are many wavelet bases, but for those typically used, the $\psi_{k\ell}(x)$ look like irregular wave packets. For the construction of such bases, we refer, e.g., to [17] (compare also Section 5.9 of [44]).

Wavelet regression proceeds in a similar manner as Fourier regression, using a training set (x_j, Y_j), $j = 1, \dots, N$. Analogously to the Fourier series (36), square integrable $m(x)$ allows for a wavelet expansion

$$m(x) = \sum_{\ell=-\infty}^{\infty} c_\ell \phi_\ell(x) + \sum_{k=0}^{\infty} \sum_{\ell=-\infty}^{\infty} d_{k\ell}\, \psi_{k\ell}(x)$$

with $c_\ell = \langle m, \phi_\ell \rangle$, $d_{k\ell} = \langle m, \psi_{k\ell} \rangle$. These coefficients can be estimated by

$$\hat{c}_\ell = \frac{1}{N} \sum_{j=1}^{N} Y_j \phi_\ell(x_j), \qquad \hat{d}_{k\ell} = \frac{1}{N} \sum_{j=1}^{N} Y_j \psi_{k\ell}(x_j),$$

using the fast wavelet transform (FWT) for less computational load. Then, we apply, e.g., hard-thresholding and set $\hat{d}^*_{k\ell} = \hat{d}_{k\ell}\, 1_{[\tau,\infty)}(|\hat{d}_{k\ell}|)$, i.e., we only keep the large coefficients. The final wavelet estimate of $m(x)$ is

$$\widehat{m}_N(x; \tau) = \sum_{\ell} \hat{c}_\ell \phi_\ell(x) + \sum_{k,\ell} \hat{d}^*_{k\ell} \psi_{k\ell}(x).$$

Note that we do not have to threshold the coefficients of the basis functions $\phi_\ell(x)$ as those typically have bounded and disjoint or at least almost disjoint supports such that, for given x, almost all of these terms vanish anyhow.

For the choice of the hyperparameter τ, there is the well-known recommendation by Donoho and Johnstone [19] to use a universal threshold, which, in our case, is $\tau_u = \sigma_\varepsilon \sqrt{2 \log N / N}$, where σ_ε^2 denotes the variance of the residuals ε_j as usual and has to be estimated from the training set.

6.2 General Sieve Estimates

To introduce sieve estimates in a general context, we consider the general regression model (2). We additionally assume that the regression function is square-integrable w.r.t. the common distribution of the input variables X_j, i.e., $\mathrm{E}\, m^2(X_1) < \infty$. As usual, we write (X_{N+1}, Y_{N+1}) for a generic random variable with the same distribution but independent of the data (X_j, Y_j), $j = 1, \dots, N$, in the training set.

From (2), we see that $m(x)$ is the conditional expectation of Y_j given that we know $X_j = x$. This immediately implies

$$\mathrm{E}\,(Y_{N+1} - m(X_{N+1}))^2 = \min_f \mathrm{E}\,(Y_{N+1} - f(X_{N+1}))^2,$$

where the minimum runs over all measurable real-valued functions f on the d-dimensional input space. Hence, $m(x)$ minimizes a **risk** $R(f)$, which is defined as the mean-squared prediction error E $(Y_{N+1} - f(X_{N+1}))^2$. Note that, in contrast to (6), we consider the expectation of the squared prediction error not only for a fixed $X_{N+1} = x$ but with respect to the randomness of Y_{N+1} and of X_{N+1}.

The idea behind **empirical risk minimization** is to replace the risk by the empirical risk on the training set

$$\widehat{R}(f) = \frac{1}{N} \sum_{j=1}^{N} (Y_j - f(X_j))^2$$

and minimize w.r.t. f. Under the general regression model, $\widehat{R}(f) \rightarrow R(f)$ for increasing N by the law of large numbers. Minimizing the empirical risk with respect to f is of course a least-squares approach but with a twist. Initially, we do not consider minimization w.r.t. a finite number of parameters but with respect to whole functions.

Allowing for arbitrary functions f, we would get interpolation, i.e., we would estimate $m(X_j)$ by Y_j, $j = 1, \ldots, N$. To get reasonably smooth estimates of m, we could regularize the loss function $\widehat{R}(f)$ by adding a roughness penalty. An example is discussed in Sect. 10.3 on spline smoothing where we exploit the same kind of idea as, e.g., in ridge regression (compare Sect. 3.2).

Here, we pursue an alternative road where we restrict the function class as in the previous section. Let $\mathscr{F}_N$ be function classes depending on the sample size N and increasing with it, i.e., $\mathscr{F}_N \subset \mathscr{F}_{N+1}$ for all N. In general, $\mathscr{F}_N$ also may depend on the data of the training set of size N, but here we consider only fixed classes. The functions in $\mathscr{F}_N$ are usually determined by a finite number of parameters, increasing with N and representing the size of the function class.

As an estimate of the regression function m, we consider the function $\widehat{m}_N$ minimizing the empirical risk over $\mathscr{F}_N$, i.e.,

$$\widehat{R}(\widehat{m}_N) = \min_{f \in \mathscr{F}_N} \frac{1}{N} \sum_{j=1}^{N} (Y_j - f(X_j))^2 \tag{39}$$

We again encounter the **bias-variance dilemma** (compare Sect. 2.3):

(a) If $\mathscr{F}_N$ is not "too large," we do not have to estimate too many parameters in solving the minimization problem (39), and the variability will be rather small. In particular,

$$\min_{f \in \mathscr{F}_N} \mathrm{E}\,(Y_{N+1} - f(X_{N+1}))^2 \approx \min_{f \in \mathscr{F}_N} \frac{1}{N} \sum_{j=1}^{N} (Y_j - f(X_j))^2$$

(b) If $\mathscr{F}_N$ is not "too small," then m may be well approximated by a function m_N from $\mathscr{F}_N$, and the bias introduced by looking only at functions from $\mathscr{F}_N$ is rather small. In particular, if m_N denotes the function in $\mathscr{F}_N$ closest to the unknown regression function m, i.e.,

$$\mathrm{E}\,(m(X_{N+1}) - m_N(X_{N+1}))^2 = \min_{f \in \mathscr{F}_N} \mathrm{E}\,(m(X_{N+1}) - f(X_{N+1}))^2, \tag{40}$$

then we have $m_N \approx m$.

If the conditions of both (a) and (b) are satisfied reasonably well, we have

$$\hat{m}_N \overset{a)}{\approx} m_N \overset{b)}{\approx} m.$$

From these considerations, we see that function classes $\mathscr{F}_N$ have to grow with N, but not too fast, such that the sieve estimate $\widehat{m}_N$ converges to m. To better understand what is going on, it is illuminating to have a look at a crucial mathematical inequality, which is also the starting point for proving that sieve estimates work under appropriate conditions:

$$\begin{aligned} &\mathrm{E}\,(\widehat{m}_N(X_{N+1}) - m(X_{N+1}))^2 \\ &\le 2 \sup_{f \in \mathscr{F}_N} \left| \frac{1}{N} \sum_{j=1}^{N} (Y_j - f(X_j))^2 - \mathrm{E}\,(Y_{N+1} - f(X_{N+1}))^2 \right| \\ &\quad + \inf_{f \in \mathscr{F}_N} \mathrm{e}(m(X_{N+1}) - f(X_{N+1}))^2. \end{aligned} \tag{41}$$

The average (w.r.t. to the randomness in the input X_{N+1}) estimation error is bound by two terms, which represent the variance and the bias. The first term on the right-hand side of the inequality represents the error component due to the fact that we only have a limited amount N of random data. It reflects the random component of the error and measures how well m_N is estimated by $\hat{m}_N$.

Typically, there is a function m_N satisfying (40). It is the best approximation of m by a function from $\mathscr{F}_N$. Then, the second term on the right-hand side of the inequality is $\mathrm{E}\,(m(X_{N+1}) - m_N(X_{N+1}))^2$. It represents the *approximation error* due to the fact that we only allow for choosing functions from the restricted class $\mathscr{F}_N$ instead of the full function class $\mathscr{F}$.

For a sieve estimate to perform well on average for suitably large size N of the training set, in particular, the approximation error has to become small. To guarantee this, we have to verify that the particular family of functions $\mathscr{F}_N$ has a **universal approximation** property, i.e., for each regression function m, satisfying the weak condition $\mathrm{E}\,m^2(X_1) < \infty$

$$\inf_{f \in \mathscr{F}_N} \mathrm{E}\left(m(X_{N+1}) - f(X_{N+1})\right)^2 \to 0 \qquad \text{for } N \to \infty. \tag{42}$$

This property is essential and has to be checked for each statistical learning method based on sieves.

Usually, verification of the universal approximation property is not hard as it suffices to show that, similarly to (42),

$$\inf_{f \in \mathscr{F}_N} \sup_x |m(x) - f(x)| \to 0 \qquad \text{for } N \to \infty$$

for m in the much smaller class of continuous functions vanishing outside of a finite interval, which are infinitely often differentiable (compare Corollary A1 in [36]). For the two examples considered in the previous section, the universal approximation property is implied by looking at the Taylor expansion for polynomial regression and by the fact that the trigonometric polynomials are a basis of L^2.

The supremum on the right-hand side of (41) converges to 0 for increasing N, intuitively by the law of large numbers. The simple version which implies the convergence of the sample mean to the expectation does not suffice here, as we need uniform convergence over increasing sets of functions. However, for function classes, satisfying certain restrictions on their size, there are suitable *uniform law of large numbers*, which imply that the supremum on the right-hand side becomes small for increasing N. The corresponding theory is rather involved (compare [36]), but it provides some hints about the size of appropriate hyperparameters for the corresponding machine learning algorithms (compare Sect. 14.2 regarding the choice of the hyperparameter H of neural networks with one hidden layer for independent resp. time series training data).

As an illustration of such theoretical results, we sketch the justification for sieve estimates by looking at a simple example. Let $\psi_1, \psi_2, \ldots$ be bounded functions on the input space with $|\psi_k(x)| \leq 1$, e.g., basis functions $e_k(x)$ from the previous section. For bounds $\beta_L > 0$, we consider the function class

$$\mathscr{F}_N = \Big\{ \sum_{k=1}^{L_N} b_k \, \psi_k(x) \;\text{ with real-valued } b_1, \ldots, b_{L_N} \text{ satisfying } \sum_{k=1}^{L_N} |b_k| \leq \beta_{L_N} \Big\}$$

and assume that it satisfies the universal approximation property. The sieve estimate $\widehat{m}_N$ w.r.t. this function class solves

$$\frac{1}{N} \sum_{j=1}^{N} \big(Y_j - \sum_{k=1}^{L_N} b_k \, \psi_k(X_j)\big)^2 = \min!$$

over all $b_1, \ldots, b_{L_N}$ with $\sum |b_k| \leq \beta_{L_N}$. If $L_N \to \infty,\ \beta_{L_N} \to \infty$ with increasing N such that

$$\frac{L_N \beta_{L_N}^4 \log \beta_{L_N}}{N} \to 0\,, \ \frac{\beta_{L_N}^4}{N^{1-\delta}} \to 0 \quad \text{for some } \delta > 0$$

then $\widehat{m}_N$ converges to m in an appropriate sense, i.e., the learning method works (Theorem 10.1 of [36]).

Note that additional to the hyperparameter L_N controlling the complexity of the functions in $\mathscr{F}_N$, i.e., the number of parameters b_k to be estimated, we have a second hyperparameter β_N, which puts a bound on the total size of parameters. Similar to the LASSO constraint (20), it guarantees a certain smoothness of the function estimate $\widehat{m}_N(x)$. From the growth condition on the two hyperparameters, which both address the same issues in keeping the variance small, we see that we have some freedom of choice. If we prefer a rather large L_N, then we have to select a more restrictive β_N and vice versa.

7 Learning from Trees

In this chapter, we consider, as an example of sieve estimates in statistical learning, the **classification and regression trees** (CART), which, since the early eighties [9], have been and still are one of the most popular learning methods. They are easy to understand and quick to calculate such that they are also quite suitable as a basis for ensemble learning methods, discussed in the next chapter, where many similar function estimates have to be calculated simultaneously.

As sieve estimates, tree-based methods are using a particularly simple and flexible class of functions for approximations, the piecewise constant functions. It is well-known from integration theory that every integrable $m(x)$ on d-dimensional space may be arbitrarily well approximated in the following sense by piecewise constant functions:

$$m(x) \approx \widehat{m}(x) = \sum_{k=1}^{M} b_k 1_{A_k}(x), \qquad \text{i.e.} \int |m(x) - \widehat{m}(x)| dx \ \text{ small} \tag{43}$$

for large enough M and suitable real-valued coefficients b_k and measurable d-dimensional sets $A_k, k = 1, \dots, M$. Recall that the indicator function $1_A(x) = 1$, if x lies in A, and $= 0$ else.

This kind of universal approximation property of the piecewise constant functions still holds, if we restrict the A_k to hyperrectangles, not necessarily finite ones, i.e., to intervals for $d = 1$, to rectangles for $d = 2$, to cuboids for $d = 3$, etc. For suitable $a_i < b_i$, such a hyperrectangle R corresponds to the set of all x satisfying $a_i \le x_i \le b_i, i = 1, \dots, d$, where we allow for $a_i = -\infty$ and $b_i = \infty$.

Note that for input vectors X_j with a continuous range of values, it does not matter if we consider $<$ or $\le$ for the inequalities as the boundary values a_i, b_i are assumed with probability 0. If discrete-valued features are present, we would have to take a bit more care, but for ease of notation, we do not consider these complications here.

Hyperrectangles are used in the piecewise constant approximation (43) due to the iterative manner in which the tree-based function $\widehat{m}(x)$ is constructed from the training set. It is defined by successive splits along coordinates, using a tree-growing algorithm of the following general structure:

(i) At each current terminal node, find the optimal split point t_i on the real line for each coordinate $i = 1, \ldots, M$, separately.
(ii) Using the results from (i), decide which coordinate i provides the best split in general.
(iii) Calculate the function values at the new terminal nodes.

Starting with a constant function $\widehat{m}(x) = b_0$, after several iterations, this results in a function of the form

$$\widehat{m}(x) = \sum_{k=1}^{M} \hat{b}_k 1_{R_k}(x) \tag{44}$$

with hyperrectangles $R_1, \ldots, R_M$ forming a partition of $\mathbb{R}^d$, i.e., $R_k = [a_{1k}, b_{1k}] \times \ldots \times [a_{dk}, b_{dk}]$, where $\infty \leq a_{ik} < b_{ik} \leq \infty, i = 1, \ldots, d, k = 1, \ldots, M$. The details depend on the application, the corresponding optimality concept, and the choice of stopping rules.

We illustrate this procedure with a simple artificial example from predictive maintenance for a crane moving heavy loads. We look at the expected time $m(x)$ for which the crane will be inoperable during the next 4 weeks due to some failures and necessary repairs. If $m(x)$ is too large, we schedule a major maintenance for the near future. As input variables, we observe, for the last 3 months, the total time x_1 of usage of the crane, the maximum load x_2 carried, and the average load x_3 per day throughout this period; hence, the input vector x is three-dimensional. As a training set, we have pairs of data X_j, Y_j from the past, where X_j are the inputs for a 3-month period and Y_j the corresponding time of inoperability.

We use a tree-based function $\widehat{m}(x)$ for approximating $m(x)$, which can be represented by the (graph-theoretic) tree in Fig. 22. It starts from some root, has a depth of 2, and ends in four **terminal nodes** or *leafs*, i.e., $M = 4$. Each of these nodes corresponds to one of the sets R_k in (44). The first split uses the first coordinate and split point t_1, and the second and the third split happen to be coordinate 3 resp. 2 with split points t_2, t_3. This results in the function of the

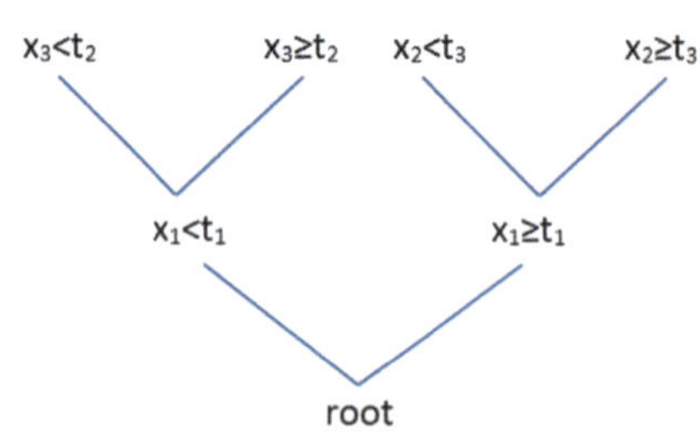

Fig. 22 Regression tree with four terminal nodes

form (44) with $\hat{b}_k$ determined from the training set as described in the next section. R_2, e.g., consists of all x for which $x_1 < t_1$ and $x_3 \geq t_2$, and R_4 of those with $x_1 \geq t_1, x_2 \geq t_3$.

In the following two sections, we consider regression and classification problems separately. For later reference, we denote by $\mathscr{T}(M)$ the set of all functions of the form (44), i.e., they are given by a regression tree with M terminal nodes.

7.1 Regression Trees

We again consider the general regression model (2) and the quadratic loss function $L(y, m) = (y-m)^2$. For a sieve estimate of the regression function $m(x)$, we choose as the corresponding function class $\mathscr{F}_N$ the class of regression trees $\mathscr{T}_N = \mathscr{T}(M_N)$, consisting of all functions of the form (44) with M_N terminal nodes. Note that this includes also all trees with $M \leq M_N$ terminal nodes, as coefficients $b_k = 0$ are allowed.

Instead of choosing a rather small M_N to avoid overfit, we also could consider larger regression trees and restrict their complexity additionally by a bound on the coefficients, i.e.,

$$\mathscr{T}_N^\beta = \{f \in \mathscr{T}(M_N); \sum_{i=1}^{M_N} |b_k| \leq \beta_N\}$$

with the same constraint as in (20). $\mathscr{T}_N^\beta$ provides more flexibility for the price of having to choose two hyperparameters (M_N, β_N) instead of only one. Here, we focus on the simpler case of $\mathscr{T}_N$.

The sieve estimate $\widehat{m}(x)$ of $m(x)$ is given as the solution of the least-squares problem

$$\sum_{j=1}^{N} (Y_j - f(X_j))^2 = \min_{f \in \mathscr{T}_N}! \quad \text{resp.} \quad \sum_{j=1}^{N} \left(Y_j - \sum_{k=1}^{M_N} b_k 1_{R_k}(X_j) \right)^2$$
$$= \min_{R_1,\ldots,R_{M_N}, b_1,\ldots,b_{M_N}} !$$

Given $R_1, \ldots, R_{M_N}$, the coefficients are immediately specified. Just note that the latter representation of the loss may be written as

$$\sum_{k=1}^{M_N} \sum_{j; X_j \in R_k} (Y_j - b_k)^2$$

Setting its partial derivatives w.r.t. b_k to 0, we get

$$\hat{b}_k = \frac{1}{N_k} \sum_{j=1}^{N} Y_j 1_{R_k}(X_j), \qquad \text{with} \quad N_k = \sum_{j=1}^{N} 1_{R_k}(X_j) = \text{no. of } X_j \text{ in } R_k,$$

i.e., $\hat{b}_k$ is just the mean of the responses Y_j from the part of the training sample with inputs $X_j \in R_k$.

Algorithm for building a regression tree:

(o) Initialize by setting $M = 1$, R equal to the whole d-dimensional input space such that

$$\widehat{m}(x) = \widehat{m}_0 = \overline{Y}_N = \frac{1}{N} \sum_{j=1}^{N} Y_j.$$

(i) For each $i = 1, \ldots, d$, split R into two parts $A_{i0}(t) = \{x; x_i < t\}$ and $A_{i1}(t) = \{x; x_i \geq t\}$. Consider the means of Y_j with X_j in $A_{i0}(t)$ resp. $A_{i1}(t)$:

$$\widehat{m}_{i0}(t) = \text{mean}\{Y_j;\, X_{ji} < t\}, \qquad \widehat{m}_{i1}(t) = \text{mean}\{Y_j;\, X_{ji} \geq t\}.$$

Choose the optimal split point t_i for the ith coordinate, which minimizes

$$d_i(t) = \sum_{j; X_j \in A_{i0}(t)} (Y_j - \widehat{m}_{i0}(t))^2 + \sum_{j; X_j \in A_{i1}(t)} (Y_j - \widehat{m}_{i1}(t))^2$$

(ii) Choose the split coordinate i_1 that minimizes $d_i(t_i)$ over $i = 1, \ldots, d$. Set $R_0 = A_{i_1 0}(t_{i_1})$, $R_1 = A_{i_1 1}(t_{i_1})$, which provides a split of R into two disjoint subsets.

Repeat steps (i) and (ii) for each subsample $\{(X_j, Y_j), X_j \in R_s\}, s = 0, 1$, resulting in four disjoint hyperrectangles, then repeat steps (i) and (ii) again for the hyperrectangles corresponding to the now four terminal nodes, etc.

This algorithm doubles the number of terminal nodes in each iteration, which leads finally to a very large tree. Furthermore, we have to decide when to stop the iteration. There are various ways to deal with those problems. We discuss two of them, where the first one avoids overfit already during the construction of the tree, whereas the second one allows for growing a large tree, which in retrospect is **pruned**, i.e., simplified by deleting unnecessary splits.

Individual Stopping

In the course of the iteration, we decide for each terminal node separately if we stop here or if the corresponding hyperrectangle R_i is split further. There are various stopping criteria that also may be combined, e.g.,

(a) Do not split a terminal node if one of the two resulting daughter nodes, i.e., the corresponding new R_k, would contain less than $n_{\min}$ data.

(b) Do not split a terminal node if this does not improve overall prediction performance much. More precisely, we choose a performance criterion D and a complexity cost s for the tree $\mathbf{T}$ with M terminal nodes fitted to the data with resulting regression function estimate (44)

$$D(\mathbf{T}) = \frac{1}{N} \sum_{k=1}^{M} \sum_{j;X_j \in R_k} (Y_j - \hat{b}_k)^2, \qquad s(\mathbf{T}) = M.$$

$D(\mathbf{T})$ is the average squared prediction error on the training set if using the prediction function $\widehat{m}(x)$, and $s(\mathbf{T})$ is the number of terminal nodes as a measure of the size of the tree.

Now we consider one additional split resulting in a tree $\mathbf{T}^*$ with $M+1$ terminal nodes, say $R_k^* = R_k, k = 1, \ldots, M-1$, and R_M is split into two subsets R_M^*, R_{M+1}^*. As the first $M-1$ hyperrectangles do not change, $\hat{b}_k^* = \hat{b}_k, i = 1, \ldots, M-1$. $\hat{b}_M^*, \hat{b}_{M+1}^*$ are the means of the Y_j with $X_j \in R_M^*$ resp. $\in R_{M+1}^*$. As we have one parameter more in $\mathbf{T}^*$ to fit to the data, the performance always improves, i.e., $D(\mathbf{T}^*) \leq D(\mathbf{T})$. For deciding if the additional split is worthwhile, we choose a parameter $\alpha > 0$ to balance between the goals of fitting the optimal prediction function $m(x)$ well enough on the one hand and avoiding overfit due to a too complex function estimate, which would impair generalization, on the other hand. Then, we consider a penalized least-squares approach as, e.g., in Sect. 3.2 and compare the complexity penalized performance $C_\alpha(\mathbf{T}) = D(\mathbf{T}) + \alpha s(\mathbf{T})$ of the old and the new tree:

$$\Delta(\mathbf{T}, \mathbf{T}^*) = C_\alpha(\mathbf{T}) - C_\alpha(\mathbf{T}^*) = \sum_{j;X_j \in R_M} (Y_j - \widehat{m}_M)^2$$

$$- \sum_{j;X_j \in R_M^*} (Y_j - \widehat{m}_M^*)^2 - \sum_{j;X_j \in R_{M+1}^*} (Y_j - \widehat{m}_{M+1}^*)^2 - \alpha.$$

We do not split this node further if the penalized gain $\Delta(\mathbf{T}, \mathbf{T}^*) < \delta$ for some chosen $\delta > 0$. Note that due to the particular choice of $s(\mathbf{T})$, the decision only depends on $\alpha + \delta$, i.e., we only have to choose one hyperparameter here.

Individual stopping has one drawback, which already is known from other bottom-up stepwise regression methods, which start from a very simple model, here a constant $\widehat{m}(x) = \widehat{m}_0$, and then consider more and more complex models by adding additional terms and parameters until some stopping criterion is satisfied. It may happen that we stop as every further split of one hyperrectangle into two parts does not result in a considerable reduction of the prediction error $D(\mathbf{T})$, i.e., the penalized gain Δ is small. However, subsequent splits, leading to a partition of some R_k into 2^r parts, $r > 1$, may lead to a considerable improvement, but the individual stopping algorithm will never take them into consideration. Therefore, the method of first constructing a very large tree, which afterward is pruned to a reasonable size, often is preferred though it is a bit more time-consuming.

Pruning
Here, we first grow a very large tree and prune it back by merging terminal nodes, i.e., by removing splits. For two trees, we write $\mathbf{T} \subset \mathbf{T}^*$, if each component R_k of the partition corresponding to $\mathbf{T}$ is the union of some partition components of $\mathbf{T}^*$ or, vice versa, $\mathbf{T}^*$ can be constructed by splitting terminal nodes of $\mathbf{T}$ further.

Let $\mathbf{T}_{\max}$ denote the maximal tree where $M = N$, and each component $R_{\max,j}$ contains exactly one data point X_j, $j = 1, \ldots, N$ from the training set. For a given $\alpha > 0$, there is a smallest subtree $\mathbf{T}_\alpha \subset \mathbf{T}_{\max}$ with

$$C_\alpha(\mathbf{T}_\alpha) = \min_{\mathbf{T} \subset \mathbf{T}_{\max}} C_\alpha(\mathbf{T}).$$

To find $\mathbf{T}_\alpha$, we prune the maximal (or a large enough) tree by removing from top to bottom unnecessary splits, e.g., using the same kind of criteria as in individual stopping, we remove a split of tree $\mathbf{T}^*$ if $\Delta(\mathbf{T}, \mathbf{T}^*) < \delta$, which results in a subtree $\mathbf{T} \subset \mathbf{T}^*$ with one split less. We proceed until for all pairs of terminal nodes, $\Delta(\mathbf{T}, \mathbf{T}^*) \geq \delta$.

Individual stopping as well as pruning algorithms depend on the hyperparameter $\alpha + \delta$. As for other nonparametric function estimates, we can choose it adaptively from the data by crossvalidation, etc.

As an example, we consider the training set of size $N = 50$ already used for Fig. 21. We fit a regression tree using the *fitrtree* algorithm of the MATLAB statistics and machine learning toolbox, which has an inbuilt stopping rule, where we used the default settings. To our experience, it shows a slight tendency to overfit, which is not bad as one easily can prune it back to compare the performance of trees with variable number of nodes. In this simple example with one-dimensional features, we can find a parsimonious function estimate $\widehat{m}(x)$ by looking at the plots of the functions. Figure 23 shows the resulting tree with six terminal nodes, and on the left-hand side of Fig. 24, the piecewise constant function estimate $\widehat{m}_6(x)$ (blue) is plotted together with the true regression function $m(x)$ (red) and the training data. The larger tree

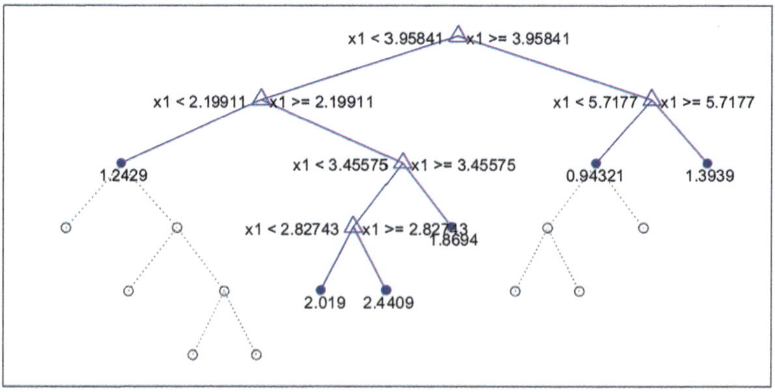

Fig. 23 Regression tree fitted to the data of Fig. 24

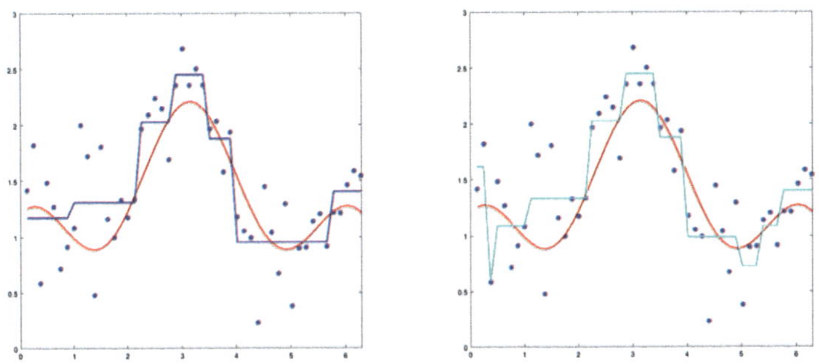

Fig. 24 Tree-based function estimates using six (blue) resp. ten (light blue) function estimates together with the training data and the true regression function (red)

with ten terminal nodes chosen by *fitrtree* is indicated in Fig. 23 by the additional dotted branches and empty nodes. It describes the function estimate $\widehat{m}_{10}(x)$ (light blue), which is shown on the right-hand side of Fig. 24 together with the data and the true function. It looks quite similar to $\widehat{m}_6(x)$ but shows some overfitting artifacts around $x = 0$ and $x = 5$, which are not related to the shape of $m(x)$.

7.2 Classification Trees

The original tree-based regression function estimates are simple and easy to calculate. However, they lead to a piecewise continuous approximation $\widehat{m}(x)$, which may not be so appealing if the true function $m(x)$ is known to be smooth. For classification purposes, however, the discreteness of the values of the function represented by a tree is rather an advantage as the targets of a data analysis, the class labels, assume finitely many values, too.

Let again (X_j, Y_j), $j = 1, \ldots, N$, be i.i.d. and, now, $Y_j \in \{0, 1\}$, i.e., we consider only two classes. We consider classification functions of the form

$$c(x) = \sum_{k=1}^{M} c_k 1_{R_k}(x)$$

given by a tree with M terminal nodes generating the partition of the input space into $R_1, \ldots, R_M$. By restricting the values of c_k to 0 and 1, we get that $c(x)$ also assumes only the values 0 and 1.

The construction of the tree proceeds analogously to regression. However, we have to choose estimates $\hat{c}_k$ of the coefficients and select loss functions for individual stopping or pruning tuned to classification problems. Let us recall the optimal Bayes classifier from (9):

$$c_B(x) = 1 \quad \text{if } \pi(x) = \text{pr}(Y_j = 1|X_j = x) > \frac{1}{2}, \qquad \text{and } c_B(x) = 0 \quad \text{else.}$$

For the tree-based classifier $c(x) = c_k, x \in R_k$, the coefficient c_k should depend on $\pi_k = \text{pr}(Y_j = 1|X_j \in R_k)$. The maximum likelihood estimate of π_k is based on relative frequencies and given by

$$\hat{\pi}_k = \frac{\text{no. of } j \text{ with } Y_j = 1, X_j \in R_k}{\text{no. of } j \text{ with } X_j \in R_k}.$$

Using the idea of the Bayes classifier, we set

$$\hat{c}_k = 1 \quad \text{if } \hat{\pi}_k > \frac{1}{2}, \qquad \text{and } \hat{c}_k = 0 \quad \text{else.}$$

The resulting classifier $\hat{c}(x) = \sum_{k=1}^{M} \hat{c}_k 1_{R_k}(x)$ corresponds to a **majority vote** in each area R_k of the sample space. If we want to classify a new object with feature vector, say, X_{N+1}, we check in which hyperrectangle R_k it is located. Then, we classify it as belonging to the same class 0 or 1 as the majority of objects from the training set lying in the area R_k, as, e.g., $\hat{\pi}_k > \frac{1}{2}$ if the number of j with $Y_j = 1, X_j \in R_k$ is greater than the number of j with $Y_j = 0, X_j \in R_k$.

For classification, splits are considered to be good if they result in rather pure terminal nodes. If, e.g., we split R_M into R_M^*, R_{M+1}^*, then the split is good if most objects with X_j lying in R_M^* belong to the same class, and the same holds for R_{M+1}^*, too. To make this idea precise, we have to specify how to measure the purity of a node.

The **impurity of the node** k is given by $\phi(\hat{\pi}_k)$ for some suitable function $\phi : [0, 1] \to [0, \infty)$ satisfying

$\phi(0) = \phi(1) = 0$ (pure nodes have impurity 0)
$\phi(\frac{1}{2}) = \max$ (maximal impurity if both classes are equally present)
$0 < \phi(p) = \phi(1 - p), 0 < p < 1,$ (symmetric treatment of both classes)

Popular choices are (for two classes, for more classes, the general forms are given in chapter 9 of [44], (9.17)):

$\phi(p) = 2p(1 - p)$	*Gini index*
$\phi(p) = -p \log p - (1 - p) \log(1 - p)$	*cross-entropy* or *deviance*
$\phi(p) = 1 - \max(p, 1 - p) = \min(1 - p, p)$	*misclassification error*

For **growing** a tree, the Gini index or cross-entropy is preferable. In case of only two classes, both impurity measures lead to similar results. Moreover, in both cases, $\phi(p)$ is differentiable, which helps in numerical optimization. For $K > 2$ classes, the Gini index frequently is preferable as it results in more homogeneous classes, whereas cross-entropy favors subsets of similar size. For subsequent **pruning**, however, frequently, the misclassification error is used, though one may stick to the other two impurity measures too (compare Section 9.2 of [44] for a more detailed discussion and an artificial example).

Algorithm for building a classification tree for two classes: Proceed as in the previous section for building a regression tree, but replace $d_i(t)$ with the total impurity of the split

$$I_i(t) = \phi(\hat{\pi}_{i0}(t)) + \phi(\hat{\pi}_{i1}(t)),$$

where, for $i = 1, \ldots, d$,

$\hat{\pi}_{i0}(t) =$ relative frequency of objects j with $Y_j = 1, X_{ji} < t$,
$\hat{\pi}_{i1}(t) =$ relative frequency of objects j with $Y_j = 1, X_{ji} \geq t$,

and replace in the pruning step $D(\mathbf{T})$ by the total impurity of the tree

$$J(\mathbf{T}) = \sum_{k=1}^{M} \phi(\hat{\pi}_k).$$

As mentioned above, it is not uncommon to use different impurity measures ϕ in $I_i(t)$ (for growing the tree) and $J(\mathbf{T})$ (for pruning the tree).

Figure 26 shows a training set with two-dimensional features, consisting of $N = 400$ observations, half of them from class 1 (blue) resp. class 2 (red). These data are part of the sample already used as example for classification with classical methods and neural networks (compare Figs. 18 and 32). We calculated a classification tree using the *fitctree* algorithm of the MATLAB statistics and machine learning toolbox with default settings and applied a bit of additional pruning. The resulting tree with

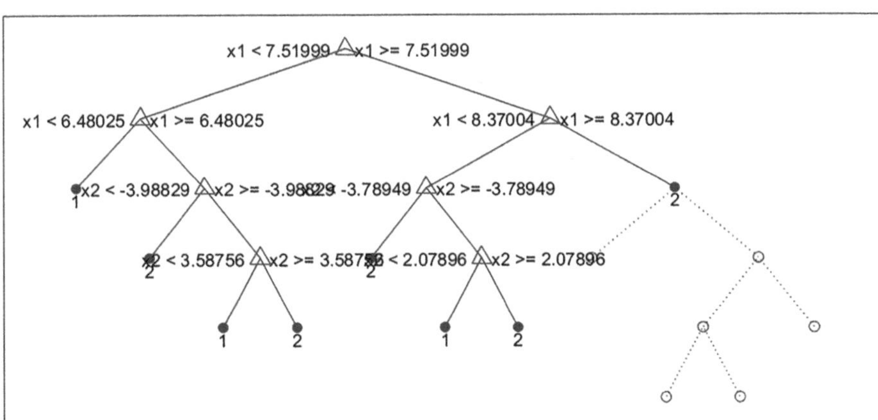

Fig. 25 Classification tree fitted to the data shown in Fig. 26

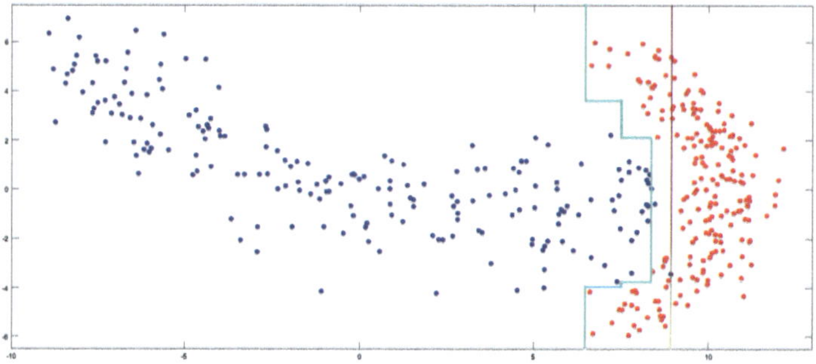

Fig. 26 Training set from two classes 1 (blue) and 2 (red) together with the decision boundary based on the classification tree of Fig. 25

Table 4 Misclassification rates in training and validation sample of Fig. 26

No. terminal nodes	8	11
Training sample	0.5 %	0.25 %
Validation sample	2.75 %	3.00 %

8 terminal nodes is shown in Fig. 25, where the dotted branches indicate the pruned parts of the original tree with 11 terminal nodes chosen by *fitctree*. Figure 26 shows the decision boundary (light blue) resulting from the classification rule $\hat{c}_8(x)$ based on the smaller tree. Here, a new observation with feature vector $X_{N+1} = x$ would be classified as 1 if x lies left of the boundary and as 2 otherwise. The rule based on the larger classification tree adds two vertical lines to the decision boundary, which are so close that they are practically indistinguishable in Fig. 26. If $X_{N+1} = x$ lies between these lines, then $\hat{c}_{11}(x) = 1$. In this manner, the larger tree achieves to additionally classify the one right-most observation from class 1 in the training set correctly, which obviously is an overfitting effect. This becomes apparent from looking at Table 4 with the misclassification rates. On the training set, the larger tree gives slightly better results but on an independent, identically generated validation set (shown on the right-hand side of Fig. 19) where $\hat{c}_8(x)$ performs slightly better than $\hat{c}_{11}(x)$.

7.2.1 Classification Performance Measures

In this section, we briefly review a few concepts for representing the performance of a classifier, e.g., one based on a classification tree. We stick to the case of two classes.

From medical diagnosis, we borrow the notions of *sensitivity* and *specificity*. In this context, the two classes are the infection with a disease (class 1) and the healthy state (class 0). Then:

Sensitivity = (conditional) probability of diagnosing a disease given that the patient suffers from it,

Specificity = (conditional) probability of diagnosing healthiness given that this is true.

Let $c(x)$ be a classification function that may be random, e.g., chosen as the estimate $\hat{c}(x)$ learned from the training data. If (X_{N+1}, Y_{N+1}) is a new pair of input and output data, independent of $c(x)$, then sensitivity is

$$\mathrm{pr}(c(X_{N+1}) = 1 | Y_{N+1} = 1) = \frac{p_{11}}{p_{11} + p_{10}}$$

and the specificity is

$$\mathrm{pr}(c(X_{N+1}) = 0 | Y_{N+1} = 0) = \frac{p_{00}}{p_{00} + p_{01}}$$

where $p_{k\ell}$ is the probability of the classifier choosing class ℓ and the true class being k. These probabilities may be estimated from a training set of size N by the relative frequencies

$$\hat{p}_{k\ell} = \frac{1}{N}\big(\text{ no. of } j \le N \text{ with } Y_j = k, c(X_j) = \ell\big), \ , \ell = 0, 1,$$

from which we also get estimates of sensitivity and specificity. These probabilities may be combined to the so-called *confusion table*

$$P = \begin{pmatrix} \hat{p}_{00} & \hat{p}_{01} \\ \hat{p}_{10} & \hat{p}_{11} \end{pmatrix}$$

There is a trade-off between sensitivity and specificity if we change the hyperparameters of the fitted classification function ($\alpha+\delta$ for the complexity cost $s(\mathbf{T}) = M$ considered above) in building a classification tree. This is visualized by the so-called *receiver operating characteristic* (ROC) *curve*, which is a plot of sensitivity against specificity. It may be used for comparing the performance of different classifiers.

The examples that we have considered up to now are with their low-dimensional feature vectors atypical for the application of regression and classification trees. Therefore, we have a look at another example with real data from the UCI machine learning repository [74]. The so-called ionosphere data consist of 34-dimensional feature vectors X_j, each representing 17 different complex-valued measurements from a radar array. Each observation belongs to one of two classes, denoted as 0="good" resp. 1="bad." This classification problem is typical for quality control or process surveillance applications in the industry.

To judge the performance of a tree-based classification rule, we split the original sample of size 351 into a training sample of size $N = 270$ and a validation set of size $M = 81$, where in both subsets, about one third of the data are bad, i.e., $Y_j = 1$. Figure 27 shows the classification tree with 16 terminal nodes fitted to the training set by using the MATLAB algorithm *fitctree* with default settings. To investigate the performance, we estimate the confusion table from the training set

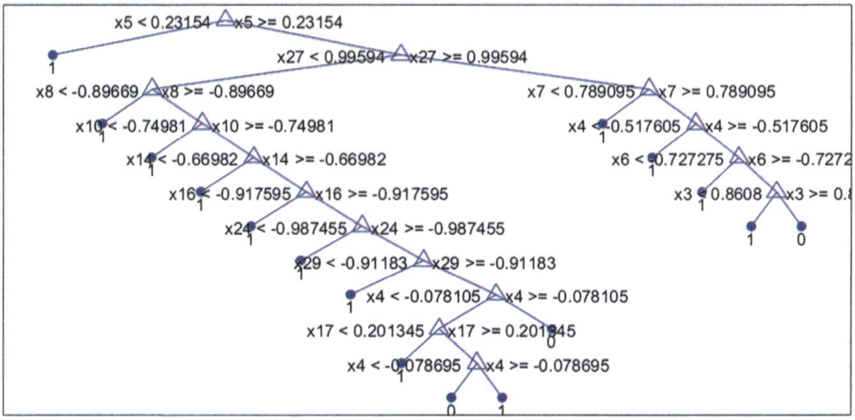

Fig. 27 Classification tree for the training sample being part of the ionosphere data

and, more importantly, from the independent validation set not used in training at all. The results are

$$P^{train} = \begin{pmatrix} 61.11\% & 1.85\% \\ 0\% & 37.04\% \end{pmatrix} \qquad P^{val} = \begin{pmatrix} 64.20\% & 3.70\% \\ 2.47\% & 29.63\% \end{pmatrix}$$

As expected, error probabilities are larger for the validation set but still reasonable regarding the small effort we have spent on finding the classification rule. Adding the off-diagonal elements, we get a misclassification rate of 1.85 % in the training set and of 6.17 % in the validation set. In the latter, both types of errors (classifying a good object as bad and vice versa) occur with similar frequencies, which often is desirable, in particular if the consequences are of similar severity for both error types. However, in practice,. there also are situations where we should avoid a particular type of error with high probability, whereas the other type of error is more tolerable. This problem is discussed in more detail in Sect. 12.4.

8 Basic Neural Networks

In the 1990s, neural networks were the basis of one of the most popular statistical learning methods, and thanks to the success in deep learning (compare Sect. 13), they are now again very popular. They also belong to the class of sieve estimates and use a special class of approximating functions for, in particular, solving regression, classification, and forecasting problems. For the sake of simplicity, we present the basics of neural network learning in this chapter only for feedforward networks with **one** hidden layer, postponing discussion of more general networks to Sect. 13. They already may produce quite good results for less involved problems with only moderate sample size and feature vector dimension, which frequently arise in an engineering context.

A neural network represents a function $f_H(x;\vartheta)$ mapping the d-dimensional input vector x to an output y, which in this section is assumed to be one-dimensional. Its complexity depends on a hyperparameter H, which is the number of neurons in the hidden layer. For applications in regression and forecasting, we use network functions of the form

$$f_H(x;\vartheta) = v_0 + \sum_{h=1}^{H} v_h \psi\big(w_{0h} + \sum_{i=1}^{d} w_{ih} x_i\big). \tag{45}$$

It depends on the choice of a fixed so-called **activation function** $\psi(u)$ and a finite number of **weights** or parameters $v_0, \ldots, v_H$ and $w_{ih}, i = 0, \ldots, d, h = 1, \ldots, H$. ϑ denotes the vector of all $(d+2)H+1$ parameters v_h, w_{ih}. We denote the class of network functions with given size $H = H_N$ of the hidden layer by

$$\mathscr{F}_N^{\text{nn}} = \big\{f_H(x;\vartheta); \vartheta \in \Theta_N\big\}. \tag{46}$$

The parameter set Θ_N is the set of admissible values for the weight vector. In some applications, it may coincide with the full space of dimension $(d+2)H+1$, but sometimes, it is advisable to use a smaller subset as discussed in the context of identifiability below.

Recall the empirical risk of a predictor function $f(x)$ from Sect. 6.2

$$\widehat{R}(f) = \frac{1}{N}\sum_{j=1}^{N}(Y_j - f(X_j))^2,$$

calculated from the training set $(X_1, Y_1), \ldots, (X_N, Y_N)$. Then, a neural network sieve estimate of the regression function $m(x)$ in (2) is $\widehat{m}_N(x) = f_H(x;\hat{\vartheta}_N)$ where the weight vector ϑ is estimated by empirical risk minimization or least squares, i.e., $\hat{\vartheta}_N$ minimizes

$$Q(\vartheta) = \sum_{j=1}^{N}\big(Y_j - f_H(X_j;\vartheta)\big)^2 \tag{47}$$

over ϑ in Θ_N. Correspondingly, the function estimate $\widehat{m}_N(x)$ minimizes the empirical risk over the function class $\mathscr{F}_N^{\text{nn}}$

$$\widehat{R}(\widehat{m}_N) = \min_{f\in\mathscr{F}_N^{\text{nn}}} \frac{1}{N}\sum_{j=1}^{N}\big(Y_j - f(X_j)\big)^2.$$

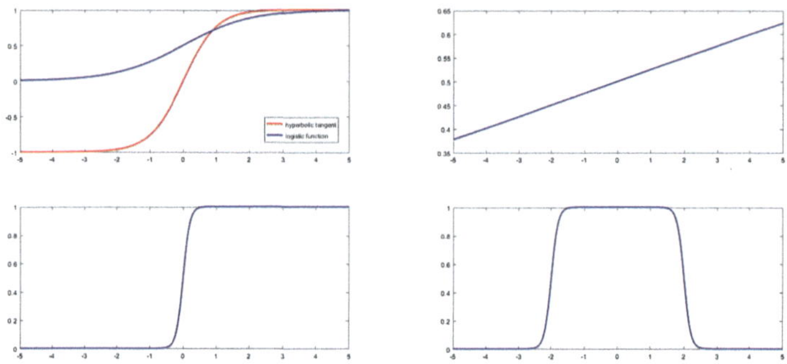

Fig. 28 Sigmoid functions and scaled versions

For regression, sigmoid, i.e., roughly S-shaped, functions are mainly chosen as activation functions. They look like and frequently are distribution functions of continuous real-valued random variables, which are symmetrically distributed around 0, e.g., the standard normal distribution function. Popular choices for those functions are the **logistic function** and the **hyperbolic tangent**

$$\psi_{\log}(u) = \frac{1}{1+e^{-u}} = \frac{e^u}{1+e^u}, \qquad \text{resp.} \quad \psi_{\tanh}(u) = \tanh(u) = \frac{e^u - e^{-u}}{e^u + e^{-u}}. \tag{48}$$

These two activation functions (compare Fig. 28, upper left) are essentially identical as $\tanh(u) = 2\psi_{\log}(2u) - 1$. They differ only by their range of values, which is [0,1] for the logistic function and $[-1, +1]$ for the hyperbolic tangent. Also, $\psi_{\log}(u) > 0$, whereas $\psi_{\tanh}(-u) = -\psi_{\tanh}(u)$, which both may be useful properties in one or the other situation.

Another popular activation function is $\psi_{\text{RBF}}(u) = e^{-u^2}$, which is the density of the standard normal distribution up to standardization. Neural networks using this function type are called **radial basis function networks**.

Intuitively, sigmoid activation functions are suitable for many regression applications, as they cover typical shapes. They are approximately linear in the central part around 0 and slope off to constants for $u \to \pm\infty$. This corresponds to the frequently observed fact that physical processes show an approximately linear behavior in a certain range of nice inputs, but are not increasing over all bounds for outlying inputs, e.g., due to saturation effects. More precisely, due to the influence of weights, the components of the neural network function (45) are scaled and shifted versions of $\psi(u)$:

$$\psi_{s\mu}(u) = \psi(\frac{u-\mu}{s}).$$

This function is centered around μ, such that it can be moved around on the real line. Due to the scaling by $s > 0$, it becomes almost linear over a widening range if s increases (compare Fig. 28, upper right, where $\psi = \psi_{\log}$ and $s = 10$), and in the other extreme, it becomes closer and closer to a pure threshold or jump function with a jump at μ and constant otherwise (compare Fig. 28, lower left, where $\psi = \psi_{\log}$ and $s = 0.1$). Figure 28 (lower right) shows $\psi_{\log}(10(x+2)) - \psi_{\log}(10(x-2))$ as an approximation of the indicator function of the interval $[-2, +2]$. In general, using two suitably shifted sigmoid functions with weights v_k of different sign and small s, we may approximate the indicator functions of arbitrary intervals.

The latter fact shows that the family $\mathscr{F}_N^{\text{nn}}, N \geq 1$ of neural network output functions has a **universal approximation** property among all integrable functions as we know from integration theory that linear combinations of interval indicator functions have this property, and they may be approximated arbitrarily closely by linear combinations of neural network output functions. A formal proof of the universal approximation property relies again on Weierstrass' theorem (compare [46]).

As we already have mentioned, the parameters or weights are **not identifiable** from the function (45), i.e., quite different sets of parameter values result in identical functions. If we are only interested in estimation or forecasting, this is usually not a problem. Problems may arise in Monte Carlo studies or in resampling approaches like the bootstrap (compare Sect. 9.1), where the variability of network-related quantities is studied by repeatedly applying the network algorithm to similar training sets. Depending on the quantities of interest, the non-identifiability of weights may cause problems. A straightforward solution to this problem is the restriction of the parameter set Θ_N in (46) from the full space to a subset where identifiability is guaranteed for almost all parameters, but the universal approximation property is retained.

Non-identifiability is caused by the arbitrariness of the enumerations of the neurons in the hidden layer, i.e., the terms of the sum (45), and by symmetry properties of the activation function. For example, $\psi_{\tanh}(u) = -\psi_{\tanh}(-u)$, i.e., the function values do not change if, for any $h \geq 1$, we multiply v_h and all w_{ih} by -1. In this case, we get identifiability for almost all ϑ, if we define Θ_N as the set of parameter vectors satisfying an order and nonnegativity condition on the output weights $v_1 \geq v_2 \geq \ldots \geq v_H \geq 0$.

As an example, we consider the training sample of size $N = 51$ and one-dimensional inputs X_j, which we already have used for Fig. 7. The true regression function $m(x)$ is estimated by feedforward neural networks (45) with activation function tanh and increasing number $H = 4, 6, 9$ of neurons. Figure 29 (left) shows the true function (red) and the estimate $f_4(x; \hat{\vartheta}_N)$ (blue), which looks quite reasonable. On the right-hand side, $f_6(x; \hat{\vartheta}_N)$ (magenta) and $f_9(x; \hat{\vartheta}_N)$ (black) are plotted together with the data. Both show considerable overfit (compare Sect. 2.3), trying to even interpolate some of the noisy data.

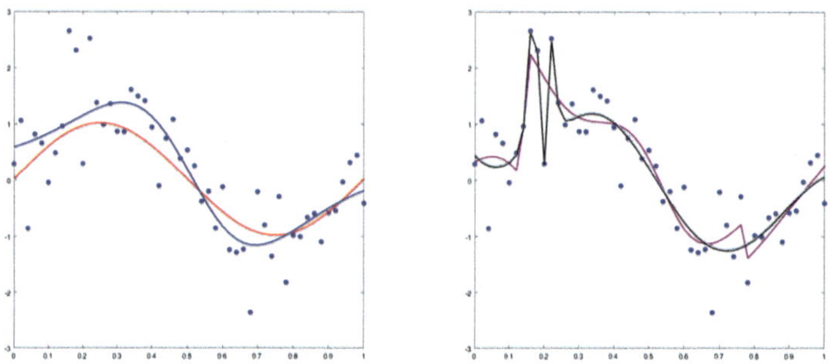

Fig. 29 Data, true function $m(x)$ (red) and neural network estimate with $H = 4$ (blue) on the left-hand side resp. $H = 6$ (magenta) and $H = 9$ (black) on the right-hand side

8.1 From Perceptron to Neural Network

The perceptron of [68] is the ancestor of neural networks. Originally, it presented a mathematical model for a biological neuron, which becomes active (fires) or not, i.e., it creates an output signal 1 or 0, depending on d sensory inputs, which are signals sent from other neural cells. Hence, it works as a classification rule with only $M = 2$ classes and can be represented by an input-output relation of the form (50) with only one neuron ($H = 1$) and, hence, only one set of input weights $w_i, i = 0, \ldots, d$, with activation function $\psi(u) = 1_{[0,\infty)}(u)$ and $v_0 = 0, v_1 = 1$. With the parameter vector $\vartheta = (w_0, \ldots, w_d)$, this particular simple version

$$c_{pc}(x; \vartheta) = \psi\big(w_0 + \sum_{i=1}^{d} w_i x_i\big) = 1_{[0,\infty)}\big(w_0 + \langle \mathbf{w}, x \rangle\big)$$

of (50) is called the perceptron function, where $\mathbf{w} = (w_1, \ldots, w_d)$ and $\langle \cdot, \cdot \rangle$ denotes the scalar product in d-dimensional input space. The perceptron function is represented by the diagram in Fig. 30, where the neuron fires if the weighted sum of the inputs lies above the threshold w_0.

If we want to learn the weights of the perceptron from a training set, we can apply empirical risk minimization as in Sect. 6.2. As, here, we try to learn a classification function directly, we use the probability of misclassification as risk function, and

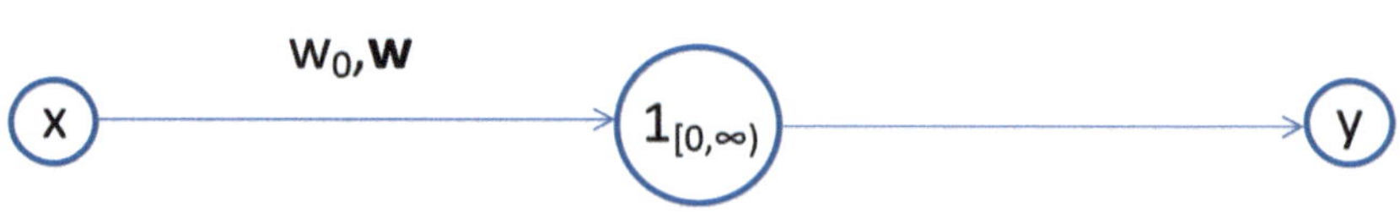

Fig. 30 The perceptron function

the corresponding empirical risk is the relative frequency of misclassifications in the training sample, depending on the weights $\vartheta = (w_0, \ldots, w_d)$:

$$\widehat{R}(\vartheta) = \frac{1}{N} \sum_{j=1}^{N} L\big(Y_j, c_{pc}(X_j; \vartheta)\big),$$

where $L(y, c) = 1$ if $y \neq c$, and $= 0$, else. We already have presented the Widrow-Hoff algorithm as a method for learning the perceptron weights from training data and for finding a perfect classification, provided there is one. As we have discussed in Sect. 5.3.1, there are many situations, where the perceptron classificator fails.

Motivated by the fact that biological neural nets consist of many neurons and in the hope of improving the ability to solve many different classification tasks, later on artificial neural networks were considered, which consist of many perceptrons. Also, to extend the applications from classification to general function estimation, the thresholding of (50) was replaced by general sigmoid functions ψ. The perceptron building blocks were arranged in several layers such that feedforward neural networks are also known as *multilayer perceptrons*. In this chapter, we only consider networks with one layer of perceptrons and one output, i.e., with output functions (45) represented by the diagram in Fig. 31.

The network consists of three layers: from left to right, the input layer containing the feature vector x, the **hidden layer** with H neurons and a constant 1, and finally the output layer with one node corresponding to the output y. As a **feedforward network**, all edges that originate from a node end up in a node of a subsequent layer, never in a node of the same or a previous layer.

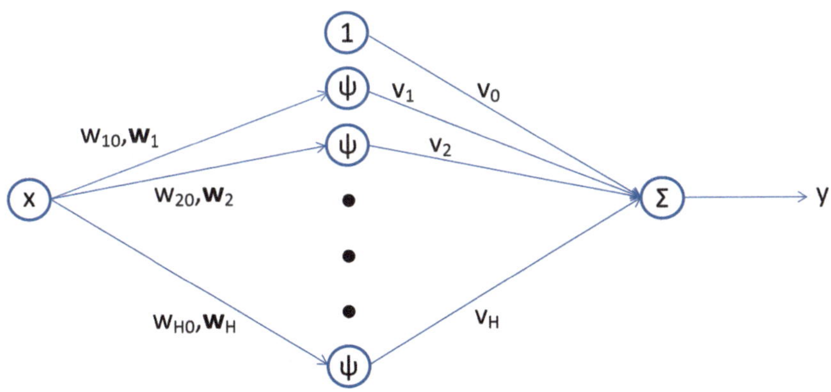

Fig. 31 Feedforward neural network with output function 45

8.2 Backpropagation

Backpropagation is a well-known learning algorithm, which allows feedforward networks to learn their weights from the training set. Mathematically, it is nothing but a special numerical procedure for solving the nonlinear least-squares problem of minimizing (47) w.r.t. to the weight vector ϑ. Backpropagation requires not much storage but may converge slowly or even not at all.

For the sake of illustration, we consider a neural network with only one neuron in the hidden layer representing the function

$$y = \psi(w_0 + \sum_{i=1}^{d} w_i x_i) = \psi(w_0 + < \mathbf{w}, x >),$$

where $w_i = w_{i1}$, $< \mathbf{w}, x >$ denotes the scalar product of the input weight vector and the input vector, and ψ is any differentiable activation function, e.g., the logistic one. For the sake of simplicity, we also fix the weight w_0 of the constant in advance such that the loss function $Q(\vartheta) = Q(\mathbf{w})$ of (47) depends only on the input weights $\mathbf{w} = (w_1, \ldots, w_d)$.

An elementary numerical procedure for minimizing Q is **gradient descent**. Starting from some weight vector $\mathbf{w}(n)$, the next approximation is calculated by proceeding into the direction where Q decreases as fast as possible:

$$\mathbf{w}(n+1) = \mathbf{w}(n) - \eta \operatorname{grad} Q(\mathbf{w}(n)),$$

$$\operatorname{grad} Q(\mathbf{w}) = -\sum_{j=1}^{N} 2\Big(Y_j - \psi(w_0 + < \mathbf{w}), X_j >)\Big)\psi'(w_0 + < \mathbf{w}, X_j >)X_j.$$

To improve the rate of convergence, the tuning constant $\eta > 0$ may be decreased in the course of the iteration. The gradient descent algorithm judges the quality of the input weights $\mathbf{w}(n)$, i.e., of the current network, by looking simultaneously at all the data in the training set. The network is applied to all inputs $X_1, \ldots, X_N$, and the weight vector is modified only afterward.

Backpropagation is kind of a gradient descent procedure, but it looks at single data from the training set one after the other. The network is applied successively to the X_j, and after each single step, the weights are modified, again in the direction of steepest descent but of the component function $Q_j(\mathbf{w}) = \big(Y_j - f_H(X_j; \vartheta)\big)^2$ instead of $Q(\vartheta)$. In our example with only one neuron and fixed w_0, this results in

$$\mathbf{w}(n+1) = \mathbf{w}(n) - \eta \operatorname{grad} Q_j(\mathbf{w}(n)), \quad j = 1, \ldots, N,$$

$$\operatorname{grad} Q_j(\mathbf{w}) = -2\Big(Y_j - \psi(w_0 + < \mathbf{w}, X_j >)\Big)\psi'(w_0 + < \mathbf{w}, X_j >)X_j.$$

After one pass through the N training data, the iteration starts over again with the first data point. N steps of backpropagation correspond roughly to one step of a batch-mode algorithm like gradient descent.

The Widrow-Hoff learning rule for the perceptron (compare Sect. 5.3.1) is essentially a backpropagation algorithm, keeping in mind that $\psi(u)$ is replaced in the definition of $Q(\mathbf{w})$ by the thresholding function $1_{[\tau,\infty)}(u)$, which is not differentiable. After presenting (X_j, Y_j) to the perceptron, the weights are modified into the direction of steepest descent of $Q_j(\mathbf{w})$ too, i.e., into the direction of X_j for $Y_j = 1$, $\psi(w_0 + < \mathbf{w}, X_j >) = 0$ and into the direction $-X_j$ for $Y_j = 0$, $\psi(w_0 + < \mathbf{w}, X_j >) = 1$. For correctly classified (X_j, Y_j), the weights remain unchanged.

Backpropagation is quite popular as it is fast and easy to implement. However, it has two essential drawbacks. If, in each pass through the sample, the training data $(X_j, Y_j), j = 1, \ldots, N$, are used in their original order of enumeration, then it may happen that the algorithm cycles through the same values and never converges. This can be avoided by using the data in a randomized order in each round of N backpropagation steps.

The other drawback is that in contrast to the usual situation in statistical estimation by minimization (e.g., least-squares) or maximization (e.g., maximum likelihood), the error introduced by numerical approximation of the solution is **not negligible**. For backpropagation, it is of the same order as the statistical error due to the randomness of the data! A precise version of this statement has been given by White [81], which will be discussed briefly in the next section. An easy remedy is to use backpropagation for calculating quick and reasonably good approximations of the optimal weights, which then are used as starting values of a batch-mode optimization like gradient descent. Usually, few or even one iteration of the latter suffices to get estimates that are statistically efficient (compare [81] for the details).

Of course, one may use any other numerical algorithm for solving nonlinear extremum problems, to determine the weights corresponding to a minimum of the function $Q(\mathbf{w})$. In particular, conjugate gradient methods seem to work well frequently in applications. Of course, all algorithms run the risk to end in a local minimum of $Q(\mathbf{w})$. If the solution shows a suboptimal performance, it may be worthwhile to run the numerical optimization algorithm with different starting values.

8.3 *Classification with Neural Networks*

To conclude the introduction to neural networks, let us remark that (45) works well in regression and forecasting. In classification, we want to estimate class probability functions $p_\ell(x)$ (compare Sect. 2.4), which only assume values between 0 and 1 and sum to 1. To guarantee that this holds also for the network output function, we additionally apply a squashing function to (45), which maps the real line to [0,1].

If we only have $M = 2$ classes, coded, e.g., as 0 and 1, the logistic function is the most popular choice. We approximate $p_1(x)$ by functions of the form

$$g_H(x; \vartheta) = \psi_{\log}\Big(v_0 + \sum_{h=1}^{H} v_h \psi\big(w_{0h} + \sum_{i=1}^{d} w_{ih} x_i\big)\Big), \tag{49}$$

and due to $p_1(x) + p_0(x) = 1$, we immediately get an estimate for the second class probability $p_0(x)$, too. If we are only interested in the final decision between classes 0 and 1, we may replace $\psi_{\log}$ by the indicator function $1_{[0,\infty)}(u)$ of the nonnegative numbers to get immediately the classification function

$$c_H(x; \vartheta) = 1_{[0,\infty)}\Big(v_0 + \sum_{h=1}^{H} v_h \psi\big(w_{0h} + \sum_{i=1}^{d} w_{ih} x_i\big)\Big). \tag{50}$$

Instead of squashing, we apply a thresholding operation to a linear combination of the outputs of the individual H neurons, and we classify an object as 1 if this linear combination is $\geq -v_0$ and as 0, else.

This approach can be generalized to $M > 2$ classes. Here, it is convenient to code the classes by dummy variables, which are unit vectors in M-dimensional space (compare Sect. 12.1). For $M = 3$, e.g., the class variables Y_j from the training set would assume values in $\{(1, 0, 0), (0, 1, 0), (0, 0, 1)\}$. The class probabilities given the d-dimensional input $X_j = x$ also are three-dimensional vectors $(p_1(x), p_2(x), p_3(x))$ with nonnegative entries summing to 1.

To estimate the class probabilities, we start with a neural network of the form (45) but with M different outputs:

$$z_\ell = v_{0\ell} + \sum_{h=1}^{H} v_{h\ell} \psi\big(w_{0h} + \sum_{i=1}^{d} w_{ih} x_i\big) = f_{H\ell}(x; \vartheta), \quad \ell = 1, \ldots, M,$$

where now the parameter vector ϑ consists of all $M + H(M + d + 1)$ weights $v_{0\ell}, v_{h\ell}, w_{ih}$.

Again, the z_ℓ are no probabilities, such that we have to apply some squashing to the linear activation functions, describing the transformation from the hidden to the output layer. We use the generalization of the logistic function from the binomial to the multinomial case, which, in the machine learning literature, is called the **softmax** activation function. The M-dimensional output function $g_H(x; \vartheta) = (g_{H1}(x; \vartheta), \ldots, g_{HM}(x; \vartheta))$ of the resulting network is given by

$$g_{H\ell}(x; \vartheta) = \frac{e^{z_\ell}}{\sum_{k=1}^{M} e^{z_k}}, \quad \ell = 1, \ldots, M. \tag{51}$$

Replacing ϑ by the estimate $\hat{\vartheta}$ from the training set, $g_{H\ell}(x; \hat{\vartheta})$ approximates the class probability $p_\ell(x)$. To get the final classification, we apply the Bayes classifier (compare Sect. 2.4), i.e., we decide for class k if $g_{Hk}(x;\vartheta) \geq g_{H\ell}(x;\vartheta), \ell = 1, \ldots, M$, or, equivalently, $z_k \geq z_\ell, \ell = 1, \ldots, M$, due to monotonicity of the softmax transformation.

To estimate the network weights ϑ from a training set $(X_j, Y_j), j = 1, \ldots, N$, it would be principally possible to again apply least squares, i.e., to minimize $Q(\vartheta) = \sum_{j=1}^{N} ||Y_j - g_H(X_j;\vartheta)||^2$ w.r.t. ϑ as in (47). However, it is more efficient to exploit the relationship between the softmax activation function and multinomial logistic regression and use a maximum likelihood approach instead (compare Sect. 5.2 for the case $M = 2$). Therefore, instead of minimizing $Q(\vartheta)$, we maximize the multinomial log-likelihood function

$$L(\vartheta) = \sum_{j=1}^{N}\sum_{\ell=1}^{M} Y_{j\ell} \log g_{H\ell}(x;\vartheta) = \max_{\vartheta \in \Theta_N}!$$

or, equivalently, to minimize the negative log-likelihood $-L(\vartheta)$, which is also called the *cross-entropy* (compare [67, Section 5.3], or [47, Section 10.2], for a more detailed discussion).

As an example, we consider the training and the validation sets used for Figs. 18 and 19 to demonstrate the application of linear and quadratic discriminant analysis. Figure 32 shows on the left-hand side the training set with two-dimensional feature vectors X_j belonging to $M = 3$ classes, which are color-coded, and on the right-hand side an identically generated but independent validation set. As in Figs. 18 and 19, we plot the relevant parts of the decision boundaries, which partition the feature space into three parts. If, e.g., a feature vector $X_{N+1} = x$ is observed in the lower left region, containing most of the observations from class 1 (blue), then it is also classified as class 1, i.e., using the dummy variable coding, $\widehat{Y}_{N+1} = (1, 0, 0)$. Here, the three decision boundaries are the sets $\{x;\, \hat{p}_k(x) = g_{Hk}(x;\hat{\vartheta}) = g_{H\ell}(x;\hat{\vartheta}) = \hat{p}_\ell(x)\}, 1 \leq k < \ell \leq 3$, where $g_{Hk}(x;\hat{\vartheta})$ is given by (51) with $\hat{\vartheta}$ maximizing $L(\vartheta)$. In Fig. 32, decision boundaries from neural network classifiers

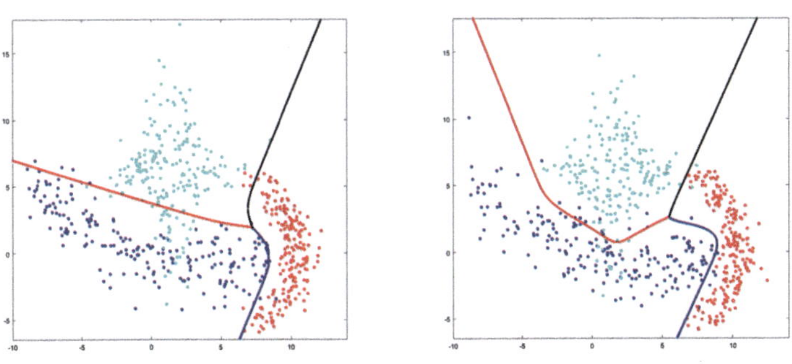

Fig. 32 Two-dimensional input data from three classes marked in different colors (left, training set; right, validation set) with decision boundaries from neural network classifiers (left, $H = 2$; right, $H = 6$)

with $H = 2$ neurons (left-hand side) and with $H = 6$ neurons (right-hand side) are shown.

Using more neurons and, hence parameters, allows for a more complex decision rule but may lead to overfit. This is rather unlikely here, as the sample size $N = 600$ is considerably larger than the number $M + H(M + d + 1) = 36$ of fitted network weights for the larger network; nevertheless, we test the capability for generalization by applying the decision rule to a validation set. Table 5 shows the relative number of misclassifications in the training set (in-sample error) and in the validation set (out-of-sample error) for networks with $H = 2, 4, 6$ neurons. Comparing the results with Table 3, $H = 2$ performs only slightly worse than quadratic discriminant analysis. For $H = 4$ and $H = 6$, we get considerably smaller error rates in-sample and out-of-sample. Maybe even larger networks may lead to further small improvements. We have not tested this, as straightforward feedforward neural networks with only one hidden layer anyhow are rarely the best machine learning approach for classification. Most successful classification examples use several layers and/or particularly structured networks. This can be easily seen from an illuminating, still rather simple example in the context of automatic digit recognition (with $M = 10$ classes), discussed by Hastie et al. [44] (Section 11.7).

Figure 33 shows the class probability estimates $g_{Hk}(x; \hat{\vartheta}), k = 1, 2, 3$, from neural network classifiers with 2 resp. 6 neurons. In training and validation set, one third each of the data belonged to each of the three classes. Looking, in particular, at the observations from class 1, we see that increasing the number of neurons H not only leads to less misclassifications but also to more clear-cut decisions. For $H = 2$, the estimated class probability $\hat{p}_1(X_j) = g_{H1}(X_j; \hat{\vartheta})$ for features from class 1 (blue) is in most cases slightly large than 0.8, leading to a correct classification. For

Table 5 Misclassification rate in training and validation sample of Fig. 32

H	2	4	6
Training sample	7.8 %	4.0 %	4.0 %
Validation sample	8.0 %	6.5 %	5.8 %

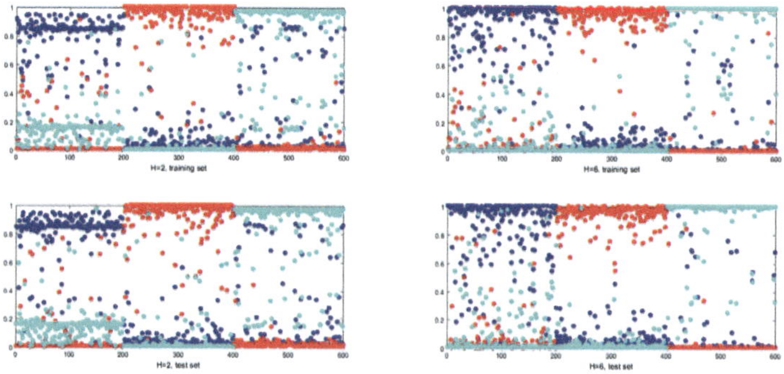

Fig. 33 Class probability estimates $g_{H1}(x; \hat{\vartheta})$ (blue), $g_{H2}(x; \hat{\vartheta})$ (red), $g_{H3}(x; \hat{\vartheta})$ (light blue), evaluated at training set (upper row) and validation set (lower row) for $H = 2$ (left) and $H = 6$ (right)

$H = 6$, however, most $\hat{p}_1(X_j)$ are close to 1 such that there remains now doubt that these data belong to class 1.

8.4 Neural Network Regression Statistics

In this section, we study the properties of neural network regression estimates $\widehat{m}_N(x) = f_H(X_j; \hat{\vartheta}_N)$ in more detail. We consider the general regression model (2) with the usual independence assumptions on inputs and residuals. We also assume that we consider only weight vectors ϑ in a subset Θ_N in which they are almost everywhere identifiable.

Let us first consider the case where the number of neurons H is fixed and $N \to \infty$. Then, $\widehat{m}_N(x)$ converges to the output function $f_H(x; \vartheta_{\text{opt}})$ of a particular neural network with H neurons. Its weight vector ϑ_{opt} provides the best fit to the unknown $m(x)$ among all network functions in $\mathscr{F}_N^{\text{nn}} = \{f_H(x; \vartheta); \vartheta \in \Theta_N\}$ in the sense

$$D_0(\vartheta_{\text{opt}}) = \min_{\vartheta \in \Theta_N} D_0(\vartheta), \qquad \text{with}$$

$$D_0(\vartheta) = \mathrm{E}\left(m(X_j) - f_H(X_j; \vartheta)\right)^2 = \int \left(m(x) - f_H(x; \vartheta)\right)^2 p(x) dx_1 \ldots dx_d,$$

where $p(x)$ is the density of $X_1, X_2, \ldots$.

A sample approximation of the distance D_0 is given by

$$D_N(\vartheta) = \frac{1}{N} \sum_{j=1}^{N} \left(m(X_j) - f_H(X_j; \vartheta)\right)^2.$$

If $D_N(\vartheta_N) = \min_\vartheta D_N(\vartheta)$, then ϑ_N is a sample approximation of the optimal ϑ_{opt}.

To describe the behavior of $\widehat{m}_N(x)$ for increasing N, [28] introduce three auxiliary quadratic matrices. $A(\vartheta)$ has the entries $\frac{\partial^2}{\partial \vartheta_k \partial \vartheta_\ell} D_0(\vartheta)$, i.e., it is the Hesse matrix of $D_0(\vartheta)$.

$$B_1(\vartheta) = 4\sigma_\varepsilon^2 \,\mathrm{E}\, G_H(X_j; \vartheta), \qquad B_2(\vartheta) = 4\,\mathrm{E}\left(m(X_j) - f_H(X_j; \vartheta)\right)^2 G_H(X_j; \vartheta),$$

where $G_H(x; \vartheta)$ denotes the quadratic matrix with entries $\frac{\partial}{\partial \vartheta_k} f_H(x; \vartheta) \frac{\partial}{\partial \vartheta_\ell} f_H(x; \vartheta)$. Under rather weak regularity conditions, in particular on $\psi(u)$, we have that not only $\hat{\vartheta}_N$ converges to ϑ_{opt} for increasing N but that the estimation error is approximately multivariate normally distributed, more precisely for large enough N

$$\sqrt{N} \begin{pmatrix} \hat{\vartheta}_N - \vartheta_N \\ \vartheta_N - \vartheta_{\text{opt}} \end{pmatrix} \text{ approximately distributed as } \mathscr{N}\left(0, \begin{pmatrix} \Sigma_1 & 0 \\ 0 & \Sigma_2 \end{pmatrix}\right), \tag{52}$$

with $\Sigma_i = A(\vartheta_{\text{opt}})^{-1} B_i(\vartheta_{\text{opt}}) A(\vartheta_{\text{opt}})^{-1}$, $i = 1, 2$ [28].

The estimation error $\hat{\vartheta}_N - \vartheta_{\text{opt}}$ decomposes into two components $\hat{\vartheta}_N - \vartheta_N$ and $\vartheta_N - \vartheta_{\text{opt}}$, which represent the variance part and the bias part of the error. They are approximately independent for large N, as the off-diagonal blocks of the covariance matrix in (52) are 0. In particular, the bias part is asymptotically negligible if $m(x) = f_H(x; \vartheta_{\text{opt}})$, i.e., the network correctly specifies the true regression function m for the given finite number H of neurons. Usually, that will not be the case, and we need to increase H with N such that the bias component of the error becomes smaller, but not too fast to keep the variance in check. This again illustrates the usual bias-variance dilemma (compare Sect. 2.3).

Note that as an immediate consequence, $\sqrt{N}(\hat{\vartheta}_N - \vartheta_{\text{opt}})$ is asymptotically normal, too, with asymptotic covariance matrix $\Sigma = \Sigma_1 + \Sigma_2$. Using this terminology, the result about the asymptotic inefficiency of backpropagation (compare [81]) can be stated as follows. Let $\hat{\vartheta}_N^{\text{bp}}$ denote the numerical approximation of the least-squares estimate $\hat{\vartheta}_N$ by the backpropagation algorithm. Then, $\sqrt{N}(\hat{\vartheta}_N^{\text{bp}} - \vartheta_{\text{opt}})$ is asymptotically normal with mean 0 and covariance matrix $\Sigma^{\text{bp}} > \Sigma$, i.e., $\Sigma^{\text{bp}} - \Sigma$ is positive definite. Therefore, the variability of $\hat{\vartheta}_N^{\text{bp}}$ is larger than that of $\hat{\vartheta}_N$ even for very large N, and, hence, a network regression function estimate is suboptimal if it is based only on backpropagation.

If the activation function $\psi(u)$ is smooth enough, then we get easily from (52) that

$$\sqrt{N}\big(\widehat{m}_N(x) - f_H(x; \vartheta_{\text{opt}})\big) \text{ is approximately normal for large } N,$$

In particular, $\widehat{m}_N(x) \rightarrow f_H(x; \vartheta_{\text{opt}})$ for all x where ϑ_{opt} depends on the fixed number H of neurons.

However, for increasing N, i.e., for an ever-growing amount of information, we would like to get better and better approximations of the regression function $m(x)$ and, hence, forecasts. We know already that, for fixed H and large enough N, $\widehat{m}_N(x) \approx f_H(x; \vartheta_{\text{opt}})$. As the family $\mathscr{F}_N^{\text{nn}}$, $N \geq 1$, of neural networks has a universal approximation property, we also have $f_H(x; \vartheta_{\text{opt}}) \approx m(x)$ for large enough H. Hence, the goal, that $\widehat{m}_N(x) \rightarrow m(x)$, is achievable if H grows with N. Again we are confronted with the bias-variance dilemma, as H and with it the number of network weights must not grow too fast to avoid overfit.

We have a brief look at the corresponding theory, similar to the discussion at the end of Sect. 6.2. We again have to add some boundedness conditions, e.g., $0 \leq \psi(u) \leq 1$, and an upper bound β on the total size of output weights such that $|f_{H_N}(x; \vartheta)| \leq \beta_N$. As this bound is allowed to grow with N, the subclasses of $\mathscr{F}_N^{\text{nn}}$ given by

$$\mathscr{F}_N^{\beta} = \{f_H(x; \vartheta); \vartheta \in \Theta_N, \sum_{h=0}^{H} |v_h| \leq \beta_N\}.$$

still have a universal approximation property.

Let $\widetilde{m}_N$ denote the sieve estimate, which minimizes the empirical risk $\widehat{R}(f)$ over the bounded neural network output functions in $\mathscr{F}_N^{\beta}$. Then (Theorem 16.1 of [36]),

$$||\widetilde{m}_N - m||^2 = \int \left(\widetilde{m}_N(x) - m(x)\right)^2 p(x)dx \to 0$$

almost surely, if $H, \beta_N \to \infty$ for $N \to \infty$ such that for some $\delta > 0$

$$\frac{H\beta_N^4 \log(H\beta_N^2)}{N} \to 0 \qquad \text{and} \qquad \frac{\beta_N^4}{N^{1-\delta}} \to 0.$$

Here, $p(x)$ again denotes the density of the inputs X_j. From the rate condition on the left-hand side, we can choose H rather large and β_N small or vice versa. Both have to be considerably smaller than N.

8.5 Network Regularization and Specification

In this section, we assume the training set (X_j, Y_j), $j = 1, \ldots, N$, and, hence, the dimension d of the input vectors given. Questions of feature selection, which also are important for the performance of learning rules, are discussed in Sect. 12.

For learning algorithms, the performance of regression estimates, classificators, and forecasts depends on the choice of hyperparameters. Sieve estimates, in particular, consist of growing classes of parametrized functions, and here, the number of parameters to be learned from the training set is crucial. For general neural networks, the number of weights depends on the number d of components of the input vector as well as on the number of neurons. However, the structure of the network is also very important. A feedforward network with only one hidden layer, in which all nodes of the input layer are connected with all the H neurons of the hidden layer, may be outperformed by networks with a different structure and considerably less parameters, if we allow for more than one, say K, hidden layers with $H_1 + \ldots + H_K \ll H$. It may also be profitable, to consider networks that are not completely connected, which means that we set certain weights to 0 in advance and such reduce the number of free parameters.

If we want to apply neural networks to a given training set, we therefore have to decide upon the structure of the network, i.e., for feedforward networks:

- How many hidden layers K should we consider?
- Of how many neurons H_k does the kth hidden layer consist?
- Which nodes of the network (inputs, hidden neurons, outputs) have to be connected?

A parsimonious network is desirable, i.e., a network prescribed by just enough, but not too many parameters to provide an adequate fit to the training data. For network specification, the same kind of ideas as for other statistical learning methods are used. We give a brief review of some of them for the running example of this chapter, a feedforward network with only one hidden layer and one output. Here, the only hyperparameter, determining the network structure, is the number H of neurons in the hidden layer.

Weight Decay

The popular method of weight decay in training a neural network is closely related to the ideas of enforcing sparsity and to avoid overfit in ridge regression (compare Sect. 3.2). Here, the choice of H, which may be rather large, does not matter so much as we restrict the overall size of parameters by applying penalized least squares. Instead of minimizing the original empirical risk, we minimize a regularized version

$$\widehat{R}_\lambda(\widehat{m}_{N,\lambda}) = \min_{f \in \mathscr{F}_N^{\text{nn}}} \frac{1}{N} \sum_{j=1}^{N} (Y_j - f(X_j))^2 + \lambda\Big(\sum_{h=0}^{H} v_h^2 + \sum_{i=0}^{d} \sum_{h=1}^{H} w_{ih}^2 \Big). \qquad (53)$$

Alternative penalties may also be used, e.g., based on the L^1-norm of the weight vector as in the LASSO (compare Sect. 3.3).

Let $\hat{\vartheta}_N(\lambda)$ denote the weight vector of the neural network output function $f_H(x;\theta)$ minimizing the penalized empirical risk in (53). Figure 34 demonstrates the effect of weight decay with $\lambda = 0.1$ and quadratic penalty. For the same data as used in Fig. 28, the left-hand side shows the data, the true function $m(x)$ (red) and the regularized estimate $f_9(x;\hat{\vartheta}_N(\lambda))$ (black). $f_4(x;\hat{\vartheta}_N(0))$ (blue), which is a reasonable estimate without weight decay, is also shown for comparison. Both estimates look quite similar, but the regularized estimate with $H = 9$ and weight decay turns to be better from comparing the plots of the estimation errors $m(x) - f_4(x;\hat{\vartheta}_N(0))$ (blue) and $m(x) - f_9(x;\hat{\vartheta}_N(\lambda))$ (black) on the right-hand side. Note that without weight decay, the estimate using $H = 9$ neurons shows considerable overfit (compare Fig. 28).

The regularization weight λ is, of course, another hyperparameter that has to be chosen somehow. From experience, it often suffices to try out a few different values

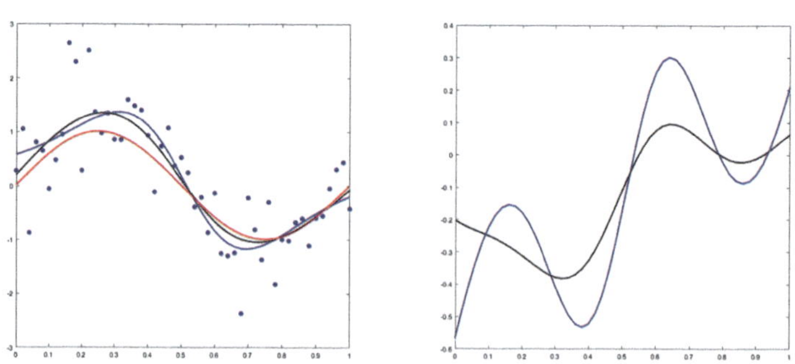

Fig. 34 Data, true function $m(x)$ (red), neural network estimates with $H = 4, \lambda = 0$ (blue) resp. $H = 9, \lambda = 0.1$ (black) on the left-hand side and estimation errors on the right-hand side

to find a neural network that performs well enough. For a more systematic approach, we could use crossvalidation or other methods for selecting this hyperparameter. In contrast to the direct choice of the discrete H, this has the advantage that λ has a continuous range. Moreover, using weight decay for multilayer networks, we do not have to worry about K hyperparameters, i.e., the sizes $H_k, k = 1, \ldots, K$, of the hidden layers and can focus in choosing the one decisive hyperparameter λ instead.

For multilayer neural networks and also for other statistical learning approaches with classifiers or predictors of a rather complex structure, focusing on the overall size of weights as a measure of complexity and, hence, the tendency of the learning algorithm toward overfit may be an oversimplification due to the interdependencies of the components of the input-output function. A more careful investigation on the complexity of the used function class, e.g., of (46), would be recommendable. For theoretical purposes, there are general measures like covering numbers or the Vapnik-Chervonenkis (VC) dimension based on the work of Vapnik [80], which are fundamental for proving the consistency of neural networks, i.e., in a regression setting, their ability to learn every reasonable function $m(x)$ for increasing training sample size N and correspondingly increasing function class (46) (compare, e.g., [36], Chapter 16). However, they are rather difficult to use in practice.

Repeated Significance Testing

The question of selecting a model of appropriate size for fitting a regression function to the training set has already a long history in classical linear regression (compare Sect. 3.1). There, one popular approach to select hyperparameters, representing the number of model parameters, is repeated significance testing, e.g., in the version of *stepwise regression*. We start with a very simple model, then add one model component after the other, and subsequently test if the component contributes significantly to explaining the data-generating mechanism. We stop once adding to the complexity of the model does not lead to further improvement.

This idea may also be used for choosing the number H of neurons in the hidden layer where some care has to be taken due to peculiarities of the neural network output functions. We start by fitting the function (45) with $H = 1$ to the training set. Then, we add one neuron after the other to the network. At the beginning of step no. $I - 1$ of the iteration, we already have $H = I - 1$ neurons and now add the next one by minimizing the empirical loss (47) w.r.t. ϑ for $H = I$. In this larger model, we test the hypothesis $v_I = 0$ against the alternative $v_I \neq 0$. If the test accepts the hypothesis, the last neuron does not really contribute to the output as its output weight is not significantly different from 0. In this case, we stop at $H = I - 1$; otherwise, we proceed with the iteration.

Testing for the non-vanishing output weight v_I is a nonstandard statistical testing problem as parameters $w_{0I}, \ldots, w_{pI}$ are not identifiable, i.e., completely arbitrary, under the hypothesis. So we have to use particular tests, e.g., Lagrange multiplier tests [21], also known as score tests, which are tuned to such a situation.

Crossvalidation and Validation

In statistical learning for regression and classification, where the training set consists of independent observations, (X_j, Y_j), crossvalidation (compare Sect. 3.4) is a popular method for choosing hyperparameters. For neural network regression, e.g., the leave-one-out approach has the following form.

For every given $\ell = 1, \ldots, N$, we remove the ℓth observation from the training set, resulting in N pseudo samples of size $N-1$ each, which have the same structure as the original data. We fit a neural network with H neurons to each of those pseudo samples by minimizing (47) resulting in estimates $\hat{\vartheta}_N^{(-\ell)}, \ell = 1, \ldots, N$. Then, for each ℓ, we apply the resulting output function $f_H(x; \hat{\vartheta}_N^{(-\ell)})$ to forecasting the left-out Y_ℓ from X_ℓ. As the latter are independent of the ℓth pseudo sample, the squared forecasting error gives a fair impression of the performance of the predictor. Averaging over ℓ, we get

$$\mathrm{CV}(H) = \frac{1}{N} \sum_{\ell=1}^{N} \left(Y_\ell - f_H(X_\ell; \hat{\vartheta}_N^{(-\ell)})\right)^2$$

as a valid approximation of the generalization error (21) as a function of H, i.e., in our case

$$\mathrm{mspe}(H) = \mathrm{E}\,\{\left(Y_{N+1} - f_H(X_{N+1}; \hat{\vartheta}_N)\right)^2 | \mathscr{T}\}, \tag{54}$$

where the expectation is computed given that the training set $\mathscr{T}$ is known. Hence, we choose $H = \widehat{H}_{\mathrm{CV}}$ as the number of neurons that minimizes $\mathrm{CV}(H)$.

As crossvalidation is widely applied to forecasting with neural networks, we want to add a word of warning. If the training data are part of a signal observed over time, i.e., a time series, then crossvalidation does not necessarily result in a good choice of H (compare Sect. 14.2). Due to the dependence structure, the observation left out still depends, and usually rather strongly, on the remaining data, i.e., the ℓth pseudo-sample, such that the above argument does not hold any longer.

The general approach of crossvalidation, i.e., approximating the generalization error by the average error over pseudo samples, can also be used with different resampling methods. For example, we can use a residual-based bootstrap method as described in Sect. 9.1.2 to generate bootstrap pseudo-samples $(X_j^b, Y_j^b), j = 1, \ldots, N$, for $b = 1, \ldots, B$ such that each of the pseudo-samples mimics the structure of the original training set. Then, we fit the neural network to each of the subsamples resulting in weight vectors $\hat{\vartheta}_N^b, b = 1, \ldots, B$, and approximate (54) by

$$\widehat{\mathrm{mspe}}_{rb}(H) = \frac{1}{B} \sum_{b=1}^{B} \frac{1}{N} \sum_{j=1}^{N} \left(Y_j - f_H(X_j; \hat{\vartheta}_N^b)\right)^2.$$

As this kind of bootstrap is known to work for neural networks [28], we may choose $H = \widehat{H}_{rb}$ as the minimizer of $\widehat{\mathrm{mspe}}_{rb}(H)$. For, e.g., $B = N$, the computational effort is comparable to crossvalidation, as fitting the neural network output function to the pseudo-samples is the most time-consuming part.

For large neural networks, crossvalidation, bootstrap, and similar resampling techniques are often computationally too expensive. Fortunately, such networks are usually considered if the number of available data is also very large. In that case, we can afford to use only part of the data $(X_1, Y_1), \ldots, (X_N, Y_N)$ as the training set from which $\hat{\vartheta}_N$ is estimated. The rest of the data, typically 10 % or 20 %, $(X_{N+1}, Y_{N+1}), \ldots, (X_{N+M}, Y_{N+M})$ serves as validation set to compare the performance of different networks. In that case, (54) is approximated by the average error over the validation set

$$\mathrm{V}(H) = \frac{1}{M} \sum_{j=N+1}^{N+M} (Y_j - f_H(X_j; \hat{\vartheta}_N))^2,$$

which then is minimized w.r.t. to H (compare Sect. 3.4).

9 The Wisdom of Many: Ensemble Learning

If applied to the same training set, different statistical learning methods, in particular similar ones using different hyperparameters, usually result in different forecasting or classification functions. Combining them in a suitable manner results in new and often better performance. Examples for such ensemble methods, using an ensemble of, e.g., regression predictors, are bagging, random forests, or boosting.

We start with bagging, where first we have to give a preliminary discussion of the bootstrap as a popular resampling method in statistics, for which the "b" in bagging stands.

9.1 Bootstrap Approximations of Prediction Errors

As an illustration, we again consider the general regression model (2)

$$Y_j = m(X_j) + \varepsilon_j, \;\; j = 1, \ldots, N, \tag{55}$$

with the usual assumptions on inputs and residuals. $\widehat{m}$ denotes some estimator for the regression function m.

A main performance measure for $\widehat{m}$ is its capability for generalization, i.e., the quality of forecasts for future, out-of-sample data. Let (X_{N+1}, Y_{N+1}) be a new pair of input-output data not used in calculating $\widehat{m}$ and independent of the training sample (X_j, Y_j), $j = 1, \ldots, N$. If we forecast Y_{N+1} by $\widehat{m}(X_{N+1})$ and use the loss function $L(y, f)$, e.g., quadratic loss $(y - f)^2$, for measuring prediction performance, then the expected loss should be small. Note that, for this purpose, only the randomness of the new data is taken into account, and the training sample and, hence, the prediction function $\widehat{m}$ are treated as given, i.e., the error which we want to keep small is

$$\mathrm{err}(\widehat{m}) = \mathrm{E}^* \; L\big(Y_{N+1}, \widehat{m}(X_{N+1})\big)$$

where E^* is an abbreviation for the (conditional) expectation given that the training sample is fixed and known. The in-sample empirical loss

$$\widehat{\mathrm{err}}_{in}(\widehat{m}) = \frac{1}{N} \sum_{j=1}^{N} L\big(Y_j, \widehat{m}(X_j)\big)$$

is unfortunately not a good approximation to the unknown quantity $\mathrm{err}(\widehat{m})$. It is systematically smaller due to overfit and the related dependence of $\widehat{m}$ on the forecasted data used for averaging as discussed in detail in Sect. 3.4.

To get a non-distorted estimate for $\mathrm{err}(\widehat{m})$, an out-of-sample empirical loss is needed. Popular methods are crossvalidation or splitting the data into training and validation set where the first one may become computationally expensive for the large samples frequently used in machine learning, and the second one does not make efficient use of the data.

Crossvalidation (compare Sect. 3.4) is a *resampling method* where from the original training set, many *pseudo samples* with similar properties are generated. In crossvalidation, these are just subsamples were one or few data are left out. Then, the estimate is calculated from each of these pseudo samples and used for predicting data, which are independent of the data used for calculating the estimate. In crossvalidation, these are the data left out from the particular subsample. Averaging over all the prediction errors provides a legitimate approximation of $\mathrm{err}(\widehat{m})$.

Another tried-and-tested resampling approach is the so-called *bootstrap*, where pseudo samples are generated by randomly drawing from the training set. In many cases, it can be proven that the empirical loss is a consistent estimate of $\mathrm{err}(\widehat{m})$ for sample size N and number of pseudo datasets growing. However, bootstrap approaches do not always work, and the underlying theory has to be investigated to be sure that one gets a valid estimate of $\mathrm{err}(\widehat{m})$ or other performance measures. We illustrate the idea with the following two bootstrap procedures applied to regression.

9.1.1 The Pairwise Bootstrap

We generate B bootstrap pseudo samples $X_j^b, Y_j^b, j = 1, \dots, N, b = 1, \dots, B$, by drawing purely randomly an observation pair from the original training set $\mathcal{T} = \{(X_j, Y_j), j = 1, \dots, N\}$. More formally, let $K_{j,b}, j = 1, \dots, N, b = 1, \dots, B$, be independent, identically distributed discrete random variables assuming with probability $\frac{1}{N}$ each of the values $1, \ldots, N$. Then, choose (X_j^b, Y_j^b) as the observation pair with random index $K_{j,b}$:

$$(X_j^b, Y_j^b) = (X_{K_{j,b}}, Y_{K_{j,b}}), \quad , j = 1, \ldots, N, b = 1, \ldots, B.$$

In other words, the (X_j^b, Y_j^b) are independent identically distributed with values in $\mathscr{T}$, and, given the training set $\mathscr{T}$, they assume each of the values $(X_j, Y_j), j = 1, \ldots, N$ with equal probability:

$$\text{pr}\{(X_j^b, Y_j^b) = (X_k, Y_k) \mid \mathscr{T}\} = \frac{1}{N}, \quad j, k = 1, \ldots, N, b = 1, \ldots, B.$$

For each of the B pseudo samples, we calculate an estimate $\widehat{m}^b(x)$ in exactly the same manner as we get $\widehat{m}(x)$ from $X_j, Y_j, j = 1, \ldots, N$. We use these prediction function estimates to forecast each of the original output data Y_j by $\widehat{m}^b(X_j)$. Then, we average over all errors:

$$\widehat{\text{err}}_{pb}(\widehat{m}) = \frac{1}{B}\frac{1}{N}\sum_{b=1}^{B}\sum_{j=1}^{N} L(Y_j, \widehat{m}^b(X_j)).$$

This is an example where a bootstrap approach does not work though pairwise bootstrapping leads to good results in other contexts. In regression, however, $\widehat{\text{err}}_{pb}(\widehat{m})$ systematically underestimates $\text{err}(\widehat{m})$ due to the same reasons as for $\widehat{\text{err}}_{in}(\widehat{m})$. Many observation pairs in the real sample and in each bootstrap pseudo sample will coincide such that the bootstrap estimates $\widehat{m}^b(x)$ of the regression function $m(x)$ depend too much on the forecasted data, and we are in an overfitting situation.

9.1.2 The Residual-Based Bootstrap

In predicting from a regression model, the input variables X_j are treated as observable, i.e., as given. The essential randomness that makes it difficult to infer the optimal prediction function m from the data is hidden in the errors or residuals ε_j, which are independent and identically distributed. Hence, it is plausible to concentrate on those and to resample from $\varepsilon_1, \ldots, \varepsilon_N$ in constructing the bootstrap pseudo data. However, we neither observe the residuals nor do we know the regression function m needed to construct the output pseudo data from inputs and pseudo residuals.

The key to solving both problems simultaneously is an initial estimate $\widetilde{m}(x)$ for the function $m(x)$. Using it, we get the usual *sample residuals*

$$\tilde{\varepsilon}_j = Y_j - \widetilde{m}(X_j), \ j = 1, \ldots, N.$$

Their empirical distribution, i.e., the discrete distribution with mass $\frac{1}{N}$ at each of those values, is usually not centered around 0 in contrast to the distribution of the true residuals ε_j, having mean 0. For bootstrapping, the pseudo data should mimic the properties of the real data. Therefore, we center the sample residuals before proceeding:

$$\hat{\varepsilon}_j = \tilde{\varepsilon}_j - \frac{1}{N}\sum_{k=1}^{N} \tilde{\varepsilon}_k, \; j = 1, \ldots, N.$$

This step is indispensable, as otherwise the bootstrap will not work even in simple regression settings (compare [29]).

We resample, i.e., we draw with replacement, the bootstrap pseudo residuals ε_j^b from the centered sample residuals $\hat{\varepsilon}_1, \ldots, \hat{\varepsilon}_N$:

$$\text{pr}^*\{\varepsilon_j^b = \hat{\varepsilon}_k\} = \frac{1}{N}, \; j, k = 1, \ldots, N, b = 1, \ldots, B, \tag{56}$$

whereas above * as an abbreviation denotes that we assume the original training set and, hence, the $\hat{\varepsilon}_j$ to be given. We can easily generate such bootstrap residuals by setting

$$\varepsilon_j^b = \hat{\varepsilon}_{K_{j,b}},$$

where the random indices $K_{j,b}$ are defined as in Sect. 9.1.1.

From the residuals, we construct the bootstrap pseudo data from the regression model, replacing the unknown m by its initial estimate

$$Y_j^b = \widetilde{m}(X_j) + \varepsilon_j^b, \; j = 1, \ldots, N, b = 1, \ldots, B.$$

Note that we choose the original inputs X_j as the inputs in the bootstrap world, i.e., we have $X_j^b = X_j$, as, as mentioned above, the inputs are observable and can be treated as fixed.

Once we have the B bootstrap pseudo samples $(X_j, Y_j^b), j = 1, \ldots, N$, we calculate the estimate $\widehat{m}^b(x)$ and use it for forecasting Y_j. For $B, N \to \infty$, the average of the errors

$$\widehat{\text{err}}_{rb}(\widehat{m}) = \frac{1}{B}\frac{1}{N}\sum_{b=1}^{B}\sum_{j=1}^{N} L(Y_j, \widehat{m}^b(X_j))$$

converges to $\text{err}(\widehat{m})$ if the initial estimate is suitably chosen. We consider two concrete situations:

(i) *parametric regression* (compare, e.g., Sect. 2.1): In this setting, $m(x) = g(x; \beta)$, where g is a known function up to the parameter vector β with fixed dimension q. Let $\tilde{\beta}$ be an initial estimate, e.g., the least-squares estimate solving

$$\sum_{j=1}^{N} \left(Y_j - g(X_j; \beta)\right)^2 = \min_{\beta}!$$

Then, the bootstrap works with the initial estimate $\widetilde{m}(x) = g(x; \tilde{\beta})$.

(ii) *spline smoothing* (compare Sect. 10.3): If we are interested in the performance of $\widehat{m}(x) = \widehat{m}_{s,\lambda}(x)$, which is the spline smoother with regularization parameter λ, then the bootstrap works if we choose $\widetilde{m}(x)$ as a spline smoother too. However, in this nonparametric setting, the bootstrap only works if $\widetilde{m}(x)$ is smoother than the estimate of interest $\widehat{m}(x)$. More precisely, we have to choose

$$\widetilde{m}(x) = \widehat{m}_{s,\tilde{\lambda}}(x) \qquad \text{such that } \frac{\tilde{\lambda}}{\lambda} \to \infty$$

for increasing N. Note that, for consistency, we need $\tilde{\lambda}, \lambda \to 0$ anyhow. In practice, the initial estimate has to use a stronger roughness penalty $\tilde{\lambda} > \lambda$, and hence $\widetilde{m}(x)$ is smoother than $\widehat{m}(x)$.

9.1.3 Mathematical Justification of the Bootstrap

Bootstrap approximations of the distributional characteristics of estimators, predictors, classificators, etc. are easy to implement. As already stated, they have to be applied with care as they do not always work. Here, mathematical theory is useful also from a practical point of view as it tells us when we may use which bootstrap algorithm. In this section, we give a brief introduction into the basics of this theory without any technical details.

Essentially, the bootstrap does not only approximate estimation errors like $\mathrm{err}(\widehat{m})$ but the whole distribution of, e.g., $\widehat{m}$. For a precise formulation, we need a measure for the closeness of distributions, i.e., a distance $d(F, G)$ between distributions F, G. An easy-to-explain example is the Kolmogorov distance for real-valued random variables (like $\widehat{m}(x)$): $d_K(F, G) = \sup_x |F(x) - G(x)|$, where F, G denote the distribution functions on the real line.

If U, V are two random variables with distributions F resp. G, then, with some abuse of notation, we write $d(U, V) = d(F, G)$ for convenience but keep in mind that d is not a distance between random variables but between probability measures.

We again consider the case of parametric regression. Let $\hat{\beta}$ denote the least-squares estimate of the parameter β. Under weak assumptions, it follows from an appropriate central limit theorem that, for large enough N, $\sqrt{N}(\hat{\beta} - \beta)$ has approximately a q-dimensional normal distribution with mean 0 and some covariance matrix Σ_β. This implies immediately by standard arguments that the scaled estimation error $\sqrt{N}\big(\widehat{m}(x)-m(x)\big) = \sqrt{N}\big(g(x;\hat{\beta})-g(x;\beta)\big)$ at x is approximately normally distributed too with mean 0 and some variance σ_x^2 depending on x.

Using the bootstrap of Sect. 9.1.2, we get another approximation for the distribution of the scaled estimation error, which, for complicated functions m and not too large samples, outperforms the approximation by a normal distribution. This is not only an empirical observation, but it can be explained by theory too (compare [39]). Under appropriate conditions, we can prove for quite a number of distances d, e.g., d_K, that

$$d(\sqrt{N}(\widehat{m}(x) - m(x)), \sqrt{N}(\widehat{m}^b(x) - \widetilde{m}(x))$$

converges to 0 for increasing N. $\widehat{m}^b(x)$ denotes the regression function estimate calculated from the bth pseudo sample for any given b. Therefore, the distribution of $\sqrt{N}\big(\widehat{m}(x) - m(x)\big)$ may be approximated by the (conditional) distribution of $\sqrt{N}(\widehat{m}^b(x) - \widetilde{m}(x))$ given that the original training set $X_j, Y_j, j = 1, \ldots, N$, is known.

The advantage of the bootstrap approximation is that $\widetilde{m}(x)$ is known. As X_j, Y_j, $j = 1, \ldots, N$ are also given, by drawing B bootstrap pseudo samples from the original training set, we can generate a large number of copies of $\widehat{m}^b(x)-\widetilde{m}(x), b = 1, \ldots, B$. Then, we can use them to numerically calculate any distributional characteristic of the scaled error $\sqrt{N}(\widehat{m}^b(x) - \widetilde{m}(x))$ by Monte Carlo simulation. An example for this approach is discussed in the next section.

9.1.4 Bootstrap Approximation of the Predictive Distribution

The bootstrap provides a method for numerically calculating the **predictive distribution**, which we have discussed in the context of Bayesian learning (compare Sect. 4.1), by Monte Carlo simulation without any specific assumptions on the distributions of the Y_j. We illustrate this approach with the derivation of confidence intervals for $m(x)$ and for $Y_{N+1} = m(x) + \varepsilon_{N+1}$ given $X_{N+1} = x$, where we use the residual-based bootstrap of Sect. 9.1.2. Recall that the forecast of Y_{N+1} is given by $\widehat{Y}_{N+1} = \widehat{m}(x)$.

For $0 < \alpha < 1$, let q_α denote the α-quantile of the distribution of the scaled estimation error $\sqrt{N}(\widehat{m}(x) - m(x))$, i.e.,

$$\alpha = \mathrm{pr}(\sqrt{N}(\widehat{m}(x) - m(x)) \le q_\alpha).$$

Then, for $0 < \gamma < 1$, we have $q_{\gamma/2} < \sqrt{N}(\widehat{m}(x) - m(x)) \le q_{1-\gamma/2}$ resp.

$$\widehat{m}(x) - \frac{1}{\sqrt{N}} q_{1-\gamma/2} \le m(x) < \widehat{m}(x) - \frac{1}{\sqrt{N}} q_{\gamma/2} \tag{57}$$

with probability $1 - \gamma$. (57) represents the confidence interval of level $1 - \gamma$ for $m(x)$. However, the quantiles involved are not known.

If the bootstrap is known to work, then, as discussed in the previous section, $q_\alpha \approx q^*_\alpha$, where the latter is the α-quantile of the distribution of the estimation error in the world of bootstrap pseudo samples, i.e.,

$$\alpha = \mathrm{pr}^*(\sqrt{N}(\widehat{m}^b(x) - \widetilde{m}(x)) \le q^*_\alpha).$$

Using Monte Carlo simulation, this probability may be approximated by the relative frequency over B bootstrap samples for which the scaled estimation error lies below q^*_α, i.e., the latter quantile may, up to a small approximation error for large enough B, be calculated by solving

$$\alpha = \frac{1}{B}\sum_{b=1}^{B} 1_{(-\infty, q^*_\alpha]}\big(\sqrt{N}(\widehat{m}^b(x) - \widetilde{m}(x))\big).$$

Replacing the unknown quantiles in (57) by their known bootstrap approximations, we get as a bootstrap confidence interval of level $1 - \gamma$ for $m(x)$:

$$\widehat{m}(x) - \frac{1}{\sqrt{N}} q^*_{1-\gamma/2} < m(x) < \widehat{m}(x) - \frac{1}{\sqrt{N}} q^*_{\gamma/2}. \tag{58}$$

For judging the precision of the forecast $\widehat{Y}_{N+1}$ of Y_{N+1}, we consider quantiles of the difference $\widehat{Y}_{N+1} - Y_{N+1}$, i.e.,

$$\alpha = \mathrm{pr}(\widehat{Y}_{N+1} - Y_{N+1} \le q^Y_\alpha) = \mathrm{pr}(\widehat{m}(x) - m(x) - \varepsilon_{N+1} \le q^Y_\alpha),$$

and a confidence interval for Y_{N+1} is given by $\widehat{Y}_{N+1} - q^Y_{1-\gamma/2} \le Y_{N+1} < \widehat{Y}_{N+1} - q^Y_{\gamma/2}$, which holds with probability $1 - \gamma$. Again, we use that $\widehat{m}(x) - m(x)$ is approximately distributed as $\widehat{m}^b(x) - \widetilde{m}(x)$. Moreover, a crucial step in showing this result is the approximability of the distribution of ε_j by the distribution of the bootstrap residual ε^b_j. Therefore, we again replace the quantile q^Y_α by its bootstrap analogue $q^{*,Y}_\alpha$

$$\alpha = \mathrm{pr}^*(\widehat{m}^b(x) - \widetilde{m}(x) - \varepsilon^b_{N+1} \le q^{*,Y}_\alpha) \approx \frac{1}{B}\sum_{b=1}^{B} 1_{(-\infty, q^{*,Y}_\alpha]}\big(\widehat{m}^b(x) - \widetilde{m}(x) - \varepsilon^b_{N+1}\big),$$

where ε^b_{N+1} is randomly generated from (56). We get as an approximate $(1 - \gamma)$-bootstrap confidence interval for Y_{N+1}

$$\widehat{Y}_{N+1} - q^{*,Y}_{1-\gamma/2} \le Y_{N+1} < \widehat{Y}_{N+1} - q^{*,Y}_{\gamma/2}. \tag{59}$$

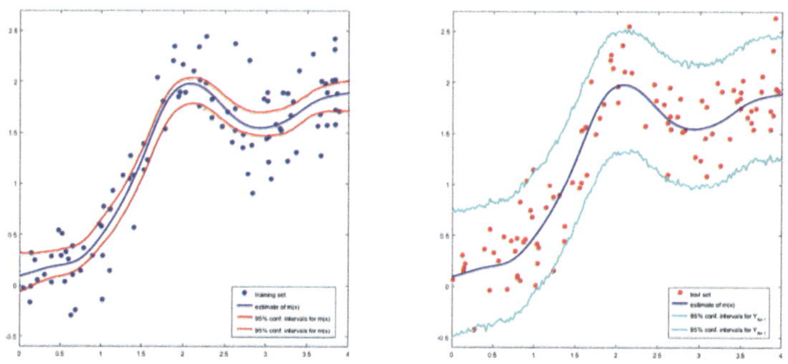

Fig. 35 Data, function estimate $m_{\mathrm{NW}}(x,h)$ (blue) and 95 % bootstrap confidence intervals (red) for $m(x)$ (left); test data, function estimate $m_{\mathrm{NW}}(x,h)$ (blue) and 95 % bootstrap confidence intervals (light blue) for Y_{N+1} (right)

We have used here a rather straightforward approach to bootstrap confidence intervals. There are more refined methods that may improve the approximation quality (compare [18]).

As an example, we consider a training sample using the same regression function as in Fig. 5 but with random inputs X_j, $j = 1, \ldots, N = 101$, which are uniformly distributed on [0,4]. We estimate the regression function $m(x)$ by a Nadaraya-Watson kernel estimate $\widehat{m}_{\mathrm{NW}}(x,h)$, described in Sect. 10.1, which is similar to the local averages shown in Fig. 7. The hyperparameter $h = 0.23$ is chosen by crossvalidation (compare Sect. 10.1.1). For generating the bootstrap data, we apply the residual-based bootstrap, using as initial estimate also a Nadaraya-Watson kernel estimate $\widetilde{m}(x) = \widehat{m}_{\mathrm{NW}}(x, h_B)$. As we have discussed at the end of Sect. 9.1.2, the initial estimate should be a bit smoother than the estimate of interest, which for this kind of function estimate is achieved by choosing $h_B = 0.30 > h$.

On the left-hand side of Fig. 35, we show the data on the estimate (blue) together with the 95 % confidence intervals (red) for $m(x)$, which we have calculated from (58) using $B = 1000$ bootstrap resamples.

To validate the approach, we also look at the 95 % bootstrap confidence intervals for Y_{N+1} given $X_{N+1} = x$ calculated from (59), which are shown as light-blue bands on the right-hand side of Fig. 35, together with a test sample that has been generated independently of and in the same manner as the training set. We see that 5 out of $N = 101$, i.e., $\approx 5\%$, of the test data lie outside the band, which is what we expect given the level of confidence of the intervals.

9.2 *Bootstrap Aggregation (Bagging)*

We consider again a training set $\mathscr{T} = \{(X_j, Y_j), j = 1\ldots, N\}$ where Y_j is real valued in regression and assumes the two values 0 and 1 in binary classification. General classification with more than two classes can be handled analogously.

We want to estimate the regression function $m(x)$ resp. the classification function $c(x)$. To stress the dependence of the estimates $\widehat{m}(x), \hat{c}(x)$ on the data, we write $\widehat{m}(x) = \widehat{m}(x, \mathscr{T}), \hat{c}(x) = \hat{c}(x, \mathscr{T})$. If, e.g., the out-of-sample input is $X_{N+1} = x$, then we predict the corresponding output Y_{N+1} by $\widehat{m}(x, \mathscr{T})$.

We now assume that we have B training sets $\mathscr{T}_b, b = 1, \ldots, B$, of size N each. Then, we also have B usually different predictions of Y_{N+1}

$$\widehat{m}_b(x) = \widehat{m}(x, \mathscr{T}_b), \quad b = 1, \ldots, B.$$

Our goal is to combine them for an overall better forecast. This is called *aggregation*. Our hope is that a combination of the knowledge of various experts (i.e., predictors or classifiers) drawing from different sources of information (i.e., the training sets) leads to an overall improvement.

In regression, we just *average* the predictions and use

$$\widehat{m}_A(x) = \frac{1}{B} \sum_{b=1}^{B} \widehat{m}_b(x)$$

whereas in classification, we rely on a *majority vote*, i.e.,

$$\hat{c}_A(x) = 1 \quad \text{if the number of } b \text{ with } \hat{c}_b(x) = 1 \text{ is greater than } \frac{N}{2},$$

and $\hat{c}_A(x) = 0$ else. In case that the same number of classifiers $\hat{c}_b(x)$ decides for class 1 resp. class 0, there is a certain ambiguity in defining the aggregated classifier.

In practice, we have to work with only one training set usually. To be able to aggregate, we artificially generate several training sets $\mathscr{T}_b, b = 1, \ldots, B$, using the bootstrap, e.g., the pairwise version, i.e., with random indices $K_{j,b}$ as in Sect. 9.1.1

$$\mathscr{T}_b = \{(X_{K_{1,b}}, Y_{K_{1,b}}), \ldots, (X_{K_{N,b}}, Y_{K_{N,b}})\}, \; b = 1, \ldots, B,$$

We just are sampling with replacement N input-output pairs from the original training set $\mathscr{T}$ to get B bootstrap pseudo samples, and then we use them for aggregation. This procedure is called *bagging*, and the resulting predictors and classificators are denoted by $\widehat{m}_{BA}(x)$ resp. $\hat{c}_{BA}(x)$.

Note that, in regression, we do not use the pairwise bootstrap for inference about the distribution of $\widehat{m}(x)$, which would not work (compare Sect. 9.1.1), but only for getting a whole set of different predictors.

In his original paper [8], Breiman applied bagging to several real datasets and considered classification as well as regression problems. For checking the performance independently, he put aside about 10 % of the data in each case as a test set $\mathscr{P}$. The rest was used as the training set $\mathscr{T}$. As a learning method, Breiman used classification resp. regression trees (compare Sect. 7) with hyperparameters selected by crossvalidation.

He then compared the performance of the plain tree-based regression or classification estimate $\widehat{m}(x, \mathscr{T})$ resp. $\hat{c}(x, \mathscr{T})$ with the bagged versions $\widehat{m}_{BA}(x), \hat{c}_{BA}(x)$. As performance measure, quantifying the potential of generalization (compare Sect. 2.2), the average squared error, resp., the percentage of misclassifications on the independent test set $\mathscr{P}$, was considered. Bagging always led to an improvement, and in the majority of applications, the prediction error, resp., misclassification rate, decreased even by 20–50%.

9.2.1 When Does Bagging Work?

Though Breiman's experiment with real data showed a considerable improvement due to bagging, this is not necessarily so. He illustrated the problem with an artificial example based on a general linear regression model similar to (13)

$$Y_j = \sum_{k=1}^{p} \beta_k X_{jk} + \varepsilon_j, \; j = 1, \ldots, N,$$

with independent identically standard normal residuals ε_j. The input vectors X_j are of moderately high dimension $p = 30$ and have a p-dimensional normal distribution with mean 0 and a covariance matrix indicating that the correlations between the components $X_{jk}, k = 1, \ldots, p$, are non-negligible in all cases and rather strong for some cases. This is a rather typical situation in the applications of statistical learning where a lot of features can be used but do not all represent completely different information about the output. In such situations, as discussed in Sect. 3.1, we are looking for sparser models resulting in regression function estimates and forecasts

$$\widehat{Y}_{N+1} = \widehat{m}(X_{N+1}, \mathscr{T}) = \sum_{i=1}^{q} \hat{\beta}_{k_i} X_{N+1,k_i}$$

where we use only a selection of $q < p$ feature values with indices $k_1, \ldots, k_q$. The latter are chosen adaptively from the data by a forward selection approach starting with a simple model and including one feature component after the other to the forecast. Usually, this is done until a stopping criterion is fulfilled, but here, we just consider the submodels for various but fixed q.

To investigate the effect of bagging, artificial data were generated from (13), where three different scenarios were considered: a sparse one, where only 3 of the 30 regression coefficients were different from 0; an almost full model, where only 3 coefficients were not vanishing and the remaining 27 were of similar size; and an intermediate one with 15 vanishing and nonvanishing β_k each. Of course, in constructing the predictor, we do not assume that we know which coefficients are 0 and, hence, which of the corresponding input variables are irrelevant.

Applying bagging in this situation, Monte Carlo simulation showed that bootstrap aggregation always improved the prediction performance, but the effect was negligible for large values of q. In particular, in the full regression scenario with many nonvanishing components of relevance, we would expect that the submodels provide a better fit for large q, and in such a situation, the minor gain in performance is not worth the considerable additional computational effort.

The idea of bagging is to add some variability to the construction of the predictor to average out a particular unfortunate data constellation in the original training set $\mathcal{T}$. Let us assume, as in the example, that we have $p = 30$ input variables $X_{j1}, \ldots, X_{jp}$, where likely only a few are relevant for predicting Y_j. This may be due to the fact that Y_j depends only weakly on some of the X_{jk}, or, in more common practice, some of the X_{jk} are strongly dependent among themselves such that they essentially contain the same information and using only one of them in the prediction of Y_j suffices. In such situations, we should fit a model with parameter dimension $q < p$ to the data, and in cases, where there is such a sparser representation of the feature vector, bagging helps.

The selection of parameters $\beta_{k_1}, \ldots, \beta_{k_q}$ and, hence of feature components, may be quite sensitive to the data, i.e., if we just modify the data slightly, we can get a completely different selection with perhaps a much better predictive performance. Bagging should ameliorate this situation by considering various similar but somewhat different dataset. This idea works for small enough q. If, however, q is close to p, then there is not much room for variability in selecting the actually used predictor variables as the model includes anyhow almost all of them. Predictors $\widehat{m}(X_{N+1}, \mathcal{T}_b), b = 1, \ldots, B$, are more or less the same, and we do not profit from aggregating them.

9.2.2 Bagging for Classification

Now, we have a closer look at the binary classification problem with two classes, labeled 0 and 1 (compare also Sect. 2.4). Let

$$p_1(x) = \text{pr}\{Y_{N+1} = 1 | X_{N+1} = x\}$$

denote the probability that an object with input value x belongs to class 1. Let $\hat{c}(x, \mathcal{T})$ denote a classificator derived from the training sample, and let

$$q(x) = \text{pr}(\hat{c}(x, \mathcal{T}) = 1),$$

i.e., the probability of observing a training set $\mathcal{T}$ for which a new object with input $X_{N+1} = x$ is classified as belonging to class 1. The probability that we observe a training set for which the resulting classificator $\hat{c}$ classifies the object correctly is

$$\text{pr}\{\hat{c}(x, \mathcal{T}) = Y_{N+1} | X_{N+1} = x\} = q(x)p_1(x) + (1 - q(x))(1 - p_1(x)) = r(x),$$

and the overall probability of correct classification is $r = \mathrm{E}\, r(X_{N+1})$. As $0 \leq q(x) \leq 1$, we also have $r(x) \leq \max(p_1(x), 1 - p_1(x))$ with equality if and only if

$$q(x) = 1 \qquad \text{for } p_1(x) = \max(p_1(x), 1 - p_1(x)), \text{ i.e. } p_1(x) \geq \frac{1}{2}.$$

and $q(x) = 0$ else. The latter situation corresponds to the optimal Bayes classifier c_B (9) based on the (unknown) true class probabilities $p_1(x)$:

$$c_B(x) = 1 \text{ if } p_1(x) \geq \frac{1}{2}, \text{ and } c_B(x) = 0, \text{ else.}$$

In this case, the probability of correct classification is

$$r_B = \mathrm{E}\, \max(p_1(X_{N+1}), 1 - p_1(X_{N+1})) \geq r,$$

i.e., r_B is an upper bound on the rate of correct classifications in general.

Looking for classificators for which r is close to the optimum r_B, we first require that the classificator $\hat{c}(x, \mathscr{T})$ satisfy a rather weak assumption that it is *order-correct*, i.e.,

$$q(x) > \frac{1}{2} \text{ if and only if } p_1(x) > \frac{1}{2},$$

i.e., if objects with input value x belong more often to class 1, then, on the average over the training sets, class 1 is chosen more frequently by the classification rule $\hat{c}(x, \mathscr{T})$ than class 0. Under this condition, bootstrap aggregation by majority vote improves the situation. This corresponds to the classification rule

$$\hat{c}_{BA}(x) = 1, \text{ if } \frac{n_1}{B} > \frac{1}{2}, \qquad \text{and } \hat{c}_{BA}(x) = 0, \text{ else,}$$

where n_1 is the number of bootstrap pseudo samples $\mathscr{T}_b, b = 1, \ldots, B$, for which $\hat{c}(x, \mathscr{T}_b) = 1$. By the law of large numbers, the condition $n_1/B > \frac{1}{2}$ is approximately identical to $q(x) > \frac{1}{2}$, and this implies, due to the assumptions of order-correctness, $p_1(x) > \frac{1}{2}$. Therefore, for large enough B, the bagging classificator $\hat{c}_{BA}(x)$ is approximately identical to the Bayes classifier at every x for which the original classification rule is order-correct at x. If this assumption is satisfied for all x, then the overall probability of correct classification r_{BA} of $\hat{c}_{BA}$ is approximately equal to r_B, i.e., the bagged classificator is almost optimal for large enough bootstrap replications B even if the original classification function $\hat{c}(x, \mathscr{T})$ is far from optimal.

Even if the classificator is not order-correct everywhere, but only at most x, then still the bagged classificator will be nearly optimal. Otherwise, however, bagging may worsen the situation in the sense that r_{BA} is even smaller than the probability of correct classification using the original single $\hat{c}(x, \mathscr{T})$. In such cases, bagging increases the likelihood of wrong classifications, in contrast to regression where the performance of the bagged predictor is always at least as good as that of the original predictor.

9.3 Boosting

On the first glance, boosting is an ensemble method similar to bagging, but the underlying principles differ considerably. In classification, e.g.:

- *bagging* is a majority vote among many classificators based on pseudo samples drawn from the training set in a purely random manner,
- *boosting* is a **weighted** majority vote among many classificators iteratively generated from the training set to give more weight to "difficult" data to focus on the task of learning their correct classification.

We again consider binary classification with a training set $\mathscr{T}$ of i.i.d. (X_j, Y_j), $j = 1, \dots, N$, where Y_j assume values 0 and 1. $\hat{c}(x) = \hat{c}(x, \mathscr{T}) \in \{0, 1\}$ is a classification rule based on the training sample. For ease of notation, we consider a transformation of class labels from $0, 1$ to $-1, +1$ and denote the transformed outputs and classifications by

$$Y_j' = 2Y_j - 1, \qquad \hat{c}'(x) = \hat{c}'(x, \mathscr{T}) = 2\hat{c}(x) - 1.$$

Then, $Y_{N+1}'\hat{c}'(X_{N+1}) = 1$ for correct and $= -1$ for incorrect classifications, and the error rate in the training sample can be written as

$$\widehat{err}_{in}(\hat{c}) = \frac{1}{N}\sum_{j=1}^{N} L(Y_j, \hat{c}(X_j)) = \frac{1}{2} - \frac{1}{2N}\sum_{j=1}^{N} Y_j'\hat{c}'(X_j),$$

where $L(y, c) = 1$ if $y \neq c$, and $= 0$, else.

Boosting starts with some given reasonable classifier, say $\hat{c}(x, \mathscr{T})$, which even may be weak in the sense that it is only a bit better than pure guessing of the classes. Then, a sequence of weak classifiers $\hat{c}_1 = \hat{c}, \hat{c}_2, \dots, \hat{c}_K$ are generated iteratively by fitting the original classification rule to repeatedly modified versions $\mathscr{T}_1 = \mathscr{T}, \mathscr{T}_2 \dots, \mathscr{T}_K$ of the original training set $\mathscr{T}$:

$$\hat{c}_1(x) = \hat{c}(x, \mathscr{T}), \hat{c}_2(x) = \hat{c}(x, \mathscr{T}_2), \dots, \hat{c}_K(x) = \hat{c}(x, \mathscr{T}_K).$$

Using the transformed classifiers, usual majority vote results in the classification function

$$\hat{c}_{MV}'(x) = +1, \quad \text{if} \sum_{k=1}^{K} \hat{c}_k'(x) > 0, \qquad \hat{c}_{MV}'(x) = -1, \text{ else}.$$

Boosting, however, uses weights α_k depending on the data

$$\hat{c}'_{BO}(x) = +1, \quad \text{if} \sum_{k=1}^{K} \alpha_k \hat{c}'_k(x) > 0, \qquad \hat{c}'_{MV}(x) = -1, \text{ else} .$$

For generating the $\mathscr{T}_k$, each of the training data (X_j, Y_j) gets a weight $w_j \geq 0$ with $w_1 + \ldots + w_K = 1$. Initially, $w_j = \frac{1}{N}$, i.e., all data in $\mathscr{T}_1 = \mathscr{T}$ get the same weight and are treated as equally important for constructing the classifier.

Now, for each iteration, those data get **more weight**, which have been previously **misclassified**. Hence, subsequent classifiers are forced to concentrate on these "difficult" parts of the sample and try to classify them correctly.

Boosting is, in particular, useful in situation with unbalanced training sets where the large majority of data belongs to one class and only a few to the other (compare Sect. 12.4). The interesting cases are underrepresented, and classification algorithms just move with the majority. In, e.g., quality control at the end of production, typically, there are only a few bad parts among many good ones.

Usual training algorithms try to treat all input data equally. As a consequence, for such unbalanced samples, they focus on the many good parts achieving very good classification results for them and hence a small overall error rate. The bad parts are, however, neglected, resulting in relatively many classification errors though they are economically more important as the reputation of the producer will be damaged if customers get too many bad products.

A first idea would be to give the cases, where a bad part erroneously is classified as good, more weight in the loss function used for training the algorithm, which, in our case, is $\widehat{err}_{in}$. Without changing the algorithm, this can be easily achieved by replicating the input-output pairs from the bad parts in the training set several times and adding them to the sample, to give those cases more weight in learning. In the individual iterations of boosting, the data are augmented in essentially this manner but more systematically. Instead of adding duplicates of "difficult" data, we just increase their weight, which has the same effect.

There are various variants of boosting differing somewhat w.r.t. to the iteration. Here, we present the original **AdaBoost** algorithm of [30]:

1. Initialization: $w_j^{(1)} = \frac{1}{N}, j = 1, \ldots, N, \mathscr{T}_1 = \mathscr{T}$.
2. For $k = 1, \ldots, K - 1$:

   (a) Denote by $\mathscr{T}_k$ the weighted data set with weights $w_j^{(k)}$ for $(X_j, Y_j), j = 1, \ldots, N$. Fit a classifier $\hat{c}_k(x)$ to $\mathscr{T}_k$, e.g., by minimizing the weighted empirical loss w.r.t. c

$$\sum_{j=1}^{N} w_j^{(k)} L(Y_j, c(X_j)).$$

(b) Compute the weighted error rate of $\hat{c}_k$

$$err_k = \sum_{j=1}^{N} w_j^{(k)} L(Y_j, \hat{c}_k(X_j)).$$

(c) Compute

$$\alpha_k = \log \frac{1 - err_k}{err_k}.$$

Note that $\alpha_k = 0$ for pure guessing (where $err_k = \frac{1}{2}$), $\alpha_k > 0$ for other classifiers, and α_k increasing with the quality of the classifier.

(d) Update the weights

$$\tilde{w}_j^{(k+1)} = w_j^{(k)} \exp\left(\alpha_k L(Y_j, \hat{c}_k(X_j))\right),$$

$$w_j^{(k+1)} = \frac{\tilde{w}_j^{(k+1)}}{\sum_{i=1}^{N} \tilde{w}_i^{(k+1)}}$$

Note that $\tilde{w}_j^{(k+1)} = w_j^{(k)}$ for correctly classified data, i.e., for $Y_j = \hat{c}_k(X_j)$, and $\tilde{w}_j^{(k+1)} = w_j^{(k)} e^{\alpha_k} > w_j^{(k)}$, else.

To illustrate the power of boosting, in Chapter 10 of [47], an artificial example with ten-dimensional inputs X_j, $j = 1, \ldots, N = 2000$, is considered, where all the feature components X_{ji}, $i = 1, \ldots, 10$, were independent standard normal random variables. The outputs only depended on the length of the input vectors, i.e., for some $\rho > 0$

$$Y_j = 1, \text{ if } ||X_j|| > \rho, \qquad Y_j = 0, \text{ else},$$

where ρ was chosen such that $\text{pr}(Y_j = 1) = \text{pr}(Y_j = 0) = \frac{1}{2}$. This is a highly nonlinear classification problem.

As a classifier, a *stump* was used, i.e., the most simple classification tree with only two terminal nodes corresponding to a classification function of the form

$$c(x) = 1, \text{ if } x_\ell > t, \qquad c(x) = 0, \text{ else}.$$

Fitting it to the training data, i.e., estimating the split coordinate ℓ and the threshold t, the error probability, calculated by Monte Carlo simulation, for the resulted learned classifier $\hat{c}$ was 46%, i.e., it is a rather weak classifier not much better than guessing (corresponding to 50% error probability). Boosting achieved 12.2% after $K = 400$ iterations, whereas fitting a large single classification tree with 400 nodes to the training set results in a considerably larger error probability of 26%.

9.4 Random Forests

Random Forests is the name of an ensemble method combining various regression resp. classification trees. Essentially, it is nothing but bagging, based on trees, with an additional randomization regarding the selection of predictor variables. Let $\mathcal{T}$ denote again the training set $\{(X_j, Y_j), j = 1 \ldots, N\}$, with d-dimensional inputs X_j. The algorithm proceeds as follows:

(1) For $b = 1, \ldots, B$,

   (a) Draw a bootstrap resample $\mathcal{T}_b$ from $\mathcal{T}$ (as in Sect. 9.2).
   (b) Fit a random forest tree to the sample $\mathcal{T}_b$, e.g., in regression resulting in $\widehat{m}_M(x, \mathcal{T}_b)$, in the following manner. Split subsequently every terminal node that does not yet satisfy the **stopping criterion** that the node contains $\leq n_{\min}$ data:

      (i) Select randomly a subset of M different input components $\{i_1, \ldots, i_M\} \subset \{1, \ldots, d\}$.
      (ii) Select the best variable and corresponding best splitting point only among $X_{ji_\ell}, \ell = 1, \ldots, M$.
      (iii) Split the node w.r.t. to the best input variable at the splitting point resulting in two daughter nodes.

   Proceed until all terminal nodes satisfy the stopping criterion.

(2) In regression, estimate the regression function $m(x)$ by

$$\widehat{m}_{RF}(x) = \frac{1}{B} \sum_{b=1}^{B} \widehat{m}_M(x, \mathcal{T}_b).$$

In classification, follow the majority vote in classifying an object with $X_{N+1} = x$. Recall $\hat{c}_M(x, \mathcal{T}_b) \in \{0, 1\}$. Set

$$\hat{c}_{RF}(x) = 1, \quad \text{if } \frac{1}{B} \sum_{b=1}^{B} \hat{c}_M(x, \mathcal{T}_b) > \frac{1}{2}, \qquad \hat{c}_{RF}(x) = 0, \quad \text{else}\,.$$

In the extreme case $M = d$, this algorithm is nothing but bagging based on the particular stopping criterion. Usually, M is chosen considerably smaller than d. Then, the main difference to bagging is step (1b), where the tree is grown not based on the all the input coordinates (which correspond to potential predictors), but in splitting each node, only an individual smaller random selection of M input coordinates is used. This results in a lot of variability between the trees represented by $\hat{c}_M(x, \mathcal{T}_b), b = 1, \ldots, B$, which is good for averaging out the effects of specific bad data constellations.

Another main feature of random forests that helps in reducing the computational load is the abandonment of pruning. Each tree $\widehat{m}_M(x, \mathscr{T}_b)$ is grown until all terminal nodes satisfy the stopping criterion, and then the trees are used as they are. Recall that pruning is a safeguard against overfit for tree-based methods. For random forests with hyperparameter M much smaller than the number of input variables d, overfit is already very unlikely without additional measures.

Choosing $M \ll d$ has the further advantage that each split is computed quite fast. On the other hand, it has a noteworthy disadvantage if only few input components $X_{ji}, i = 1, \ldots, d$, are relevant for predicting Y_j. In that case, for small M, the chance that the selection $X_{ji_\ell}, \ell = 1, \ldots, M$, contains none of the informative features is pretty high.

Therefore, random forests work well if many X_{ji} contribute appreciably to the optimal forecast $m(X_j)$ of Y_j. More precisely, let $\tilde{m}(X_{ji_1}, \ldots, X_{ji_M})$ denote the optimal forecast of Y_j using only the corresponding subset of feature components. Then, random forests perform well if for any $M \ll d$, all selections $X_{ji_\ell}, \ell = 1, \ldots, M$, on the average produce a much worse forecast of Y_j than the full feature vector X_j, i.e.,

$$\mathrm{E}\left(Y_j - m(X_j)\right)^2 \ll \mathrm{E}\left(Y_j - \tilde{m}(X_{ji_1}, \ldots, X_{ji_M})\right)^2.$$

For the rest of the section, we focus on regression problems and try to get more insight when random forests work. For a data-dependent tree-based prediction function $\widehat{m}(x, \mathscr{T})$, we want to predict out-of-sample new observations well, i.e., we strive for a small mean-squared forecasting error for new data (X_{N+1}, Y_{N+1}), which represents the *generalization error*:

$$\mathrm{E}\left\{\left(Y_{N+1} - \widehat{m}(X_{N+1}, \mathscr{T})\right)^2 \middle| \mathscr{T}\right\} = ge(\text{tree}).$$

For random forests, the analogous generalization error is, using $\mathscr{T}_1$ as a representative of the bootstrap pseudo-samples $\mathscr{T}_b$:

$$\begin{aligned}\mathrm{E}\left\{\left(Y_{N+1} - \widehat{m}_{RF}(X_{N+1})\right)^2 \middle| \mathscr{T}\right\} \rightarrow \quad & \mathrm{E}\left\{\left(Y_{N+1} - \mathrm{E}^{*,M}\, \widehat{m}_M(X_{N+1}, \mathscr{T}_1)\right)^2 \middle| \mathscr{T}\right\} \\ &= ge(\text{forest}),\end{aligned}$$

for $B \to \infty$, where $\mathrm{E}^{*,M}$ denotes the expectation w.r.t. to the randomness in the construction of the tree $\mathscr{T}_1$ from the training set $\mathscr{T}$ and the random selections of M input components in the calculation of $\widehat{m}_M(., \mathscr{T}_1)$.

To get a better understanding when random forests are better than simple bagging, let us pretend for a moment that the regression function estimate $\widehat{m}(x) = \widehat{m}(x, \mathscr{T})$ satisfies the weak unbiasedness condition $\mathrm{E}\ m(X_{N+1}) = \mathrm{E}\ \{\widehat{m}(X_{N+1})|\mathscr{T}\}$, i.e., it predicts on the average the same values as the true regression function $m(x)$. Then, it can be shown that

$$ge(\text{forest}) \leq \overline{\rho}\ ge(\text{tree}),$$

where $\overline{\rho}$ is a weighted correlation, which is small, if predictors $\widehat{m}_M(X_{N+1}, \mathcal{T}_i)$, $i = 1, 2$, for Y_{N+1} based on different trees $\mathcal{T}_1$, $\mathcal{T}_2$ are weakly dependent. If additionally the generalization error $ge(\text{tree})$ for single trees is small, then the random forest algorithms performs well. In comparison with bagging, for $M \ll d$, it in particular decreases the correlation component $\overline{\rho}$ in the upper bound of the generalization error due to the additional inbuilt randomization.

10 Learning by Local Smoothing

Sieve estimates like neural networks, series expansions or regression trees learn the function of interest by approximating it **globally** on the whole input space. In regression, e.g., the function estimate $\widehat{m}(x)$ at x depends on the whole training set. In this chapter, we consider another class of estimates, which instead use principles of **local** approximation. Here, $\widehat{m}(x)$ mainly depends only on those data (X_j, Y_j) for which the feature value X_j lies in a neighborhood of x.

We have already considered an example, the regression estimate (8), where $\widehat{m}(x)$ is the mean of all Y_j for which $|X_j - x| \leq h$. This estimate worked well for suitably chosen hyperparameter h, but we already noted a certain wigglyness of $\widehat{m}(x)$, which is not so pleasing to the eye, in particular if one knows that the true underlying function $m(x)$ will be rather smooth. Technically, this results from the discontinuity of $1_{[-h,+h]}(z-x)$ as a function of z and can be solved by replacing it by a continuous approximation. The plain average of output values Y_j is replaced by a weighted average that is intuitively appealing, as those observations with larger distance $|X_j - x|$ to x get less weight, i.e., less influence on the function estimate at x.

In the first section, we discuss the particular, still rather general class of weights derived from a kernel function. Then, we consider the related class of nearest neighbor estimates. The popular spline smoothers, which generalize the use of spline functions for interpolation, also belong to this family of local estimates though it is not obvious on the first glance. All these methods work well for low-dimensional feature vectors X_j, but it is rather obvious from their construction that they cause problems in higher dimensions. In the last section, we present an approach on how local smoothers can be useful for estimating functions on high-dimensional feature spaces nevertheless.

In the whole chapter, we consider the general regression model

$$Y_j = m(X_j) + \varepsilon_j, \ \ j = 1, \dots, N,$$

with independent and identically distributed zero-mean residuals independent of $X_1, \dots, X_N$ having finite variance σ_ε^2.

10.1 Kernel Smoothers

For ease of notation, we consider only one-dimensional feature vectors X_j, though kernel smoothers are also popular tools in dimension 2 and 3, in particular for denoising 2d- or 3d images. Kernel estimates of regression functions are locally weighted averages of output values where the weights are generated from a **kernel function** $K(u)$. This is a bounded, usually continuous function on the real line, integrating to 1, which is rescaled by the so-called **bandwidth** $h > 0$, which acts as the hyperparameter in this statistical learning method:

$$K_h(u) = \frac{1}{h} K(\frac{u}{h}).$$

Typical kernels are:

Gauss-kernel $K(u) = \frac{1}{\sqrt{2\pi}} e^{-\frac{1}{2}u^2}$

Epanechnikov (or *Bartlett-Priestley) kernel* $K(u) = \frac{3}{4}(1 - u^2)$ for $|u| \leq 1$ and $= 0$, else.

rectangular kernel $K(u) = \frac{1}{2} 1_{(-1,+1)}(u)$.

Note that the rectangular kernel $K_h(u)$ produces constant weights as long as $|u| < h$, such that the corresponding estimate is a plain local average of outputs. It is not continuous and suboptimal and performs worse in practice and theory, but it often allows for a better understanding why such a type of local smoother works at all.

We start with considering the special case of a deterministic design where the input values X_j are equidistant grid points on a given interval, say [0,1], where $X_j = x_j = j/N, j = 1, \ldots, N$. Then, we may estimate the regression function $m(x), 0 \leq x \leq 1$, by the *Priestley-Chao kernel estimate*

$$\hat{m}_{\mathrm{PC}}(x, h) = \frac{1}{N} \sum_{j=1}^{N} K_h(x - x_j) Y_j = \frac{1}{Nh} \sum_{j=1}^{N} K(\frac{x - x_j}{h}) Y_j. \tag{60}$$

For the rectangular kernel, this is just the local average (8). Under some not too strong additional conditions on $m(x)$ and $K(u)$, the main one being twice continuously differentiability of $m(x)$ around the location x, we have a full understanding of the performance of this kernel smoother for large enough sample size N. For example, we have for $N \to \infty, h \to 0$ such that $Nh^3 \to \infty$

$$\text{mse}\ (\hat{m}_{\mathrm{PC}}(x, h)) = \mathrm{E}\ (\hat{m}_{\mathrm{PC}}(x, h) - m(x))^2 = \frac{\sigma_\varepsilon^2}{Nh} Q_K + \big(\frac{m''(x)}{2} V_K\big)^2 h^4 + r_{N,h} \tag{61}$$

with constants, depending only on K,

$$Q_K = \int K^2(u)du, \qquad V_K = \int u^2 K(u)du.$$

The remainder term $r_{N,h}$ is negligible compared to the other two terms on the right-hand side. It follows immediately that $\hat{m}(x,h) \to m(x)$ in the sense of vanishing mean-squared error, i.e., the kernel smoother is a consistent estimate for growing N. With increasing N, the bandwidth h becomes smaller, as the feature values x_j become denser and smaller neighborhoods of x contain enough observations such that the local average approximates $m(x)$ well.

The two dominating terms on the right-hand side of (61) illustrate well the bias-variance dilemma (compare Sect. 2.3). The first term describes the behavior of the variance of the kernel smoother. For fixed N, it becomes small if h is large. The second term, involving $m''(x)$, corresponds to the systematic error, i.e., the bias E $\hat{m}_{\mathrm{PC}}(x,h) - m(x)$, which becomes small if h is the small. Again we are confronted with the problem of choosing the hyperparameter in a manner that involves balancing the tendencies to overfit and underfit.

Moreover, from the dependence of the bias term on $m''(x)$, we learn that the systematic error is particular large at locations x where the function $m(x)$ has a large curvature. In particular at peaks, $\hat{m}_{\mathrm{PC}}(x,h)$ has a systematic tendency to underestimate $m(x)$.

If $h \to 0$ like $cN^{-1/5}$ for some constant c, then the mean-squared error (61) converges to 0 with the rate $N^{-4/5}$, and the function estimate $\hat{m}_{\mathrm{PC}}(x,h) \to m(x)$ with rate $N^{-2/5}$. Under our smoothness assumptions on $m(x)$, this is more or less the optimal achievable rate of convergence. It is a bit slower than the usual convergence rate $1/\sqrt{N}$ of parameter estimates in classical statistics where we assume that our model for the data-generating process is known up to only a finite number of parameters.

This is a typical feature of all machine learning approaches, and it is the price we have to pay for the ability to learn rather arbitrary functions $m(x)$ from the data. In practice, this means that we need **larger training sets** to achieve the same kind of precision as in a well-specified white-box setting with a given finite set of parameters (compare Sect. 2.6). As a compensation, we are assured that our method will result in reasonable approximations of $m(x)$, whereas a misspecified white-box model may fail severely.

Often, it would be too cumbersome to look for a good hyperparameter $h(x)$ at every x of interest. Usually, a fixed h will be used on the whole range of input values [0,1]. In that case, the mean-squared error of $\hat{m}_{\mathrm{PC}}(x,h)$ is replaced by the *mean integrated squared error* (mise) as a performance measure. Analogously to (61), we have

$$\begin{aligned}\mathrm{mise}(\hat{m}_{\mathrm{PC}}, h)) &= \mathrm{E} \int_0^1 (\hat{m}_{\mathrm{PC}}(x,h) - m(x))^2 dx \qquad (62)\\ &= \frac{\sigma_\varepsilon^2}{Nh} Q_K + \int_0^1 (\frac{m''(x)}{2})^2 dx V_K^2 h^4 + R_{N,h},\end{aligned}$$

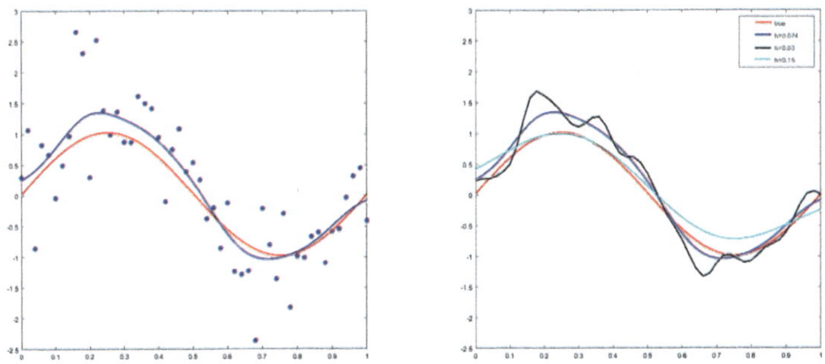

Fig. 36 Data, true function (red) and $\widehat{m}(x; h_{CV})$ (blue) on the left-hand side; additionally, $\widehat{m}(x; 0.03)$ (black) and $\widehat{m}(x; 0.15)$ (light blue) on the right-hand side

i.e., we get the same kind of behavior as for a local bandwidth $h(x)$, except that bias component depending on x is replaced by its integral.

As an example, we consider the data already used in Fig. 7. We again apply a local average with bandwidth h, but we now prefer a weighted average, more precisely a Priestley-Chao kernel estimate with Gauss-kernel. The performance of the estimate $\widehat{m}(x; h)$ depends on the hyperparameter h. For a suitably chosen bandwidth $h = 0.074$, the corresponding estimate (blue) is shown together with the true function $m(x)$ (red) and the data on the left-hand side of Fig. 36. The right-hand side additionally shows for comparison an underfitted ($h = 0.15$) and an overfitted ($h = 0.03$) estimate.

We now turn to the more common case where the regressors are random. We additionally assume that the input variables $X_1, \ldots, X_N$ are independent, identically distributed with probability density $p(x)$. In this case, the weights of the kernel smoother have to be adapted to the expected local denseness of inputs in the neighborhood of x, which is measured by $p(x)$. As the latter is not known, it has to be estimated, for which we use also a kernel approach. This complication is not due to randomness but due to the fact that we do no longer have equidistant input values.

The resulting *Nadaraya-Watson kernel estimate* of $m(x)$ is

$$\widehat{m}_{NW}(x, h) = \frac{\frac{1}{N}\sum_{j=1}^{N} K_h(x - X_j)Y_j}{\hat{p}(x, h)}, \tag{63}$$

where, in the denominator, the *Rosenblatt-Parzen* (RP) *kernel density estimate* of $p(x)$ is given as

$$\hat{p}(x, h) = \frac{1}{N}\sum_{j=1}^{N} K_h(x - X_j). \tag{64}$$

Except for the additional weights for coping with the varying denseness of data in the input space, this estimate behaves very similar to the Priestley-Chao estimate. Under appropriate assumptions, in particular twice differentiability of $m(x)$ and $p(x)$ around x, the mean-squared error is up to negligible terms $\sigma_x^2 \frac{1}{Nh} + b_x^2 h^4$ with the variance and bias terms

$$\sigma_x^2 = \frac{\sigma_\varepsilon^2}{p(x)} Q_K, \qquad b_x = \left(\frac{1}{2} m''(x) + m'(x) \frac{p'(x)}{p(x)}\right) V_K.$$

From asymptotic theory, we know that for large N and optimal rate $h \to 0$ like $cN^{-1/5}$, the scaled estimation error $\sqrt{Nh}(\widehat{m}_{\mathrm{NW}}(x,h) - m(x))$ is approximately normally distributed with mean b_x and variance σ_x^2. This property may be used to approximate the predictive distribution (compare Sect. 4.1), i.e., the distribution of $\widehat{Y}_{N+1} = \widehat{m}_{\mathrm{NW}}(X_{N+1}, h)$, if the unknown quantities in b_x, σ_x^2 are replaced by some rough and ready preliminary estimates. Moreover, the local approach has the advantage that for large enough N, the estimates $\widehat{m}_{\mathrm{NW}}(x_1, h), \widehat{m}_{\mathrm{NW}}(x_2, h)$ at two different locations $x_1 \neq x_2$ are nearly independent such that joint confidence intervals may be calculated independently of each other.

Kernel smoothers are suffering from a particular kind of **boundary effect**, i.e., they are systematically distorted for x close to the boundary of the range of input values in the training set. For illustration, we consider a kernel with finite support [-1,+1] like the Epanechnikov kernel for which this problem is easy to understand. Then, the support of $K_h(x - \bullet)$ is $[x-h, x+h]$. If, e.g., all inputs $X_j > 0$ and $x < h$, then the support of the smoothing kernel extends into a region $\{z;\ z < 0\}$ where no data are available. In this case, the local averages of $\widehat{m}_{\mathrm{PC}}$ or $\widehat{m}_{\mathrm{NW}}$ at x do not include enough terms, and an additional bias is introduced. The estimate will be too small close to the boundary if $m(0) > 0$ and too large else. The boundary bias does not even vanish for increasing sample size. However, as it is restricted to x in the shrinking interval $[0, h]$, this might not be a serious problem for large enough N, i.e., small enough h.

If, for a particular application, the boundary effect is important, then a remedy would be to use specific **boundary kernels** close to $x = 0$. They are asymmetric in contrast to the usually symmetric original kernel K and scaled such that the boundary bias is reduced (compare Section 4.4 of [37]).

10.1.1 Bandwidth Selection

As for all statistical learning methods in regression, the performance of a kernel smoother depends on a good choice for the hyperparameter controlling the smoothness of the final function estimate, which here is the bandwidth h. We look for a **global** bandwidth that is used for estimating $m(x)$ at all x in the range of interest.

We may apply the generic crossvalidation method as in Sect. 3.4. Let, e.g., $\widehat{m}_{\mathrm{NW}}^{(-i)}(x,h)$ denote the estimate calculated from the training set $X_j, Y_j, j \neq i$, leaving the ith observation out. Then, due to the independence of X_i, Y_i and $\widehat{m}_{\mathrm{NW}}^{(-i)}(x,h)$,

$$CV(h) = \frac{1}{N}\sum_{i=1}^{N}\left(Y_i - \widehat{m}_{\mathrm{NW}}^{(-i)}(X_i,h)\right)^2$$

again approximates the mean-squared prediction error (21) well. The crossvalidatory choice of bandwidth $\widehat{h}_{\mathrm{CV}}$, which minimizes $CV(h)$, has the same kind of approximate optimality for large N as in the example of Sect. 3.4.

An alternative method, specific to kernel smoothers, can be derived from the mean-squared error approximation, e.g., (61), for the Priestley-Chao estimate. Minimizing the two dominant terms on the right-hand side, an approximately (for large N) mse-optimal choice for the hyperparameter h, satisfies

$$h_{\mathrm{opt}}^5(x) = \frac{\sigma_\varepsilon^2 Q_K}{V_K^2 (m''(x))^2}\,\frac{1}{N},$$

depending on the location x at which we want to estimate $m(x)$. This expression does not provide an immediate method how to choose the **local** bandwidth as we do not know σ_ε^2 and even more the second derivative of the unknown function $m(x)$. Nevertheless, this formula provides some insight in the choice of good bandwidths. For example, at narrow peaks, where $(m''(x))^2$ is large, a smaller bandwidth is preferable, as, there, the bias is rather large. Moreover, with increasing sample size N, $h_{\mathrm{opt}}(x) \to 0$ like $N^{-1/5}$, i.e., if we have found a good bandwidth h_1 for some N and we increase the sample size by a factor 2, then $2^{-1/5}h_1$ will be a good bandwidth for the new larger sample, too.

Correspondingly, an approximately optimal global bandwidth satisfies

$$h_{\mathrm{opt}}^5 = \frac{\sigma_\varepsilon^2 Q_K}{V_K^2 \int_0^1 (m''(x))^2 dx}\,\frac{1}{N}.$$

This may be used for automatic selection of a good bandwidth, using a **plug-in** approach, i.e., σ_ε^2 and $\int (m''(x))^2 dx$ are replaced by some preliminary estimates, which are also calculated from the training set (for a recent review of the related bandwidth selection problem for kernel density estimates, compare [45]).

For the Priestley-Chao estimate on the left-hand side of Fig. 36, we have selected $h = h_{\mathrm{CV}} = 0.074$ by leave-one-out crossvalidation, i.e., by minimizing

$$\mathrm{CV}_{\mathrm{PC}}(h) = \widehat{\mathrm{mspe}}(\widehat{m}) = \frac{1}{N}\sum_{i=1}^{N}\left(Y_i - \widehat{m}^{(-i)}(X_i;h)\right)^2$$

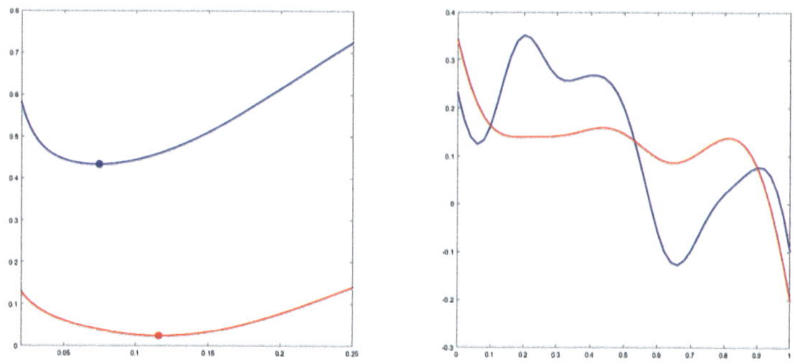

Fig. 37 $\mathrm{CV}_{\mathrm{PC}}(h)$ (blue) and ase(h) with the minimums marked (left); estimation errors for bandwidth h_{CV} (blue) and h_{opt} (red)

w.r.t. h, where $\widehat{m}^{(-i)}$ denotes the estimate based on the training set with the ith observation left out.

Figure 37 shows $\mathrm{CV}_{\mathrm{PC}}(h)$ (blue) on the left-hand side. As for this artificial data, we know the true function $m(x)$ we can compare $\mathrm{CV}_{\mathrm{PC}}(h)$ with the average squared estimation error ase$(h) = \frac{1}{N}\sum_{j=1}^{N}\left(m(X_j) - \widehat{m}(X_j; h)\right)^2$ (red), which up to a constant is a good approximation to mspe$(\widehat{m})$. ase(h) assumes its minimum at $h_{opt} = 0.116$. Note that, quite typically, the average squared error is quite flat as a function of h such that the rather large difference between h_{CV} and h_{opt} nevertheless results in a good performance of the estimate $\widehat{m}(x; h_{\mathrm{CV}})$. The slightly worse estimation quality of the crossvalidatory choice of bandwidth is due to the three data with large positive residuals around $x = 0.2$, which draw the local estimate $\widehat{m}(x; h_{\mathrm{CV}})$ upward, and the particularly large negative residual at $x = 0.68$, which draws the estimate downward. Such effects due to the particular training set cannot be avoided, but the influence of a few extreme observations becomes smaller with increasing sample size.

10.2 Nearest-Neighbor Estimates

Kernel smoothers are calculated by the intuitive (weighted) local averaging of all outputs Y_j in a neighborhood of x, e.g., in the interval $[x - h, x + h]$ for the Epanechnikov kernel. The number of training data X_j, Y_j having an influence on $\widehat{m}_{\mathrm{NW}}(x, h)$ at a given x varies with the denseness of observed input values around x, which has an influence on the reliability of the estimate and on the average depends on the probability density $p(x)$ of the X_j. To avoid this kind of variation, one would like to average over a fixed number, say k, of Y_j, i.e., one would consider the k observations for which the X_j are nearest to x, independently of their actual distance to x.

To get a precise formulation of this idea, let $d_1(x) \leq d_2(x) \leq \ldots \leq d_N(x)$ be the ordered distances $|x - X_j|,\ j = 1, \ldots, N$. If $|x - X_j| = d_k(x)$, then X_j is called the k-**nearest neighbor** of x. For a kernel function K as in the previous section,

$$\widehat{m}_{\text{nn}}(x, k) = \frac{1}{N\, d_k(x)} \sum_{j=1}^{N} K\left(\frac{x - X_j}{d_k(x)}\right) Y_j / \hat{p}_{\text{nn}}(x, k) = \widehat{m}_{\text{NW}}(x, d_k(x))$$

is called a k-nearest neighbor estimate for the regression function $m(x)$. Here,

$$\hat{p}_{\text{nn}}(x, k) = \frac{1}{N\, d_k(x)} \sum_{j=1}^{N} K\left(\frac{x - X_j}{d_k(x)}\right) = \frac{1}{N} \sum_{j=1}^{N} K_{d_k(x)}(x - X_j)$$

is the corresponding k-nearest neighbor estimate of the density $p(x)$. $\widehat{m}_{\text{nn}}(x, k)$ and $\hat{p}_{\text{nn}}(x, k)$ obviously are Nadaraya-Watson resp. Rosenblatt-Parzen estimates with variable bandwidth $d_k(x)$. The latter depends on the hyperparameter k, on x, and on all input values X_j from the training set.

If k increases, then $d_k(x)$ increases for all x, and $\widehat{m}_{\text{nn}}(x, k)$ becomes smoother. Therefore, k controls the global smoothness of the nearest neighbor regression estimate. Its local smoothness is governed by $d_k(x)$ and adapts itself to the denseness of the input values $X_1, \ldots, X_N$ around x. Similarly to kernel smoothers, we get $\widehat{m}_{\text{nn}}(x, k) \to m(x)$ if $N \to \infty$ and the hyperparameter k changes appropriately. Here we need $k \to \infty$ but slower than N, i.e., $k/N \to 0$.

For the advantage of adaptivity, a price has to be paid: $d_k(x)$ is a piecewise linear function of x and not differentiable at $(X_i + X_j)/2$, i.e., at midpoints between observations. Therefore, $\hat{p}_{\text{nn}}(x, k)$, and, hence, $\widehat{m}_{\text{nn}}(x, k)$ are not differentiable, even if $p(x)$, $m(x)$, and the kernel $K(u)$ are very smooth functions.

For the rectangular kernel $K(u) = \frac{1}{2} 1_{(-1,+1)}(u)$, we get a particular simple form, as, here, $\hat{p}_{\text{nn}}(x, k) = (k - 1)/(2N\, d_k(x))$

$$\widehat{m}_{\text{nn}}(x, k) = \frac{1}{k - 1} \sum_{j=1}^{N} 1_{(-1,+1)}\left(\frac{x - X_j}{d_k(x)}\right) Y_j = \frac{1}{k - 1} \sum_{j;|x - X_j| < d_k(x)} Y_j.$$

In this case, $\widehat{m}_{\text{nn}}(x, k)$ is just the average of the Y_j observed at the $k - 1$ nearest neighbors X_j of x.

For nearest-neighbor estimates and suitable choices of the kernel K, fast algorithms for calculating the function estimate at N locations are available. For example, for the rectangular kernel, with $X_{(1)} \leq X_{(2)} \leq \ldots \leq X_{(N)}$ denoting the ordered inputs, the following recursion may be used

$$\widehat{m}_{\text{nn}}(X_{(i+1)}, k) = \widehat{m}_{\text{nn}}(X_{(i)}, k) + \frac{1}{k - 1}\left(\tilde{Y}_{(i+[\frac{k}{2}]+1)} - \tilde{Y}_{(i-[\frac{k}{2}])}\right),$$

where $\tilde{Y}_{(\ell)}$ denotes the output corresponding to the input $X_{(\ell)}$. This algorithm allows calculation of $\hat{m}_k(X_j)$, $j = 1, \ldots, N$, with order of N operations, whereas the calculation of kernel smoothers, e.g., $\widehat{m}_{\text{NW}}(X_j, h))$, $j = 1, \ldots, N$, requires a number of operations of order $N^2 h$ if the kernel K has a compact support.

10.3 Spline Smoothers

In numerical interpolation, we start from some unknown smooth function observed at only a finite number of points, the so-called **knots** $x_1 < \ldots < x_N$. If $y_1, \ldots, y_N$ are the observed function values, then we are looking for a smooth interpolating function $g(x)$ matching the observations exactly at the knots, i.e., $y_j = g(x_j)$, $j = 1, \ldots, N$. A popular interpolation method with nice properties is the so-called spline interpolation. It approximates the unknown function between adjacent knots by simple smooth spline functions. Here, we consider only **cubic spline functions** g_s, which between the knots are polynomials of degree 3. They are chosen such that

(1) $y_j = g_s(x_j)$, $j = 1, \ldots, N$.
(2) g_s is twice continuously differentiable for all x,
(3) $g_s(x) = \sum_{k=0}^{3} c_{jk} x^k \quad$ for $x_j \le x \le x_{j+1}$, $j = 1, \ldots, N-1$.

There are functions satisfying these three conditions. The solution g_s becomes unique if we additionally impose some constraints at the two boundary points x_1, x_N. Cubic spline smoothers have an interpretable property that singles them out among all interpolating functions. They minimize the **total curvature**

$$\int_{x_1}^{x_N} \left(g''(x)\right)^2 dx$$

among all twice continuously differentiable functions g with $y_j = g(x_j)$, $j = 1, \ldots, N$.

In statistical learning, the output values Y_j are noisy and only are close to the function values $m(X_j)$ of interest up to the random residual ε_j. Therefore, interpolation makes no sense. We only require that, for a function approximation $\widehat{m}$, the deviation $|Y_j - \widehat{m}(X_j)|$ is small for all j. Assuming that the true prediction function m is smooth, i.e., in the spirit of cubic spline interpolation, twice continuously differentiable, we instead look at a least-squares fit. Without any constraints, this would lead to interpolation again and a very rough function estimate as it ignores the observation errors ε_j. To avoid this effect, we impose a constraint on the total curvature or, by the Lagrange multiplier approach, equivalently a roughness penalty. This is another example for a penalized least-squares method (compare Sects. 3.2 and 3.3).

We call $\widehat{m}_s(x)$ a **spline smoother** (or smoothing spline) with regularization parameter $\lambda > 0$ if it minimizes

$$S_\lambda(g) = \sum_{j=1}^{N} \left(Y_j - g(X_j)\right)^2 + \lambda \int \left(g''(x)\right)^2 dx$$

over all twice continuously differentiable functions g. $\lambda > 0$ is the hyperparameter of this learning method.

The minimization problem has an explicit solution that indeed is a cubic spline function with knots $X_1, \ldots, X_N$, though not an interpolating one, satisfying

(a) $\widehat{m}_s(x)$ is twice continuously differentiable for all x,
(b) $\widehat{m}_s(x)$ is a polynomial of degree 3 in any interval $[X_{(j)}, X_{(j+1)}]$, $1 \leq j < N$,
(c) $\widehat{m}_s''(x) = 0$ for $x < X_{(1)}$ and $x > X_{(N)}$,
(d) At the boundaries, $\widehat{m}_s'(X_{(1)}) = \widehat{m}_s'(X_{(N)}) = 0$.

Here, $X_{(1)} \leq \ldots \leq X_{(N)}$ denote the ordered input data. The coefficients of the pieces of cubic polynomials, which combine to $\widehat{m}_s(x)$, can be calculated as solutions of a (large) system of linear equations.

Though it is by no means obvious, spline smoothers are closely related to kernel estimates. If $N \to \infty$ and $\lambda \to 0$ not too fast we have

$$\widehat{m}_s(x) \approx \frac{1}{N} \sum_{j=1}^{N} \frac{1}{p(X_k)} \frac{1}{h_\lambda(X_k)} K^{sp}\left(\frac{x - X_k}{h_\lambda(X_k)}\right) \tag{65}$$

$$\text{with bandwidth } h_\lambda(z) = \sqrt[4]{\frac{\lambda}{Np(z)}}$$

where $p(x)$ again is the common probability density of the inputs X_j. The Silverman spline kernel

$$K^{sp}(u) = \frac{1}{2} e^{-|u|/\sqrt{2}} \sin\left(\frac{|u|}{\sqrt{2}} + \frac{\pi}{4}\right)$$

is a specific symmetric kernel with $\int u^2 K^{sp}(u) du = 0$, i.e., it assumes also negative values.

Using that, for increasing N, the range of values where $K^{sp}(u)$ is not almost 0 shrinks, we see that the spline smoother behaves like a Nadaraya-Watson kernel estimate with a bandwidth adapting to the local denseness of input values around x, i.e., to $p(x)$.

As in many similar learning approaches, the spline smoother $\widehat{m}_s(x)$ is determined by minimizing a certain loss function, here $S_\lambda(g)$, which consists of two parts. The first term on the right-hand side is the so-called **data term**, which measures the fit of the function $g(x)$ to the data, i.e., how well the Y_j of the training set are approximated by $g(X_j)$. The second term is the **smoothness term** or **roughness penalty**, which is a safeguard against overfit and guarantees solutions that are not too irregular functions. The weight λ provides the balance between the tendencies to overfit resp. underfit (compare Sect. 2.3).

In spline smoothing, we consider a quadratic data term, motivated by the popular least-squares approach in regression, and the total curvature as roughness measure of the curve, which is physically intuitive. Both have the mathematical advantage of being quadratic in g, leading to an analytic form for the minimizer of $S_\lambda(g)$.

There are other popular regularization methods, using different data and smoothness terms and, as a consequence, frequently also different function spaces. For example, we may allow for continuous functions g, which almost everywhere have a derivative g', which is integrable over an interval that contains the training inputs X_j. A suitable roughness penalty is the **total variation**

$$TV(g) = \int |g'(x)| dx.$$

This smoothness term is related to the LASSO of Sect. 3.3 and the *Tikhonov regularization* used, e.g., in numerical analysis. In image analysis, where we consider functions on two-dimensional space and the inputs x_j correspond to the pixel locations, this approach is slightly modified by allowing also discontinuities of the function g corresponding to the boundaries of objects which are part of the image. Here, the roughness is again measured by an integral of the first derivative, but it also takes into account the discontinuities of g. The smoothness term is of the form

$$T_{MS}(g) = \int \min(|g'(x)|^2, \beta) dx$$

for some $\beta > 0$, which replaces λ the hyperparameter under our control. This approach is called *Mumford-Shah regularization.*

The data term can also be changed to get estimates that are more robust against outliers in the data, which, e.g., appear more frequently when the distribution of the errors is heavy-tailed. In image analysis, this effect is called *pepper-and-salt noise* where the isolated outlying pixels are almost pure white or black and stick out of the image of moderate grayscale values. Here, using the L^1-data term

$$\sum_{j=1}^{N} |Y_j - g(X_j)|$$

is advantageous. To get more flexibility, we could use instead

$$\sum_{j=1}^{N} \rho(Y_j - g(X_j))$$

where, for some $c > 0$, we choose $\rho(u) = u^2$ for $|u| \leq c$ and $\rho(u) = 2c|u| - c^2$, else. This approach combines the advantages of the quadratic distance u^2 for small deviations of Y_j from $g(X_j)$ with the robustness of the absolute deviation $|u|$ for large values. It is motivated by Huber's M-estimates, which have been first studied in parametric robust statistics, but the idea can be generalized to statistical learning.

10.4 Local Least Squares and Local Linear Estimates

We have introduced kernel regression estimates of $m(x)$ as (weighted) local averages of the output values Y_j with X_j close to x. For a smooth regression function, this mean will be close to $m(x)$ by the law of large numbers. Alternatively, this learning method can be motivated by recalling the familiar least-squares approach. In the previous section, we have mentioned in the context of smoothing splines that a global least-squares approach does not work in statistical learning without any restriction on the function estimate. Instead of adding roughness penalties, a different idea is to approximate the function $m(x)$ locally by a simple parametric function and estimate its parameters by local least squares. In the simplest case, the local approximation is just a constant, and the function m at x is estimated by the solution of the localized LS problem

$$\sum_{j=1}^{N} (Y_j - c)^2 K_h(x - X_j) = \min_{c}!, \tag{66}$$

where, as above, we use a scaled kernel K_h to achieve that the summation essentially uses only data observed in a neighborhood of x.

The solution of this minimization problem is easily calculated by setting the derivative of the criterion function (66) w.r.t c to 0 and solving for c. The solution is

$$\hat{c} = \frac{\sum_{j=1}^{N} Y_j K_h(x - X_j)}{\sum_{j=1}^{N} K_h(x - X_j)} = \widehat{m}_{\mathrm{NW}}(x, h),$$

i.e., it coincides with the Nadaraya-Watson kernel estimate, which therefore is a local least-squares estimate.

Instead of approximating Y_j locally by a constant, we could use a polynomial function of low degree. Here, we consider only a function $c_0 + c_1(x - X_j)$, which depends linearly on the distance $x - X_j$. The corresponding local least-squares loss function is

$$\sum_{j=1}^{N} (Y_j - c_0 - c_1(x - X_j))^2 K_h(x - X_j) = \min_{c_0, c_1}!$$

Again the solution $\hat{c}_0, \hat{c}_1$ = can be calculated analytically by setting the partial derivatives w.r.t. c_0, c_1 to 0. Then, the local linear estimate is given by

$$\widehat{m}_{\ell\ell}(x, h) = \hat{c}_0,$$

and with $\widehat{m}'_{\ell\ell}(x, h) = \hat{c}_1$, we simultaneously get an estimate of the derivative $m'(x)$ of the regression function at x.

Comparing the local constant estimate $\widehat{m}_{\mathrm{NW}}(x, h)$ for $m(x)$ with the local linear estimate $\widehat{m}_{\ell\ell}(x, h)$, the latter shows a smaller bias, in particular if the local denseness of the inputs varies a lot over the range of values $X_1, \ldots, X_N$. This bias reduction is in particular striking at the boundaries, such that, there, local linear estimates usually look more plausible and explicit boundary corrections are not needed. However, not surprising, the price which we have to pay for a smaller bias is an increase in variance. Local linear estimates usually look a bit rougher, and the random error component at the boundaries is larger.

As an example, we consider data, generated as for Fig. 16, but with larger sample size $N = 101$. We calculate the Nadaraya-Watson kernel estimate with bandwidth $h_{\mathrm{CV}} = 0.23$, chosen by leave-one-out crossvalidation, and the local linear estimate, again with bandwidth chosen by crossvalidation, which, in this case, happens to be 0.23, too. Both estimates use the Gauss-kernel. On the left-hand side of Fig. 38, the data, the true function $m(x)$ (red) and the two estimates $\widehat{m}_{\mathrm{NW}}(x, h_{\mathrm{CV}})$ (blue) and $\widehat{m}_{\ell\ell}(x, h_{\mathrm{CV}})$ (black) are shown. Both estimates are almost identical with the local linear estimate slightly closer to $m(x)$ in the major part of the interval. At the boundaries, however, the larger variability of $\widehat{m}_{\ell\ell}(x, h_{\mathrm{CV}})$ leads to a greater deviation from the true function.

On the right-hand side of Fig. 38, the local linear estimate $\widehat{m}'_{\ell\ell}(x, h_{\mathrm{CV}})$ (black) of the derivative $m'(x)$ is shown. It looks reasonable in spite of the rather small sample size for derivative estimation and the not optimal bandwidth, which we have taken from $\widehat{m}_{\ell\ell}(x, h_{\mathrm{CV}})$. From Stone's Theorem ([76], compare also [37, Section 4.1]),

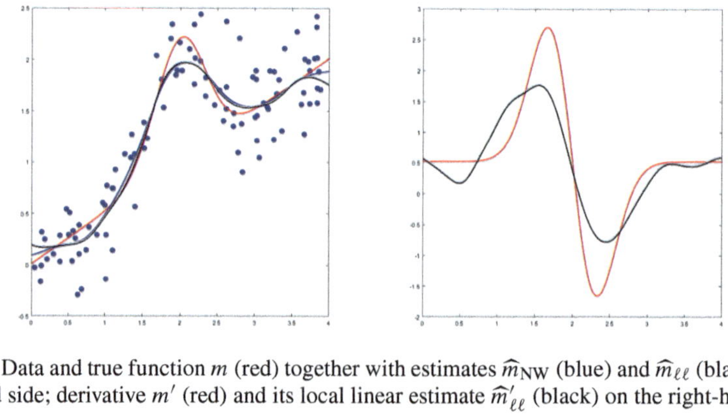

Fig. 38 Data and true function m (red) together with estimates $\widehat{m}_{\mathrm{NW}}$ (blue) and $\widehat{m}_{\ell\ell}$ (black) on the left-hand side; derivative m' (red) and its local linear estimate $\widehat{m}'_{\ell\ell}$ (black) on the right-hand side

we know that for achieving a similar performance of an estimate of a derivative, we need a considerably larger sample size N than for estimating the function itself.

10.5 The Curse of Dimensionality and Generalized Additive Models

In statistical learning, we want to estimate a function on the input space for regression, forecasting, or classification. The quality of the result, of course, depends on the amount N of data available for training. The more information we have about the data-generating process, the better solutions can be found. The dimension d of the feature vectors X_j, however, is also important. The larger d, the more difficult becomes the estimation problem for a given sample size N. This *curse of dimensionality* is particularly easy to understand in the context of local smoothing.

For kernel smoothers and local linear estimates, it is obvious that the function estimate $\widehat{m}(x)$ essentially only depends on those X_j, which are close to x. If the kernel K (now on d-dimensional space) has a compact support, e.g., $K(u) \neq 0$ only if $|u_i| \leq 1, i = 1, \ldots, d$; then only those inputs have an influence on the estimate at x for which $|x_i - X_{ji}| \leq 1, i = 1, \ldots, d$. For given N, the number of observations from the training set, satisfying this condition, will rapidly decrease with dimension d. In the end, for most locations x, e.g., $\widehat{m}_{\text{NW}}(x, h)$ will depend only on one observations or even on none. If we increase the bandwidth h, such that enough data are involved in the local averaging, then h has to be very large in high dimensions, and we will suffer from the effect discussed for nearest neighbor estimates in the next paragraph. Vice versa, if we want to have the same denseness of data around x in higher dimensions than in dimension 1 and hence the same quality of our function estimate, then roughly we have to increase the sample size from N to N^d, which soon becomes prohibitively large.

Nearest-neighbor estimates adapt their bandwidth to the local denseness of input values. This disguises the problems, as they always provide reasonably looking functions $\widehat{m}_{\text{nn}}(x, k)$. In high dimensions for given N, however, even the nearest neighbor X_ℓ of x will usually be located quite far away as the inputs will be thinly spread out over the large available space. $\widehat{m}_{\text{nn}}(x, k)$ will not be 0 or depending only on one observation as for kernel smoothers, but it is determined by observations far apart, for which $m(X_j)$ will not be close to the value $m(x)$ of interest.

Spline smoothers behave similarly to kernel smoothers due to the relationship (65). They adapt their bandwidths in a certain sense to the denseness of input values, but this does not really help against the curse of dimensionality.

Local smoothing nevertheless has meaningful applications for high-dimensional training sets, too. For this purpose, we restrict the general regression model (2) to the **generalized additive model** (GAM) where the regression function

$$m(x) = \mu + \sum_{i=1}^{d} f_i(x_i)$$

decomposes into a global mean and a sum of components, which are functions of the one-dimensional input coordinates only, with unknown functions $f_i, i = 1, \dots, d$.

For estimating $m(x)$ in a GAM, we use one-dimensional estimation methods for estimating the functions $f_i(x_i)$ on the real line where we have the choice among a lot of different methods, e.g., kernel smoothers, nearest-neighbor estimates, or spline smoothers. The general ideas for handling GAMs are the same anyhow, and we write $\hat{f}_i$ for the estimates of f_i without referring to a specific estimation method.

10.5.1 The Backfitting Algorithm

In fitting a GAM to the training set $X_j, Y_j, j = 1, \dots, N$, we simultaneously have to learn the d component functions f_i. For identifiability of the solution, we assume $\sum_{j=1}^{N} f_i(X_{ji}) = 0, i = 1, \dots, d$. We apply the iterative backfitting algorithm, which is related to the *Gauss-Seidel algorithm* for solving a system of linear equations $Ax = b$ for some $n \times n$ matrix A and some vector b. It is based on the following observation.

If μ and, for some given ℓ, $f_i(x), i \neq \ell$, would be known, we only have to estimate $f_\ell(x)$. Setting

$$Z_j = Y_j - \mu - \sum_{i \neq \ell} f_i(X_{ji}),\ U_j = X_{j\ell},\ j = 1, \dots, N,$$

we have the univariate regression problem

$$Z_j = f_\ell(U_j) + \varepsilon_j,\ j = 1, \dots, N.$$

To estimate f_ℓ, we may use any one-dimensional function estimate of the form

$$\hat{f}_\ell(u) = F(u; Z_j, U_j, j = 1, \dots, N)$$

depending on the training set $(U_j, Z_j), j = 1, \dots, N$, with one-dimensional inputs U_j. F may be a kernel estimate or a spline smoother or any other kind of estimate. The backfitting algorithm uses this representation. In each step of the iteration, all but one component estimates $\hat{f}_i l$ are fixed, and the remaining one is updated until some stopping criterion is fulfilled.

Backfitting

1. *Initialization*: $\hat{\mu} = \frac{1}{N}\sum_{j=1}^{N} Y_j,\ \hat{f}_i = 0, k = 1, \ldots, d.$
2. *Iteration*: For $\ell = 1, \ldots, d, 1, \ldots, d, 1, 2, \ldots$

$$\text{Set}\quad \widehat{Z}_j = Y_j - \hat{\mu} - \sum_{i \neq \ell} \hat{f}_i(X_{ji}),$$

$$\tilde{f}_\ell(u) = F(u; \widehat{Z}_j, X_{j\ell}, j = 1, \ldots, N)$$

$$\hat{f}_\ell(u) = \tilde{f}_\ell(u) - \frac{1}{N}\sum_{j=1}^{N} \tilde{f}_\ell(X_{j\ell})$$

3. *Stopping criterion*: After each full cycle from $1, \ldots, d$ check for $\hat{f}_\ell^{\text{old}}, \hat{f}_\ell^{\text{new}}$ denoting the function estimates at the beginning and the end of this cycle, and some $\delta > 0$, if

$$\frac{1}{N}\sum_{j=1}^{N}\left(\hat{f}_\ell^{\text{old}}(X_{j\ell}) - \hat{f}_\ell^{\text{new}}(X_{j\ell})\right)^2 \leq \delta \qquad \text{for all } \ell = 1, \ldots, d.$$

Of course, other stopping criteria may be used, e.g., the maximal number of iterations. The last step of the iteration guarantees that $\hat{f}_\ell$ satisfies the identifiability condition to safeguard against numerical errors.

In many cases, the backfitting algorithm may be shown to converge. If, in particular, F is the cubic spline smoother, the limit is

$$\widehat{m}_s(x) = \hat{\mu} + \sum_{i=1}^{d} \hat{f}_{si}(x_i)$$

which solves the following minimization problem w.r.t. all real-valued μ and twice continuously differentiable $f_1, \ldots, f_d$

$$\sum_{j=1}^{N}\left(Y_j - \mu - \sum_{i=1}^{d} f_i(X_{ji})\right)^2 + \sum_{i=1}^{d} \lambda_i \int (f_i''(u))^2 du = \text{min!}$$

where $\lambda_1, \ldots, \lambda_d$ are the weights of the roughness penalties used in the d individual univariate spline smoothers in the course of the iteration.

10.5.2 Projection Pursuit Regression

Generalized additive models replace the arbitrary regression function on d-dimensional space by a sum of functions on the real line. For this purpose, they just use dependence on the given coordinate directions x_i. However, those may not be the directions that provide the most informative univariate approximations of the whole regression function. Projection pursuit instead looks also for the most informative one-dimensional projections of the feature vectors X_j, i.e., they approximate $m(x)$ by functions of the type

$$\widehat{m}(x) = \hat{\mu} + \sum_{k=1}^{p} \hat{f}_k\Big(\sum_{i=1}^{d} \hat{b}_{ki} x_i\Big).$$

The one-dimensional argument of f_k is the projection of the input vector x onto the direction $\hat{b}_k = (b_{k1}, \ldots, b_{kd})$. For $p = d$ and the parameter vectors $\hat{b}_k$ equal to the d unit vectors of the input space, we get the original GAM back.

The number of terms p acts as a hyperparameter additional to the hyperparameters of the local smoothers, which provide the estimates $\hat{f}_k(u)$ of the component functions. Usually, we get a good approximation of the data with $p \ll d$, i.e., projection pursuit regression has kind of an inbuilt dimension reduction mechanism by looking for informative directions in the feature space. Similar ideas are used in general dimension reduction techniques like principal components analysis, too (compare Sect. 12.2). Note that $\widehat{m}(x)$ is quite similar to the output function of a feedforward neural network (45) with one hidden layer consisting of p neurons. However, the summands are not just of the form $f_k(u) = v_k \psi(u)$ for some given activation function ψ and some weights v_k, but they are completely arbitrary. Projection pursuit regression may therefore be interpreted as a kind of neural networks where not only the weights but also the activation functions are estimated adaptively from the data.

In contrast to the original GAM, we not only have to estimate the component functions by local smoothing, but we also have to learn the projection vectors $\hat{b}_k$ from the data. Based on similar ideas as backfitting, there are algorithms (e.g., [31]) that allow for solving this task. As they require a lot of calculations of functions smoothers, it is advantageous to use a computationally efficient method like, e.g., nearest-neighbor estimates.

11 Support Vector Machines

Support vector machines are one of the most popular machine learning tools for classification. For ease of notation, as in Sect. 5.3, we again consider only $M = 2$ classes with class indicators $Y_j \in \{-1, +1\}$. We briefly refer to some notions and results introduced in that section.

Let $H = \{x;\ R(x) = \langle b, x\rangle + b_0 = 0\}$ be a hyperplane in d-dimensional feature space. The corresponding linear classifier is given by $C(x) = \operatorname{sgn} R(x)$, and an object is misclassified if $Y_j C(X_j) = -1$. If H is a *separating hyperplane*, then classification on the training set is perfect, i.e., $Y_j R(X_j) > 0,\ j = 1, \ldots, N$.

In that case, there are many separating hyperplanes. We single out the *maximum margin hyperplane* $\widehat{H}$ for which

$$\min_{j=1,\ldots,N} \operatorname{dist}(X_j, H) = \min_{j=1,\ldots,N} \min_{x \in H} ||x - X_j||$$

is maximized over all separating hyperplanes. An object from the training set with feature values X_s is called a **support vector** if it lies on the boundary of the margin, i.e.,

$$\operatorname{dist}(X_s, L) = \min_{j=1,\ldots,N} \operatorname{dist}(X_j, \widehat{H}).$$

From Sect. 5.3, we have for the parameters of $\widehat{H} = \{x;\ \widehat{R}(x) = \langle \hat{b}, x\rangle + \hat{b}_0 = 0\}$

$$\hat{b}_i = \sum_{j=1}^{N} \hat{\alpha}_j Y_j X_{ji},\ i = 1, \ldots, d,$$

where the Lagrange multipliers $\hat{\alpha}_1, \ldots, \hat{\alpha}_N \geq 0$ solve

$$\sum_{j=1}^{N} \alpha_j - \frac{1}{2} \sum_{i,j=1}^{N} \alpha_i \alpha_j Y_i Y_j \langle X_i, X_j\rangle = \max_{\alpha \in \mathbb{R}^N}! \tag{67}$$

$$\text{under the constraint} \quad \sum_{j=1}^{N} \alpha_j Y_j = 0.$$

The support vectors X_s are characterized by a non-vanishing $\hat{\alpha}_s > 0$, and $\hat{b}_0$ can be calculated from the relationship $Y_s(\langle \hat{b}, X_s\rangle + \hat{b}_0) = 1$ holding for any support vector X_s.

Usually, the training set $(X_j, Y_j),\ j = 1, \ldots, N$, will not be linearly separable, i.e., the subsets $\{X_j;\ Y_j = +1\}$ and $\{X_j;\ Y_j = -1\}$ of objects in the training set, corresponding to the two classes, overlap in the d-dimensional feature space. To overcome the resulting nonexistence of separable hyperplanes and corresponding linear decision rules, we recall the original formulation (32) of finding the optimal separating hyperplane as a maximal margin problem:

$$\max_{b, b_0, ||b||=1} C! \qquad \text{under the constraints} \quad Y_j(\langle b, X_j\rangle + b_0) \geq C,\ j = 1, \ldots, N.$$

We allow for some points to lie on the wrong side of the decision boundary by introducing **slack variables** $\xi_j \geq 0$ and replacing the constraints by

$$Y_j(\langle b, X_j \rangle + b_0) \geq C(1 - \xi_j).$$

$\xi_j > 1$ is allowed; in that case X_j lies on the wrong side of the boundary, and the corresponding object is misclassified. For $0 < \xi_j < 1$, it lies on the right side, but inside the band of width $2C$ around the boundary.

The amount of overlap is bounded by enforcing $\sum_{j=1}^{N} \xi_j \leq K_s$ for some constant K_s. For integer K_s, we may have at most K_s misclassifications in the training set.

We now reformulate the optimization problem as in Sect. 5.3. We drop the condition $||b|| = 1$, set $C = ||b||^{-1}$, and redefine b_0, ξ_j accordingly. Then the equivalent formulation of (32) with additional slack variables is

$$\begin{aligned} &\frac{1}{2}||b||^2 = \min_b! \qquad \text{under the constraints} \\ &Y_j(\langle b, X_j \rangle + b_0) \geq 1 - \xi_j, \; j = 1, \ldots, N, \\ &\text{for some } \xi_j \geq 0 \quad \text{with} \quad \sum_{j=1}^{N} \xi_j \leq K_s. \end{aligned} \tag{68}$$

This is the most common formulation of the *support vector machine problem.*

11.1 Calculation of the Support Vector Classifier

As in Sect. 5.3.2, we use standard optimization techniques to solve (68). Introducing nonnegative Lagrange multipliers $\gamma, \alpha_j, j = 1, \ldots, N,$, we get the dual form of the support vector machine problem

$$\begin{aligned} &\sum_{j=1}^{N} \alpha_j - \frac{1}{2} \sum_{i,j=1}^{N} \alpha_i \alpha_j Y_i Y_j \langle X_i, X_j \rangle = \max_{\alpha_1, \ldots, \alpha_N}! \qquad \text{under the constraints} \\ &\sum_{j=1}^{N} \alpha_j Y_j = 0, \; 0 \leq \alpha_j \leq \gamma, \; j = 1, \ldots, N, \end{aligned} \tag{69}$$

and the Karush-Kuhn-Tucker conditions for $j = 1, \ldots, N$,

$$\alpha_j \big(Y_j(\langle b, X_j \rangle + b_0) - (1 - \xi_j) \big) = 0 \tag{70}$$

$$(\gamma - \alpha_j)\xi_j = 0 \tag{71}$$

$$Y_j(\langle b, X_j \rangle + b_0) - (1 - \xi_j) \geq 0. \tag{72}$$

γ is the Lagrange multiplier corresponding to the constraint $\sum_j \xi_j \leq K_s$, i.e., to the upper bound of slack we are allowing for. Both are exchangeable as tuning parameters of the learning algorithm, and as we solve the dual problem, it is more convenient to fix γ as an upper bound on the coefficients α_j.

In terms of the solution $\hat{\alpha}_j$, $j = 1, \ldots, N$, of the dual optimization problem, the solution of the original support vector machine problem (68) is given by

$$\hat{b} = \sum_{j=1}^{N} \hat{\alpha}_j Y_j X_j \qquad \text{with} \qquad \hat{\alpha}_j \neq 0 \text{ only if } Y_j(\langle \hat{b}, X_j \rangle + \hat{b}_0) = 1 - \xi_j. \tag{73}$$

The X_s with $\hat{\alpha}_s \neq 0$ are called the **support vectors**. Some of them will lie on the boundary of the margin, i.e., $\xi_s = 0$. They may be used to calculate the remaining parameter of the hyperplane used for classification as, for them,

$$\hat{b}_0 = Y_s - \langle \hat{b}, X_j \rangle.$$

Finally, we get the **support vector classifier**

$$\hat{c}(x) = \text{sgn}\widehat{R}(x) = \text{sgn}(\langle \hat{b}, x \rangle + \hat{b}_0).$$

11.2 Kernelization or the Kernel Trick

Real-world classification problems are often not linearly separable, even not approximately enough to handle them by the introduction of slack variables. So, linear classifiers as described in the last section are frequently not good, in particular for larger training samples where the nonlinearity of the data structure could be used to get a considerably improved classification performance. The main reason for the nevertheless widespread use of support vector classifiers is a computationally feasible transformation of the original nonlinear problem into a nearly linear problem. This is the so-called *kernel trick*. It uses the otherwise detrimental curse of dimensionality (compare Sect. 10.5) for profit by noting that, in high dimensions, there is a lot of empty space between the data. Therefore, it is more likely that a separating hyperplane fits in between the training data from the two classes.

To follow this idea, our first step is to map the original d-dimensional feature space $\mathbb{R}^d$ into a really high-dimensional space $\mathbb{R}^D$, using a mapping ϕ, and consider the transformed input vectors

$$Z_j = \phi(X_j), \ j = 1, \ldots, N,$$

such that $\{Z_j; Y_j = +1\}$ and $\{Z_j; Y_j = -1\}$ become linearly separable in the new feature space $\mathbb{R}^D$.

In general, calculations like solving the support vector machine problem in $\mathbb{R}^D$ become computationally extremely expensive or are even not feasible at all. The kernel trick relies on special mappings ϕ for which

$$\langle \phi(x), \phi(x') \rangle = K(x, x'), \quad x, x' \in \mathbb{R}^d \tag{74}$$

for some **kernel** K. Note that for solving (69), the feature vectors enter the calculations only via their scalar products. Therefore, the kernel trick reduces the solution of the support vector machine problem or the subsequent application of the support vector classifier in the high-dimensional space $\mathbb{R}^D$ to calculations in $\mathbb{R}^d$, e.g., we use

$$\langle Z_i, Z_j \rangle = \langle \phi(X_i), \phi(X_j) \rangle = K(X_i, X_j).$$

Note that we also do not have to evaluate ϕ. We only have to work with the kernel K.

Let us consider the example introduced at the beginning of Sect. 5, where $d = 2$ and $Y_j = +1$ for $||X_j|| > 1$ and $Y_j = -1$ for $||X_j|| < 1$, i.e., the two classes are separated by the boundary of the unit circle. For these $X_j \in \mathbb{R}^2$, it suffices to move from $d = 2$ to $D = 3$ to linearize the problem with the kernel trick. We use the mapping

$$z = \phi(x) = (x_1^2, \sqrt{2}x_1x_2, x_2^2) \qquad \text{with corresponding kernel} \qquad K(x, \tilde{x}) = \langle x, \tilde{x} \rangle^2.$$

A straightforward calculation shows $\langle \phi(x), \phi(\tilde{x}) \rangle = (x_1\tilde{x}_1 + x_2\tilde{x}_2)^2 = K(x, \tilde{x})$. The hyperplane $H = \{z; \langle b, z \rangle - 1 = 0\}$ in $\mathbb{R}^3$ is a separating hyperplane for $b = (1, 0, 1)$ as $\langle b, z \rangle - 1 = x_1^2 + x_2^2 - 1$, which implies $c(X_j) = +1$ for $||X_j|| > 1$ and $c(X_j) = -1$ for $||X_j|| < 1$.

In the example, $D = 3$ was not really large, but the dimension, necessary for linear separability, increases fast. If we suspect that the class of an object is influenced by up to 4th powers $x_i x_j x_k x_\ell$, $1 \leq i, j, k, \ell \leq d$, of the features, then, e.g., for $d = 256$, we would get $D \approx 2 \cdot 10^8$. Nevertheless, thanks to the kernel trick avoiding direct calculations in $\mathbb{R}^D$, support vector classifiers remain feasible even then.

To summarize, *kernelization* is a successful device for machine learning applications, which are based on tried-and-tested linear procedures. It consists in finding a mapping $\phi : \mathbb{R}^d \to \mathbb{R}^D$ such that

(a) $D \gg d$ to have a better chance for the existence of good linear classifiers,
(b) classifiers require only the calculation of scalar products,
(c) there is a kernel K with $K(x, \tilde{x}) = \langle \phi(x), \phi(\tilde{x}) \rangle$,
(d) replacing $\langle z, \tilde{z} \rangle$ by $K(x, \tilde{x})$ everywhere in the algorithm leads to feasible computations.

This idea is not restricted to support vector machines but can in principle be used for extending other linear classification methods. However, this particular machine learning approach has the advantage that the resulting classifier explicitly depends only on the rather small number of support vectors.

We have already mentioned that explicit evaluation of the mapping ϕ is not necessary. In practice, therefore, one starts immediately with prescribing a kernel. To judge its appropriateness, we have to know which functions $K(x, \tilde{x})$ can be represented as $\langle \phi(x), \phi(\tilde{x}) \rangle$ for a suitable ϕ. The answer is given by *Mercer's Theorem*: Yes, there exists such a representation if K is **positive definite**, i.e.,

$$\int \int K(x, \tilde{x}) f(x) f(\tilde{x}) dx d\tilde{x} \geq 0 \text{ for all square-integrable } f,$$
$$= 0 \text{ if and almost if } f = 0 \text{ almost everywhere.}$$

Popular kernels are

$$K(x, \tilde{x}) = (a + \langle x, \tilde{x} \rangle)^q \qquad \text{(polynomial of degree } q\text{)}$$
$$K(x, \tilde{x}) = \tanh(a \langle x, \tilde{x} \rangle) + b) \qquad \text{(sigmoid)}$$
$$K(x, \tilde{x}) = \exp\Big(- \frac{1}{2} ||x - \tilde{x}||^2 \Big) \qquad \text{(Gaussian or radial basis)}$$

11.3 Support Vector Machine Classifiers

In this section, we summarize the essentials about support vector machines for classification. They are the results of combining the linear support vector classifier from Sect. 11.1 with the kernel trick of Sect. 11.2. We start with choosing an appropriate kernel K having the representation (74) with some function ϕ on the original feature space.

Replacing in (69) the scalar products $\langle Z_i, Z_j \rangle = \langle \phi(X_i), \phi(X_j) \rangle$ in the new D-dimensional feature space by the kernel, we get

$$\sum_{j=1}^{N} \alpha_j - \frac{1}{2} \sum_{i,j=1}^{N} \alpha_i \alpha_j Y_i Y_j K(X_i, X_j) = \max_{\alpha_1, \ldots, \alpha_N} ! \qquad \text{under the constraints} \tag{75}$$

$$\sum_{j=1}^{N} \alpha_j Y_j = 0, \; 0 \leq \alpha_j \leq \gamma, \; j = 1, \ldots, N,$$

The solution of this convex programming problem results in the estimate $\hat{\alpha}$, and the corresponding $\hat{b}$ is given by (73). Then, from Sect. 11.1, the classification function is, with $z = \phi(x)$,

$$\widehat{R}(x) = \langle z, \hat{b} \rangle + \hat{b}_0 = \sum_{j=1}^{N} \hat{\alpha}_j Y_j \langle z, Z_j \rangle + \hat{b}_0 = \sum_{j=1}^{N} \hat{\alpha}_j Y_j K(x, X_j) + \hat{b}_0,$$

where for any fixed support vector X_s

$$\hat{b}_0 = -\sum_{j=1}^{N} \hat{\alpha}_j Y_j K(X_s, X_j).$$

Therefore, the final form of the support vector machine classifier is

$$\hat{c}(x) = \operatorname{sgn} \widehat{R}(x) \qquad \text{with} \qquad \widehat{R}(x) = \sum_{j=1}^{N} \hat{\alpha}_j Y_j \Big(K(x, X_j) - K(X_s, X_j) \Big).$$

$\hat{\alpha}_j$ are given by (75). There, γ is a hyperparameter for tuning the learning algorithm.

For an example, we use the same training set of size $N = 400$ with two-dimensional feature vectors as in Fig. 26. On the left-hand side of Fig. 39, the training data are shown, where the two classes 1 (blue) and 2 (red) are color-coded. We calculate the support vector machine classifier by solving the optimization problem (75). Then we apply the classification algorithm to the training set and to an independent, identically generated validation set shown on the right-hand side of Fig. 39. For various values of the hyperparameter γ, we get the misclassification rates shown in Table 6. Regarding the relative small error probability estimate of 1.75 % in the relevant validation set, the choice of $\gamma = 175$ seems to be reasonable. As for other machine learning procedures, e.g., local smoothers discussed in

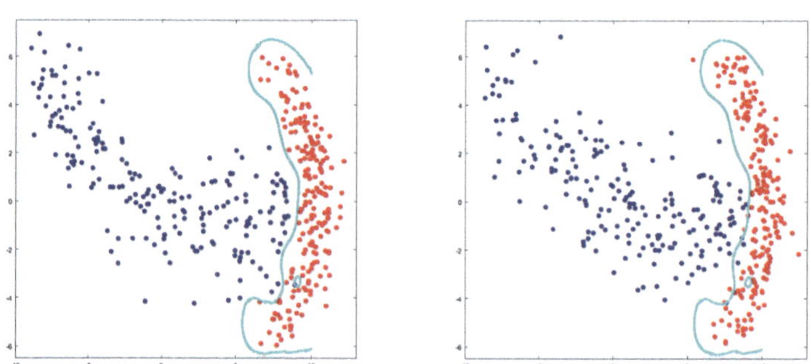

Fig. 39 Training set (left) and validation set (right) together with the decision boundary (light blue) of the support vector machine classifier using $\gamma = 175$

Table 6 Misclassification rate in training and validation sample of Fig. 39

γ	75	100	125	175	200
Training sample	11.75 %	0.25 %	0.25 %	0.25 %	0.25 %
Validation sample	12.00 %	3.75 %	2.50 %	1.75 %	2.00 %

the previous chapters, the performance fortunately is rather invariant around its optimum, such that the exact value of the hyperparameter has only a small impact as long as it lies in the right range.

Figure 39 shows the resulting decision boundary $\{x;\, \widehat{R}(x) = 0\}$ together with the training resp. validation data. A new observation with feature value $X_{N+1} = x$ is classified as 1 resp. 2, if it lies to the left resp. to the right of the boundary. Note that $\gamma = 175$ produces a small artifact from overfitting by enclosing the right-most observation from class 1 in the training set by a small ellipse.

11.4 Support Vector Machines as Penalized Function Estimators

In this section, we investigate the relationship between support vector machines and regularization methods in regression (compare Sects. 3.2, 3.3, and 10.3). Recall, e.g., the regularization approach of ridge regression and the LASSO. To avoid overfit, the residual sum of squares as target function is augmented by a penalty P for the complexity of the fitted function, which is directly related to its roughness:

$$\sum_{j=1}^{N} \left(Y_j - m(X_j)\right)^2 + \lambda P(m) = \min! \tag{76}$$

The function to be minimized consists of a *data term*, which measures how well $m(x)$ fits the data in the training set, and a *smoothness term*, which enforces a certain amount of smoothness of the function. The influence of both terms is controlled by the hyperparameter $\lambda \geq 0$, which should be chosen to balance between overfit and underfit.

In general, **penalized regression** considers only functions from some function class $\mathscr{F}$ and an arbitrary loss function $L(y, \hat{y}) \geq 0$ and penalty $P(m) \geq 0$. The regularized function estimate, then, is the solution of the minimization problem

$$\sum_{j=1}^{N} L(Y_j, m(X_j)) + \lambda P(m) = \min_{m \in \mathscr{F}}! \tag{77}$$

A special case is *penalized maximum likelihood estimates* where $L(Y_j, m(X_j))$ is, up to constants, the negative conditional log likelihood of Y_j given X_j. The resulting estimates are (conditional) maximum likelihood estimates with a regularization to avoid overfit. Note that the common quadratic loss function of (76) is of this form too, assuming that the residuals $\varepsilon_j = Y_j - m(X_j)$ of (2) are independent, identically distributed Gaussian random variables.

It turns out that support vector machines can be interpreted as a special case of the general penalized regression framework. Starting from the linear support vector classification problem in its original form (68), reformulating it in terms of Lagrange multipliers similar to Sect. 5.3.2, and applying the Karush-Kuhn-Tucker conditions (70)–(72), we get that b is determined by

$$\sum_{j=1}^{N} \left(1 - Y_j R(X_j)\right)_+ + \lambda ||b||^2 = \min_{b \in \mathbb{R}^d}!,$$

where $(z)_+ = \max(z, 0)$ denotes the positive part of z, and we use the abbreviation $R(x) = \langle b, x \rangle + b_0$. This corresponds to (77), where the function class $\mathscr{F}$ consists, as in in ridge regression, of the affine linear functions $R(x) = \langle b, x \rangle + b_0$, which provide the decision boundary for classification. For general support vector machines including kernelization, we get the same penalized regression problem; only the function class changes to functions of the form $R(x) = \langle b, \phi(x) \rangle + b_0$.

The loss function $L(y, \hat{y}) = (1 - y\hat{y})_+$ is quite different to the quadratic loss $(y - \hat{y})^2$ most common in regression. However, it shows a lot of similarity to the negative log likelihood of logistic regression (compare Sect. 5.2). This explains why classifications based on support vector machines and on logistic regression frequently lead to almost the same results.

The framework of penalized regression also includes the extension of support vector machines from classification to regression. This particular learning method has two distinctive features: it is based on the solution of a convex optimization problem, which allows for using corresponding numerical algorithms and guarantees uniqueness of the minimum, and it results in a sparse representation in terms of the input variables where the minimizing function $\widehat{R}(x)$ only depends on the support vectors, which are a small part of the training set. For regression in the general setting with kernelization, these two goals are achieved, e.g., by solving

$$\sum_{j=1}^{N} L_\varepsilon\left(Y_j - m(X_j)\right) + \lambda ||b||^2 = \min_{b \in \mathbb{R}^d}! \quad \text{with} \quad m(x) = \langle b, \phi(x) \rangle + b_0,$$

where $L_\varepsilon(u) = \max(0, |u| - \varepsilon)$ is the ε-insensitive loss function introduced by Vapnik [79] for support vector regression. The resulting regression function estimate $\widehat{m}(x) = \langle \hat{b}, \phi(x) \rangle + \hat{b}_0$ depends only on the input vectors X_j, for which $|Y_j - \widehat{m}(X_j)| \geq \varepsilon$, which here are the support vectors.

12 Preprocessing, Input Dimension Reduction, and Experimental Design

Before starting a machine learning algorithm, it is usually expedient to apply simple transformations to the raw measured feature components X^o_{jk} first, to make their values more suitable for further processing. For example, the components of the feature vectors should be standardized in one way or the other, e.g., (17), such that they cover similar ranges of values.

On the first glance, it may not be obvious why such a simple transformation is useful. In a linear regression model (13), a multiplication of the input values $X_{jk}, j = 1, \ldots, N$, with a common factor α_k just leads to a corresponding rescaling of the corresponding parameter b_k without changing the essential regression relationship. However, already in this simple setting, standardization of the input values may help. In calculating the least-squares estimate (15), a matrix inversion has to be performed, and this step becomes usually numerically more stable if the input components have similar ranges of values.

Another simple transformation is the *Box-Cox transformation* (compare [6]) for nonnegative feature components

$$X_{jk} = \frac{(X^o_{jk})^{\beta_k} - 1}{\beta_k}, \quad j = 1, \ldots, N,$$

depending on the parameter $\beta_k \neq 0$. For $\beta_k = 0$, the natural continuation of this definition is the log-transform, i.e., $X_{jk} = \log X^o_{jk}, \ j = 1, \ldots, N$. Statistical package routines, which perform this transformation, usually include a method for choosing an appropriate value for β_k data adaptively.

Many features in engineering, e.g., lifetime distributions of equipment, assume values that have a right-skewed, heavy-tailed distribution, e.g., from the flexible family of Weibull distributions, instead of the common almost Gaussian one. The Box-Cox transformation with data-adaptive parameter is a normalizing transformation, i.e., it makes the data more similar to Gaussian random variables and hedges the influence of outliers, i.e., of few extreme observations. This again frequently leads to more stable numerical optimization in the training of statistical learning systems. Due to the same reasoning, it may also be advisable to apply a Box-Cox transformation to the output values of the training set.

As an example, we consider $N = 100$ data from a Weibull distribution with parameters (2, 2), i.e., $X_j = Z_j^2$, where the Z_j are exponentially distributed with parameter $\lambda = 2$. Figure 40 shows on the left the data as circles on the horizontal axis, the true Weibull probability density (blue), and the kernel density estimate (64) based on the data (red), using a Gaussian kernel and bandwidth $h = 0.2$. Note the rather large percentage of extreme observations far to the right of the bulk of the data.

After applying the Box-Cox transformation with near-optimal exponent $\beta = 1/6$, the sample looks much more Gaussian. Figure 40 shows on the right the

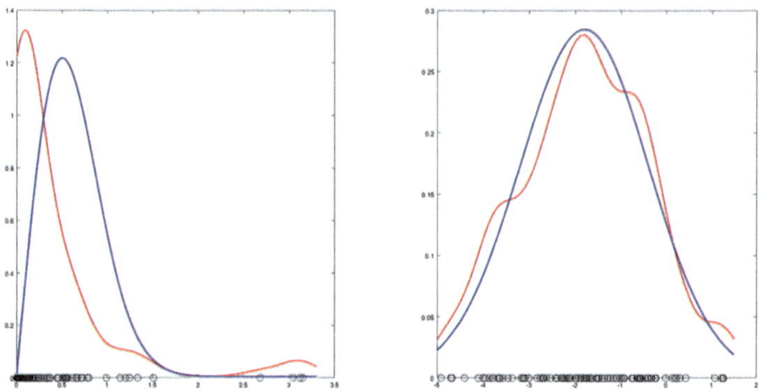

Fig. 40 Sample of Weibull-distributed data, their density, and a density estimate before (left) and after (right) a normalizing Box-Cox transformation

transformed data as circles, the kernel density estimate (64) using now bandwidth $h = 0.4$ and for comparison the normal density (blue) with mean and variance equal to the sample mean and variance of the transformed data.

Scaling and normalizing transformations apply to each single feature component and can be done easily. They may somewhat improve the performance of a learning procedure. The latter, however, crucially depends on looking at the often high-dimensional feature vector as a whole before training begins. In practice, frequently, many quantities are observed in each single experiment, which, due to the opinion of engineering experts, all may be relevant for forecasting or classification. The data analyst, then, is confronted with a high-dimensional raw input vector. Often, many input components are informative for the output on their own but quantify more or less the same kind of information such that not all of them are really needed. A typical example is the maximal, the minimal, and the average temperature of the surroundings during a step in production, which all are summary statistics derived from the same temperature curve. Typically, there are also raw input components, which on a first glance seem to be unrelated but still are strongly dependent on each other.

The task of the data analyst is to discover such hidden dependencies using statistical methods discussed below. The goal is to find a smaller set of "informative" input components out of a large set of candidates, which essentially suffice for forecasting or classification. This so-called **feature selection** step decides on the dimension d of the final input vector used for learning the requested forecasting or classification rule.

The finally used features may not be a selection from the original ones but rather are functions of them. Due to dependencies, the raw d^o-dimensional feature vectors seemingly are high dimensional, but often they are really more or less concentrated on a low-dimensional subset—not necessarily a linear subspace but in general a manifold in d^o-dimensional space. In feature selection, we try to learn this subset from the data. Then, we can transform the original inputs X_j^o of dimension d^o to a much shorter X_j having only $d \ll d^o$ components but containing almost the same information about our target function. Appropriate methods are discussed in the second and third section of this chapter.

12.1 Qualitative Inputs

Scaling and normalizing transformations are preprocessing steps, which apply to input variables, which genuinely are real numbers. However, in many machine learning applications, we also have to deal with qualitative information, e.g., the verbal description or the type number of a machine component in production. As a preprocessing step, we have to code the values of such features as numbers first.

If the values of the feature are hierarchical, i.e., ordered in a natural manner like grades attributed to a product by customers, then we can use appropriately ordered real numbers as codes and have to take care that the output of the algorithm is more or less invariant against changes in the particular coding. Here, statistical learning can help too, as we first can add an initial monotone transformation to the corresponding feature before feeding it into, e.g., a neural network, and try to learn this transformation too.

If the feature values are not ordered, then a common approach to use them for a quantitative learning system is to replace them by *dummy variables*, which are vectors of 0s and 1s of length, say L_d. This method has already been used in the context of classical linear models (compare Sect. 3.1). For example, we can set L_d equal to the number L_f of different values of the feature, the dummy variables consist of 0s except for one location, where a 1 indicates which particular values the observed feature has assumed. If, e.g., the feature values are A, B, C, then we would replace then by the dummy variables $(1, 0, 0)$, $(0, 1, 0)$, $(0, 0, 1)$ resp. In the machine learning literature, this method is often called *one-hot encoding* (compare [47, Section 10.2]) and, in particular, used in the context of classification problems. If L_f is quite large, then coding the feature values by binary numbers would reduce L_d to make computations more feasible. For example, $L_f = 4$ feature values could be coded by the dummy variables $(0, 0)$, $(0, 1)$, $(1, 0)$, $(1, 1)$, i.e., $L_d = 2$.

For really large L_f, which are encountered, e.g., in the processing of text documents where the feature values are words, the method of dummy variables reaches its limit. For such situations, which are not uncommon in modern machine learning, methods have been developed for *embedding* the feature values into L_d-dimensional vectors of real numbers, where L_d is large but still manageable. These embeddings are chosen such that the representatives of, e.g., words with similar meanings should be rather close together. There are algorithms for finding an appropriate embedding simultaneously with training the regression or classification system, or alternatively, there are open sources where one can find good embeddings at least for the English language (compare Section 10.5.1 of [47] for details). As the latter are meant for generic texts, finding a specific embedding might be useful if the feature values consist of technical terminology.

Finally, let us mention that, in particular in psychometrics and marketing research, there is a well-tested toolbox for quantifying qualitative data available. If, e.g., some information about the similarity of feature values is given, *multidimensional scaling* methods are available for transforming vectors of qualitative data into usually smaller vectors of real numbers such that it also incorporates an element of dimension reduction like the approaches described in the following sections. The principles of finding such representations could also be used for constructing similar statistical learning methods for handling large data. A review from the viewpoint of machine learning can be found in [23].

12.2 Principal Component Analysis (PCA)

The main goal of this section is to provide some intuition for methods that search for new features as functions of the original ones, resulting in less but more impactful input components. We first have a look at the classical linear situation, which we have studied in Sect. 3.1 for regression and in Sect. 5 for classification.

In both cases, we have to deal with high-dimensional features $X_1, \ldots, X_N$, which we model as independent, identically distributed random vectors with values in $\mathbb{R}^d$. We combine them to a data matrix $\mathbf{X}$:

objects	values of d features			
1	X_{11}	X_{12}	$\ldots$	X_{1d}
2	X_{21}	X_{22}	$\ldots$	X_{2d}
$\vdots$	$\vdots$	$\vdots$		$\vdots$
N	X_{N1}	X_{N2}	$\ldots$	X_{Nd}

The feature vector $X_j = (X_{j1}, \ldots, X_{jd})$ from the jth experiment shows up as the jth row of $\mathbf{X}$.

We are looking for relevant new features Z_j, which have the same explanatory power than the old ones but lie in a feature space $\mathbb{R}^q$ of considerably smaller dimension $q \ll d$. For this purpose, we first pretend that the original feature vectors X_j have a multivariate normal distribution with mean vector μ and covariance matrix Σ, i.e.,

$$\mu_i = \mathrm{E}\, X_{ji}, \qquad \Sigma_{i\ell} = \mathrm{cov}\,(X_{ji}, X_{j\ell}), \;\; i, \ell = 1, \ldots, d.$$

The corresponding probability density is of the form (1). Note that the d diagonal elements of Σ are the variances var $X_{ji}, i = 1, \ldots, d$, and the off-diagonal elements, the covariances, measure the "linear" dependence between different components (compare Sect. 1.2).

Similar to Sect. 5.1, we estimate the parameters μ, Σ of the feature distribution by their sample versions

$$\hat{\mu} = \frac{1}{N}\sum_{j=1}^{N} X_j, \qquad \mathbf{S}_{i\ell} = \frac{1}{N-1}\sum_{j=1}^{N}(X_{ji} - \hat{\mu}_i)(X_{j\ell} - \hat{\mu}_\ell)$$

The resulting sample covariance matrix $\mathbf{S}$ as an estimate of Σ shares with it the essential property that they both are symmetric and nonnegative definite matrices, i.e., all eigenvalues are ≥ 0. Let $g_1 \geq g_2 \geq \ldots \geq g_d \geq 0$ be the **eigenvalues** of $\mathbf{S}$. Then, we can decompose

$$\mathbf{S} = \mathbf{O}\mathbf{G}\mathbf{O}^T, \qquad \text{with the diagonal matrix } \mathbf{G} = \operatorname{diag}(g_1, \ldots, g_d)$$

and a suitable orthogonal matrix $\mathbf{O}$. This corresponds to a basis transformation in the feature space, where the columns of $\mathbf{O}$ form an orthogonal basis of **eigenvectors** of $\mathbf{S}$. Analogously, if $\gamma_1 \geq \gamma_2 \geq \ldots \geq \gamma_d \geq 0$ denote the eigenvalues of Σ, we have the decomposition

$$\Sigma = \Omega\Gamma\Omega^T, \qquad \text{with } \Gamma = \operatorname{diag}(\gamma_1, \ldots, \gamma_d) = \Omega^T\Sigma\Omega,$$

and the columns of Ω form an orthogonal basis of eigenvectors of Σ.

Exchanging the original basis of the feature space against a basis of eigenvectors related to the dependencies between the feature components leads to a corresponding new representation of the input vectors, which, due to the ordering of the eigenvalues by size, has components ordered by relevance. Before presenting the practically realizable sample version of this idea, we have a look at the theoretical version using the unknown mean and covariances. As before, we use X_{N+1} as an independent representative of the input vectors $X_1, \ldots, X_N$, which is not part of the training set.

The *principal component transformation* consists of first centering the feature vector around 0 by subtracting the mean and, then, applying the eigenbasis transformation. X_{N+1} is mapped onto $W = (X_{N+1} - \mu)\Omega$, and its component

$$W_\ell = \langle X_{N+1} - \mu, \eta_\ell \rangle$$

is called the ℓth **principal component** of X_{N+1}, where $\eta_1, \ldots, \eta_d$ are the normed eigenvectors of Σ, i.e., the columns of Ω. $W = \sum_{i=1}^{d} W_i \eta_i$ is the representation of $X_{N+1} - \mu$ in terms of the eigenbasis of Σ. The principal components have the following properties:

$$\mathrm{E}\, W_i = 0, \qquad \operatorname{var} W_i = \gamma_i, \qquad \operatorname{cov}(W_i, W_\ell) = 0,\ i \neq \ell$$

and, due to the ordering of the eigenvalues γ_i, $\operatorname{var} W_1 \geq \operatorname{var} W_2 \geq \ldots \geq \operatorname{var} W_d$.

As a linear combination of the $X_{N+1,i}$, the vector W of principal components has a d-dimensional normal distribution with mean vector 0 and diagonal covariance matrix Γ, if X_{N+1} is a Gaussian random vector. In particular, then, the principal components $W_1, \ldots, W_d$ are not only uncorrelated but **independent** (compare Sect. 1.2)!

The eigenvector η_1, corresponding to the first principal component W_1, denotes the direction in which the random variable X_{N+1} shows the largest variability and, therefore, is most informative for the structure of the data. More precisely,

$$\text{var } W_1 = \max\{\text{var } U;\ U = \sum_{i=1}^{d} a_i X_{N+1,i},\ a_1, \ldots, a_d \text{ real-valued}, \sum_{i=1}^{d} a_i^2 = 1\}.$$

η_2 is, then, the direction with largest variation among all directions in feature space, which are orthogonal to η_1, etc. The total proportion of variability of X_{N+1} explained by the first q principal components $W_1, \ldots, W_q$ is correspondingly $\gamma_1 + \ldots + \gamma_q$.

We have transformed the original feature vector X_{N+1} into a new feature vector W, which has uncorrelated and, in the Gaussian case, even independent components. Additionally, the latter are ordered w.r.t. to the information they provide about the structure of the input.

Recall that one major goal of **principal component analysis** is input dimension reduction. We therefore look for $q \ll d$ linear combinations of $X_{N+1,1}, \ldots, X_{N+1,d}$, which explain a large percentage of the variability in the feature vector. The answer is to choose just the first q principal components $W_1, \ldots, W_q$. An open question is how to select q, which we will discuss in the context of empirical principal component analysis based on a training set instead of knowledge of μ and Σ.

In practice, we have to use the estimates $\hat{\mu}$ of the mean vector and $\mathbf{S} = \mathbf{OGO}^T$ of the covariance matrix, which we have calculated from the training set $X_1, \ldots, X_N$ of input data. The *empirical principal component transformation* represents the centered feature vectors $X_j - \hat{\mu}$ in terms of an eigenbasis of the sample covariance matrix $\mathbf{S}$ resulting in $Z_j = (X_j - \hat{\mu})\mathbf{O}$. The components

$$Z_{j\ell} = \langle X_j - \hat{\mu}, e_\ell \rangle$$

are the (empirical) principal components of the feature vector X_j with $e_1, \ldots, e_d$ denoting the normed eigenvectors of $\mathbf{S}$, i.e., the columns of $\mathbf{O}$. Finally, we retain only the q most important components of Z_j and use the new feature vectors $Z_j^{[q]} = (Z_{j1}, \ldots, Z_{jq})$, replacing X_j as input to our statistical learning algorithm for regression or classification.

There is still the question of how to choose the number q of the principal components to be used in the further data analysis. Recall that $g_1 \geq g_2 \geq \ldots \geq g_d$ denote the eigenvalues of **S**, which estimates Σ. Correspondingly, $g_1 + \ldots + g_q$ estimates the total proportion of variability of the X_j explained by the components of $Z_j^{[q]}$.

A traditional way to choose q is by looking at the **scree graph**, i.e., a plot of g_ℓ or $\frac{g_\ell}{\sum_{i=1}^d g_i}$ against ℓ. This plot is decreasing, but from a certain point onward, it will look rather flat with rather small values g_ℓ. q is chosen visually as the component number where this flat behavior starts. As, frequently, the choice is not so obvious, there are various rules of thumb like:

(a) Choose q as small as possible but such that 90% of total variability is explained, i.e.,

$$q = \min\{\ell \leq d;\, g_1 + \ldots + g_\ell \geq 0.9 \sum_{i=1}^{d} g_i\}.$$

(b) (Kaiser's rule) Consider only principal components with above average variance, i.e.,

$$q = \max\{\ell \leq d;\;\; g_\ell \geq \frac{1}{d} \sum_{i=1}^{d} g_i\}.$$

To illustrate principal component analysis, we consider an example that is typical for judging the quality of materials like steel, concrete, etc. Usually, a lot of quantitative features are measured and are used to predict the durability. Typically, several of the available features describe similar kind of information such that the dimension of the feature vector is larger than necessary. We consider wooden pit-props cut from Corsican pine where the quality measure Y_j of interest is the maximum compressive strength. For predicting it, for each specimen, 13 observations are provided, which can be measured in a nondestructive manner. The full $N = 180$ data are described in [49], but for simplicity, we consider here only a selection of eight features:

1: Top diameter	5: No. of annual rings at top
2: Length	6: No. of knot whorls
3: Moisture content	7: Length of clear prop from top
4: Specific gravity	8: Average no. of knots per whorl

First we calculate the sample correlation matrix **C** from the data that is the standardized version of the covariance matrix **S**, resulting in (rounded to two digits):

$$\begin{pmatrix} 1.00 & 0.95 & 0.36 & 0.34 & 0.31 & 0.55 & 0.08 & -0.02 \\ 0.95 & 1.00 & 0.30 & 0.28 & 0.29 & 0.57 & 0.08 & -0.04 \\ 0.36 & 0.30 & 1.00 & 0.88 & 0.15 & -0.08 & 0.16 & 0.22 \\ 0.34 & 0.28 & 0.88 & 1.00 & 0.38 & -0.01 & 0.10 & 0.17 \\ 0.31 & 0.29 & 0.15 & 0.38 & 1.00 & 0.27 & -0.04 & 0.02 \\ 0.55 & 0.57 & -0.08 & -0.01 & 0.27 & 1.00 & -0.32 & -0.37 \\ 0.08 & 0.08 & 0.16 & 0.10 & -0.04 & -0.32 & 1.00 & 0.03 \\ -0.02 & -0.04 & 0.22 & 0.17 & 0.02 & -0.37 & 0.03 & 1.00 \end{pmatrix}$$

Then, we decompose $\mathbf{C} = \mathbf{OGO}^T$ where **G** is the diagonal matrix given by the eigenvalues of $\mathbf{C}$: $g = (2.95,\ 1.93,\ 1.06,\ 0.87,\ 0.79,\ 0.26,\ 0.08,\ 0.04)$. **O** is the orthogonal matrix given by (again rounded to two digits)

$$\begin{pmatrix} 0.52 & 0.16 & 0.23 & 0.24 & -0.06 & -0.28 & -0.07 & 0.71 \\ 0.50 & 0.20 & 0.25 & 0.27 & -0.05 & -0.27 & 0.14 & -0.69 \\ 0.36 & -0.47 & -0.08 & -0.19 & -0.36 & 0.10 & -0.67 & -0.13 \\ 0.39 & -0.43 & -0.24 & -0.30 & -0.13 & 0.08 & 0.70 & 0.08 \\ 0.31 & 0.02 & -0.37 & -0.19 & 0.82 & -0.11 & -0.19 & -0.04 \\ 0.31 & 0.53 & -0.13 & -0.04 & -0.09 & 0.78 & -0.02 & 0.01 \\ 0.04 & -0.29 & 0.79 & -0.18 & 0.36 & 0.35 & 0.02 & 0.01 \\ 0.01 & -0.41 & -0.18 & 0.82 & 0.17 & 0.31 & 0.03 & 0.01 \end{pmatrix},$$

whose columns are the eigenvectors $e_1, \ldots, e_8$ of **C**. For choosing the number of principal components q to retain, we look at the scree graph in Fig. 41 where the most likely candidate is $q = 5$ as, for larger k, the eigenvalues g_k are much smaller. Applying the rules of thumbs, discussed above, results in $q = 5$ (rule a) and $q = 3$ (Kaiser's rule b), respectively. So we would select $q = 5$, if we do not risk a loss of relevant information, and $q = 3$, if we are mainly interested in a considerable reduction of the dimension of the feature vector.

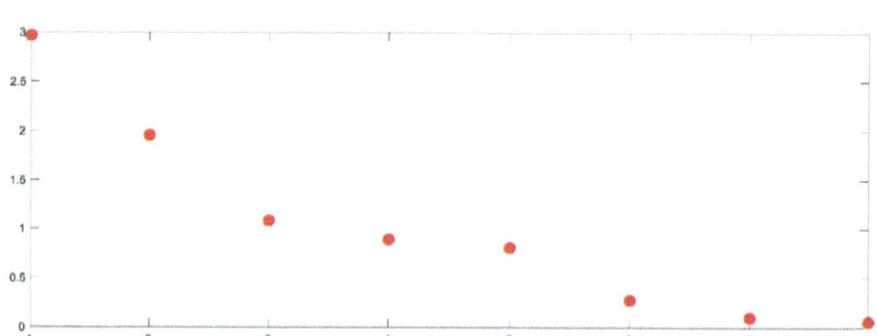

Fig. 41 Scree graph of pit-prop data

Looking at the columns of **O**, which contain the weights of the representation of the new features $Z_{j\ell}$ as a linear combination of the original features X_{jk}, one can try to interpret the new features, looking only at the dominant weights. Note that for further use as, e.g., inputs to a neural network, the absolute size and signs of the weights are not important but only their relative size and signs. Z_{j1} as the main measure of stability of the specimen essentially is a weighted average of the first six old features with slightly larger weights for X_{j1}, X_{j2}, the latter describing the overall size of the pit-prop. Z_{j2} is mainly an average of X_{j3}, X_{j4}, X_{j8} and, with the opposite sign, of X_{j6}, which has not a straightforward interpretation. Z_{j3}, Z_{j4}, Z_{j5} are mainly determined by only one of the original features, i.e., X_{j7}, X_{j8} resp. X_{j5}. Using these five new features, the dimension of the feature vector is considerably reduced with only a small loss of information, as the first five principal components still explain about 95 % of the variability of the original eight input variables.

12.3 *The Auto-Encoder*

An auto-encoder is a machine learning algorithm that tries to encode the feature vectors efficiently by using them as input as well as output of a learning system. It first reduces the usually high-dimensional input vector to a low-dimensional representation from which, then, it tries to recover the original features as well as possible. Some early applications really had a sparse encoding of information in mind, i.e., finding a representation of a signal, which requires a rather small capacity for storage or transmission but allows to reconstruct the full original signal up to small deviations.

Nowadays, auto-encoders are often applied to the common situation where the input vectors consist of many features proposed by experts as likely relevant for the data-analytic task. Often, several features describe similar information about the output of, e.g., a classification or forecasting system, though this usually is not obvious. The real information is part of a lower-dimensional structure, and auto-encoders help discover it and exploit this knowledge.

A first simple linear auto-encoder is given by principal component analysis. To illustrate the idea, we consider the principal component analysis w.r.t. the know mean vector and covariance matrix of the feature vectors, and we assume $\mu = 0$ for ease of notation. $\eta_1, \ldots, \eta_d$ denote the corresponding eigenbasis of the feature space, introduced in the previous section. Then, we have the expansion

$$X_{N+1} = \sum_{i=1}^{d} \langle X_{N+1}, \eta_i \rangle \eta_i = \sum_{i=1}^{d} W_i \eta_i$$

of a typical feature vector in this basis. By dropping the last $d - q$ principal components, we get a low-dimensional representation of X_{N+1} with mean-squared approximation error

$$\mathrm{E}\left\|X_{N+1} - \sum_{i=1}^{q} W_i \eta_i\right\|^2 = \mathrm{E}\left\|\sum_{i=q+1}^{d} W_i \eta_i\right\|^2 = \sum_{i=q+1}^{d} \operatorname{var} W_i,$$

using the orthogonality of the η_i. In principal component analysis, q is chosen such that this error is small, e.g., less than 10 % using the rule (a) of the previous section.

As principal component analysis focuses on correlations, i.e., on linear dependencies, between features, it cannot discover essentially nonlinear dependencies. General auto-encoders use, therefore, nonlinear mappings: $Z_j = g_H(X_j)$, which reduces the d-dimensional feature vector X_j to a new H-dimensional feature vector Z_j with $H \ll d$. In the second step, we try to recover the original inputs from Z_j using $\widehat{X}_j = G(Z_j)$. The goal is to choose the hyperparameter H and the mappings g_H, G such that the information loss is small, i.e., $\widehat{X}_j \approx X_j$.

For choosing the reduction and recovery functions, we may use any suitable statistical learning approach. A common choice is feedforward neural networks using the X_j as inputs and outputs for training. As a concrete example, we briefly sketch such a network with only one hidden layer. The auto-encoder is similar to the simple neural network with output function (49), except that the output is not one-dimensional but d-dimensional. The components of Z_j are the values of the neurons in the hidden layer, i.e., the hth component of the vector-valued function g_H is given by

$$Z_{jh} = \psi\big(u_{0h} + \sum_{i=1}^{d} u_{ih} X_{ji}\big), \ h = 1, \ldots, H,$$

where $u_{0h}, \ldots, u_{dh}$ are the network weights and ψ is the known activation function. The recovery step looks similar such that

$$\widehat{X}_{j\ell} = G_\ell(Z_j) = v_{0\ell} + v_{1\ell}\psi\big(w_{0\ell} + \sum_{h=1}^{H} w_{h\ell} Z_{j\ell}\big), \ \ell = 1, \ldots, d.$$

For learning all the network weights $u_{ih}, w_{h\ell}, v_{0\ell}, v_{1\ell}$, we minimize a loss function measuring the distance between the X_j and their reconstructions $\widehat{X}_j$, e.g., the usual empirical mean-squared loss $\frac{1}{N}\sum_{j=1}^{N} \|X_j - \widehat{X}_j\|^2$.

If we use linear neurons, i.e., the identity $\psi(t) = t$ as activation function, for large N the auto-encoder will approximately project the X_j onto Z_j in the H-dimensional subspace of the feature space spanned by the first principal component vectors $\eta_1, \ldots, \eta_H$. For nonlinear neurons, it will recover low-dimensional manifolds, i.e., curved subsets of intrinsic dimension H of the feature space, which characterize the essential information content of the input data (compare, e.g., Chapter 14 in [34]). As the auto-encoder is important for feature-dimension reduction as well as for recent machine learning approaches like deep learning or generative models, there are lots of variations and extensions, e.g., using other loss functions, weight decay, and other forms of regularization (compare Sect. 3.2), more than one hidden layer, etc. (compare [34] for a recent overview).

12.4 Design of Experiments (DoE) for Statistical Learning

Up to now, we have discussed how to transform input data that were already available. In situations where the training set results from measurements made in planned experiments, experimental design starts at an earlier stage. Usually, there are cost and time constraints on the final number N of data, such that we would like to have observations at carefully chosen values of the feature vectors X_j to get as much of information out of the subsequent data analysis as possible.

Successful applications of machine learning often are based on massive datasets covering a large region of the input space (and often given in advance) such that questions of experimental design are not really relevant. Nevertheless, it is worthwhile to remember some rules for good designs in interpreting the outputs of such systems, too. In particular, they may be biased if the underlying training set is unbalanced or not representative in other respects. We discuss this aspect below in more detail. In engineering applications, however, frequently, data are scarce, and experiments for generating them are expensive. Therefore, it is crucial to strategically choose the feature values at which measurements are taken well.

The extensive classical literature on design of experiments from a statistical point of view mainly is concerned with linear models like those presented in Sects. 3.1 and 5. The main goal considered is, given N, to choose $X_1, \ldots, X_N$ in a prescribed finite range of feature space in such a manner to optimize the performance of a subsequent regression or classification procedure. Some specific cases of nonlinear models have also been investigated thoroughly as they have a considerable impact in industry. [70], e.g., considered specific parametric nonlinear regression models that were successfully applied to optimize processes in the pharmaceutical industry. These were, however, based on a good white-box model (compare Sect. 2.6).

A review of the basic ideas of classical experimental design, in particular for nonlinear models, is given in Sect. 2.2 of the chapter *Optimal Experimental Design Supported by Machine Learning Regression Models* in this book, so we will not go into the details here. There are also dynamic nonlinear state space models discussed that are in particular of interest in process engineering.

Advanced statistical learning has its strength in nonlinear situations where additionally we do not believe that the models used are correctly specified. If we, e.g., use a neural network function f_H for forecasting in a regression setting, then we are aware that it is at best a useful approximation to the true underlying relationship, described by the regression function m in (13), between input and output, but often this approach provides a reasonable solution to the forecasting or classification task of interest. Therefore, we are working in a framework that is quite different to that discussed in the bulk of the elder literature on design of experiments.

Nevertheless, there are a few useful principles for choosing $X_1, \ldots, X_N$, which we can learn from the decades of experience with experimental design in linear systems. Classical design of experiments tries to get as much as possible out of a given number of experiments if the sample size is fixed. However, this might lead to rather extreme designs, which are dangerous in other respects. We illustrate this situation with a simple example. We consider the linear regression model $Y_j = m(X_j) + \varepsilon_j = b_0 + b_1 X_j + \varepsilon_j$ as in (3) with $p = 1$, believing that the relationship between input and output is described by a straight line. We want to perform experiments in the interval [0,1], and we are allowed to make $N = 51$ experiments to learn the parameters b_0, b_1. The optimal design minimizing estimation error, then, consists of making about half the observations at each of the boundary points 0 resp. 1. Intuitively, this is obvious as a straight line is completely determined by the two values $m(0), m(1)$. They can be estimated most precisely if we get as much observations as possible at those two locations and consider the two sample means of those Y_j with $X_j = 0$ resp. 1 as estimates of the values of $m(0), m(1)$. On the left-hand side, Fig. 42 shows such an optimal design for fitting a straight line together with the estimated function $\widehat{m}_{opt}(x) = \hat{b}_0 + \hat{b}_1 x$ (blue) and the—for this particular data-generating mechanism—optimal straight line (red).

This reasoning for finding an optimal design only holds if we know with absolute certainty that the input-output relationship is a straight line up to the errors ε_j. Usually, this will not be the case, and we only believe that $m(x) \approx b_0 + b_1 x$ up to small deviations, which are negligible for our data analysis. As a safeguard against failure of this assumption, it is recommendable to put a fraction of the measurements at other locations spread out over the whole range of interest. By this device, we do not lose much in terms of estimation error, but we get a warning if our regression model is grossly misspecified and has to be changed. Figure 42 shows on the right-hand side the true underlying regression function $m(x)$ (red), which is the same as

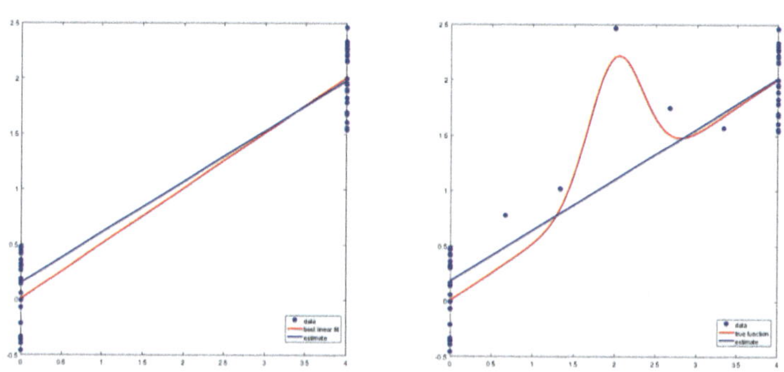

Fig. 42 Optimal design for fitting a straight line (left-hand side) and modified design for safeguarding against model misspecification (right-hand side)

the one used for Fig. 6. Here, we spread out five data, i.e., about 10 % of the sample, over the range of interest [0,4] as a safeguard against model misspecification. The fitted line $\widehat{m}_{mod}(x)$ (blue) based on this modified sample is also shown and does not differ much from $\widehat{m}_{opt}(x)$ on the left-hand side. Hence, we do not lose much by considering a slightly suboptimal design if $m(x)$ really would be a straight line. In our case, however, this simple model does not capture essential shape features of the true $m(x)$. With the modified sample, we get an according warning in terms of the very large difference between the observation at $x = 2$ and the estimate $\widehat{m}_{mod}(2)$ in the middle of the interval. Of course, this does not necessarily mean model misspecification, but it could also be caused by an outlier, i.e., an atypical observation. In any case, we are now aware that here is a problem and that we should have a closer look at the data and the training method.

Analogously, in experimental design for regression in more advanced settings, we also should try to get information at many different feature values, which are **representative** for the whole relevant part of the feature space. If the X_j are concentrated in small regions, then the learning algorithm will focus on them and will estimate the forecasting function $m(x)$ there quite well. Other regions are neglected, feature vectors X_{N+1} from those locations likely are not predicted well, and important properties of the data-generating process might be overlooked.

In classification, we have the related requirement that the training set should be **balanced**, i.e., observations from the various classes are contributing in roughly equal numbers to it. Otherwise, the learning algorithm will focus on getting the classification of those objects right, which belong to classes dominating the training set.

In the simple case of final inspection, we have only two classes, labeled 1 and 0, which consist of bad products with defects resp. of good ones. If we take the manually classified training data from the production line, we usually would have 95 % or even more data from class 0. Then, using the trivial classification rule that assigns all (!) objects to class 0 would have an in-sample classification error, i.e., the proportion of falsely classified objects from the training set, of 5 % or less. Applying this rule to a test set not used for training but generated in the same manner as the training set would result in a similar small out-of-sample error, as the test set would show the same kind of unbalancedness in favor of good products. Applying a statistical learning method to this classification problem would result in a rule quite similar to the trivial one.

Of course, in that simple case, it would be pretty obvious that the learned rule is not doing the job, i.e., finding the bad products. However, unbalancedness of the training set will result in the same kind of behavior also in more complicated situations where the problem is not immediately recognized. Classification algorithms, or, more precisely, the standard loss functions used in their training, expect roughly equal proportions of data from each of the classes under consideration, and they may not work well in grossly unbalanced situations.

As a remedy, we could use a different loss function, e.g., a weighted in-sample classification error, where false assignments made on objects from underrepresented classes get a larger weight. This would, however, require changes in the software. In

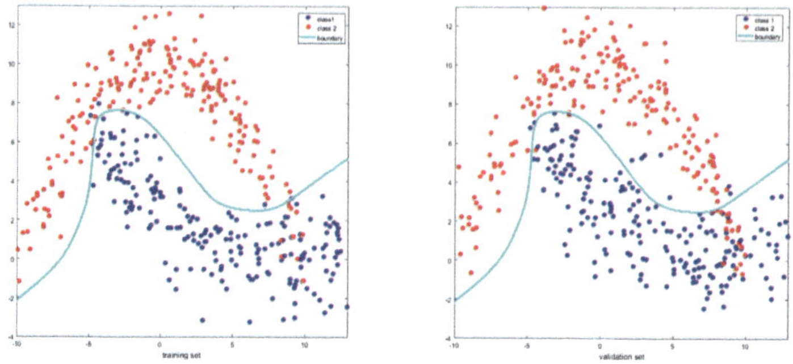

Fig. 43 Training (left) and validation (right) sample together with decision boundary from neural network classifier—balanced training set of size $N = 400$

practice, it usually is easier to use *data augmentation* instead where we artificially enlarge the part of the training set that corresponds to the underrepresented classes. We already have discussed this technique in Sect. 9.3, but there are more refined methods where the relevant data are not just duplicated (or given more weight), but where we augment the training set with some artificial data which are similar, but not equal to the underrepresented real data. James et al. [47] (Section 10.3.4), e.g., discuss this for the problem of image classification.

As an example, we consider a classification problem with two-dimensional features X_j and two classes. We start with balanced training set of size $N = 400$ with 200 observations from each class 1 (blue) and 2 (red) shown on the left-hand side of Fig. 43. We train a classification algorithm based on fitting a neural network with one hidden layer, consisting of $H = 6$ neurons, as described in Sect. 8.3. Figure 43 shows the decision boundary (light blue) between both classes, i.e., if a newly observed feature vector $X_{N+1} = x$ lies below the boundary, it is classified as belonging to class 1 and, otherwise, to class 2. On the right-hand side, we show the decision boundary together with an independent validation set that has been generated in the same manner as the training set.

To quantify the performance of this classification algorithm, we calculate the confusion tables (compare Sect. 7.2.1) P^{train} for the training set and P^{val} for the validation set, resulting in

$$P^{train} = \begin{pmatrix} 48.75\% & 1.25\% \\ 2.50\% & 47.50\% \end{pmatrix} \qquad P^{val} = \begin{pmatrix} 47.50\% & 2.50\% \\ 5.25\% & 44.75\% \end{pmatrix}$$

Recall that for $k, \ell = 1, 2$, e.g., $P^{train}_{k\ell}$ estimates the probability $p_{k\ell}$ for classifying an object from class k as belonging to class ℓ. Correspondingly, the probability for making an error is $p_{12} + p_{21}$, which is estimated by the misclassification rate 3.75 % from the training set resp. 7.75 % from the validation set. Note that both error rates P^{train}_{12} and P^{train}_{21} are roughly of the same order in the training set as well as in the validation set (with the difference partly due to the randomness of the training

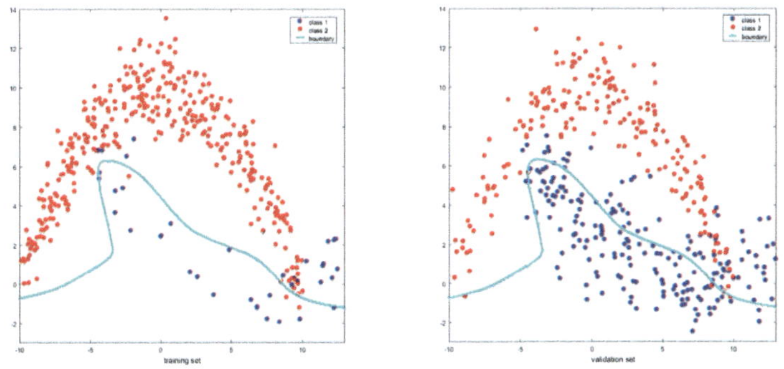

Fig. 44 Training (left) and validation (right) sample together with decision boundary from neural network classifier: unbalanced training set of size $N = 400$

set and partly due to the particular structure of the feature distributions for the two classes).

To illustrate the deteriorating effect of an unbalanced design, we look at the same kind of data, but now with a strongly unbalanced training set with only 40, i.e., 10 % observations from class 1 (blue) and the remaining 360 (red) from class 2. Figure 44 shows on the left-hand side the data and the decision boundary. The learned classification rule now is strongly in favor of class 2. The decision boundary shows some similarity with that of Fig. 43, but it is shifted downward such that considerable more features are attributed to class 2. To check the performance of the classification rule on independent data, we use the same balanced validation set as in Fig. 43. It is shown on the right-hand side of Fig. 44 together with the decision boundary calculated from the unbalanced training set. In this case, we get as confusion tables from training resp. validation set

$$Q^{train} = \begin{pmatrix} 5.25\% & 4.75\% \\ 0.75\% & 89.25\% \end{pmatrix} \qquad Q^{val} = \begin{pmatrix} 30.75\% & 19.25\% \\ 0.75\% & 49.25\% \end{pmatrix}$$

Looking at the misclassification rates, we get 5.50 % errors in the training set, which is not much larger than for the balanced design. However, from the first line of the confusion table Q^{train}, we see that almost every second observation from class 1 is falsely classified as belonging to class 2. This only has not a strong effect on the overall misclassification rate as there are so few data from class 1 in the training set. On the other hand, comparing Q^{train} and P^{train}, observations from class 2 in the training set are now more often classified correctly, which is to be expected as they constitute the large majority of training data.

The confusion table Q^{val} shows how bad the learned classification rule really is if we apply it to new data where both classes are represented equally. The strong favoritism for class 2 leads to a large relative proportion of errors (roughly 40 %) for data from class 1, leading to an overall misclassification rate of 20 %. Comparing Q^{val}_{22} and P^{val}_{22}, observations from class 2 however are much more

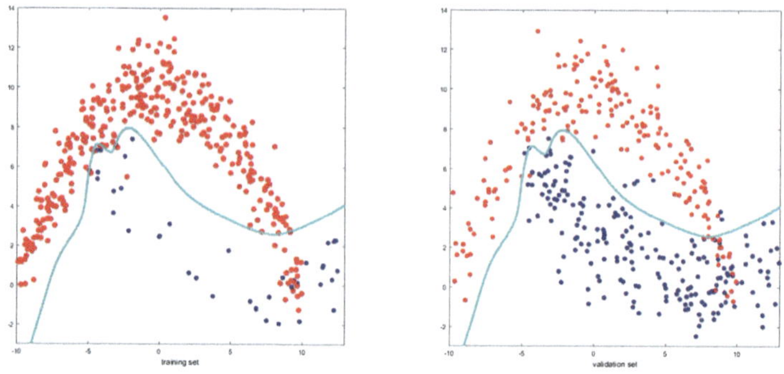

Fig. 45 Training (left) and validation (right) sample each together with decision boundary from neural network classifier: augmented training set of size $N = 720$

frequently classified correctly than by the fair classification rule learnt from the balanced sample.

The example illustrates the effect that different proportions of the classes of interest in the training sample can have on the resulting classification rule. Usually, the preference of strongly represented classes is undesirable, and ideally, the probabilities for misclassification should be roughly the same for objects from all classes. For example, during development of a product with ergonomics in mind, users are asked if, for them, the handling of the product is comfortable or not. Here, the classification does not only depend on the design parameters of the product but also on the characteristics of the specific customer. If the potential users include a large variety of people (w.r.t. sex, age, size, etc.), they all should be well represented in the training group, such that later on, a high proportion of all kinds of customers will be satisfied by the final product. In practice, however, it is often difficult to get such ideal data. For this situation, there are learning techniques like boosting (compare Sect. 9.3) that give more weight to data from underrepresented classes in the training process. Alternatively, we could change the target function, which we try to optimize in the learning stage. The latter is, however, usually more cumbersome for the user of statistical learning algorithm as he usually relies on ready-made software.

We continue the example with a look at the effect of a quick-and-dirty method for giving the underrepresented class 1 more weight. We add 8 exact copies of each of the 40 data from class 1 to the training set, such that the resulting augmented sample consists of 360 observations from each of the two classes. Then, again, we train the neural network classifier on the now $N = 720$ data. Figure 45 shows the result for the original training set (left) and the same test set of size $N = 400$ we have used before. The crude data augmentation essentially removed the effect of unbalancedness, and the classification performance is quite similar to Fig. 43. More precisely, the confusion tables based on the augmented training set are

$$\tilde{Q}^{train} = \begin{pmatrix} 48.75\% & 1.25\% \\ 3.75\% & 46.25\% \end{pmatrix} \qquad \tilde{Q}^{val} = \begin{pmatrix} 47.50\% & 2.50\% \\ 5.25\% & 44.75\% \end{pmatrix}$$

In particular, $\tilde{Q}^{val} = P^{val}$, i.e., in view of the generalization to new independent data, the performance of the classification algorithm is practically the same for the augmented training set as for a genuine balanced training set though, the decision boundaries are slightly different. As already mentioned, boosting provides a more refined approach to such data augmentation and is preferable in more complicated situations.

Let us finish this discussion with remarking that there also are situations in practice where we are interested in an unfair treatment of the classes as the two type of errors (classify class 1 erroneously as class 2 and vice versa) induce quite different costs. For example, in automatic surveillance of a component of a production plant, based on sensor outputs, we have to decide if there is a problem or not. If the machine learning system gives alarm, then a human operator checks the component, i.e., in case of a false alarm, the costs are a bit of manpower and perhaps a brief stop of the production line. If, on the other hand, a defect of the component is overlooked, then for a longer period, unsatisfactory goods are produced, and, perhaps, even more severe damage to the whole plant may be caused in the long run. In such cases, it is advisable to train the classification system such that it reduces in particular the chance of the second type of error. Unfortunately, for training data, typically, the interesting class is underrepresented. For example, we have lots of data from the production line while it runs smoothly, but defects, which we want to detect, are rarely observed. In such a case, we should use all observations from the class of interest (in the example, from defective components) and only a selection of data from the other class to get a minimum of balancedness in the training set. If the latter becomes too small for learning a satisfactory classification rule, we additionally can apply weighting techniques such that the algorithm prioritizes the goal of achieving a small probability for the important type of misclassification.

Returning to regression, research on experimental design for typical statistical learning methods is just beginning. Practitioners often just use **randomized designs**, i.e., they choose the feature vectors X_j, for which the outputs Y_j will subsequently be observed in an experiment, in a purely random manner, from a prescribed bounded region of the feature space. This may not be efficient, but good enough, and, in particular, seems to be safe, concerning the issues caused by data, which are not representative. Unfortunately, this is only true without reservation for feature vectors of smaller dimension d. We briefly sketch the problems and solutions discussed by Morgan and Rubin [59, 60].

We consider experimental design for a basic two-sample situation: A total sample of $N = 2n$ test objects (persons in bioscience, customers or machines in industrial production, etc.) is given. This sample has to be partitioned into two subsamples—the **treatment** and the **control** group—each of the same size n. The final goal is to perform an experiment with each object and to use the results for testing if there is a treatment effect and, additionally, for assessing its size. Each test object no. j is characterized by d covariates or features, which we combine to the feature vector X_j.

To avoid confounding the treatment effect of interest with the influence of the covariates, experimental design should provide a partitioning guaranteeing balance between the two groups, i.e., regarding the feature values, both parts of the training sample are similar. A simple example would be the requirement that the sample means of all d features are approximately equal between the two groups:

$$\hat{\mu}_{1i} = \frac{1}{n}\sum_{j=1}^{n} X_{ji} \approx \frac{1}{n}\sum_{j=n+1}^{2n} X_{ji} = \hat{\mu}_{2i}, \quad i = 1, \ldots, d.$$

If balance is neglected, then a significant test may not indicate an effect of the treatment but only be due to large differences in the feature values of both groups. Classical design of experiments has a lot of different tools for providing balance, but they do not work well for larger d. An alternative is a purely random assignment of half of the objects to each of the treatment and control group to avoid any experimenter bias. For low d, this leads with high probability to a more a less balanced split of the test objects into two groups. For already $d = 10$, randomization however may result in largely unbalanced subsamples. Morgan and Rubin [59] discuss an example where pure randomization leads with rather large probability to a partitioning of the data, where the differences between the covariate means in treatment and control sample are so large that at least one of them will be tested to be significantly different from 0 on the 5 % level.

The solution to the unbalancedness of randomized designs is repeated **rerandomization** until a certain balance condition is fulfilled. Morgan and Rubin [59] provide some theoretical basis, which is also summarized in [60]. The latter is a compact review of the original approach including a real data example and some extensions, and it also discusses how to incorporate expert knowledge that some of the features are more important than others for the outcome of the experiment.

One possible balance condition is based on the familiar *Mahalanobis distance*, which here is $D_M = \frac{N}{4}(\hat{\mu}_{1i} - \hat{\mu}_{2i})^T \widehat{\Sigma}^{-1}(\hat{\mu}_{1i} - \hat{\mu}_{2i}C)^T$, where $\widehat{\Sigma}$ denotes the sample covariance matrix of all the feature vectors $X_j, j = 1, \ldots, N$ (compare Sect. 5.1). The DoE algorithm with rerandomization now is:

1. For each test object of the total sample, collect the feature data.
2. Choose a balance criterion, e.g., $D_M \leq c$ for some constant $c > 0$.
3. Randomly assign $n = N/2$ objects to the treatment group.
4. Check covariate balance. If $D_M > c$, return to 3.
5. Conduct the experiment.
6. Test if there is a treatment effect, preferably by a randomization test that follows steps 2–4 (compare Section 2.2 of [59] for the details).

The stricter the balance requirement (in the example, the smaller the threshold c), the larger the computational cost of repeated rerandomization and of performing the randomization test will be, as more randomizations will be rejected. With rerandomization, standard tests for treatment effects like the t-test become too conservative, i.e., they require stronger evidence for a treatment effect than necessary, and it is harder to detect the treatment effect than necessary. If, in a practical application, they already detect a difference between treatment and control group, then we may finish the data analysis. Otherwise, it is worthwhile to apply the computationally more expensive randomization tests, which are related to other resampling methods like the bootstrap (compare Sect. 9.1) as they have a better power to detect a treatment effect.

Zhang et al. [86] discuss how to reduce feature dimension using principal components analysis before starting with experimental design based on rerandomization. Zhou et al. [87] consider the case of a sequential design where the test objects are not all given in advance but some are observed later and have then to be assigned to treatment and control group as soon as they become available.

Finding a good design for a nonlinear model requires up-to-date numerical optimization algorithms. Even then, with even moderately large input dimension, the computational costs may become impractical (compare Sect. 4.2 of the chapter *Optimal Experimental Design Supported by Machine Learning Regression Models* in this book for an example from chemical process engineering with $d = 7$). To speed up the calculations, machine learning methods (Gaussian process regression; compare Sect. 4.2) may be used for finding an efficient grid for the numerical algorithms. Details and applications are given in the chapter *Optimal Experimental Design Supported by Machine Learning Regression Models*.

13 Advanced Neural Networks and Deep Learning

Neural networks were quite popular during the late 1980s and 1990s where, due to computational restrictions, successful applications relied mainly on small or single-layer networks like (45) and (49). Then, for some time, developers' attention focused more on other learning methods like support vector machines or advanced tree-based algorithms. That changed again with the arrival of deep learning, which in many applications relies on some kind of neural networks, however much more complex ones. We start with a closer look at multilayer perceptrons, i.e., classical neural networks with more than one hidden layer.

Recall the single-layer network with output function $y = f_H(x; \vartheta)$ of (45). We rewrite it in two steps:

$$x_h^{(1)} = \psi\big(w_{0h} + \sum_{i=1}^{d} w_{ih} x_i\big) = \psi\big(w_{0h} + \langle w_{\bullet h}, x\rangle\big), \quad h = 1, \dots, H, \tag{78}$$

$$y = f_H(x; \vartheta) = v_0 + \sum_{h=1}^{H} v_h x_h^{(1)},$$

where $\langle w_{\bullet h}, x\rangle$ denotes the scalar product between the input vector x and the weight vector $w_{\bullet h} = (w_{1h}, \dots, w_{dh})$. In this representation of the function, given by the network, we separate the whole input-output-mapping into the transformation of the d-dimensional input feature vector x to the intermediate H-dimensional vector $x^{(1)}$, which represents the values attributed to each neuron in the hidden layer, and the final transformation from the hidden layer to the output layer, i.e., here, to the one-dimensional output y.

Multilayer feedforward neural networks are generated by iterating the first step, using the neuron values of the previous hidden layer as inputs to the next hidden layer. If we consider m hidden layers with $H(1), \ldots, H(m)$ neurons and set $x^{(0)} = x$, $H(0) = d$, then the output y is given by

$$x_h^{(\ell)} = \psi\big(w_{0h}^{(\ell)} + \sum_{i=1}^{H(\ell-1)} w_{ih}^{(\ell)} x_i^{(\ell-1)}\big), \quad h = 1, \ldots, H(\ell), \ \ell = 1, \ldots, m,$$

$$y = f_H(x; \vartheta) = v_0 + \sum_{h=1}^{H(m)} v_h x_h^{(m)},$$

where the parameter vector ϑ consists of the output weights $v_0, \ldots, v_{H(m)}$ and all the weights connecting input and hidden layers $w_{ih}^{(\ell)}, i = 0, \ldots, H(\ell - 1), h = 1, \ldots, H(\ell), \ \ell = 1, \ldots, m$.

A genuine multilayer network seems to have many parameters for which we would need large training sets or expect problems with massive overfitting. However, multilayer networks must not necessarily be large to be successful. By considering them, we get more structural flexibility compared to looking only at single-layer networks. Thus, training a small network, in the sense of a small number of hidden neurons, can result in a better prediction or classification performance compared to a really large single-layer network, depending on the underlying unknown input-output relation.

However, about 25 years ago, the computational facilities were not really up to the task of fitting larger multilayer perceptrons to data. This did not only concern the widely accessible hardware but also the size of available datasets for parameter estimation as well as the numerical optimization routines used for training the algorithms. If they converged at all, they frequently ended in local minima of the loss function, which were not giving satisfactory prediction or classification results, a problem which at that time was not so severe for single-layer networks.

There were some exceptions, e.g., LeCun's convolutional neural network LeNet-5 with 6 hidden layers and about 60,000 trainable parameters, which was very efficient in automatic reading of handwritten characters. It profited from the stability of CNNs (compare Sect. 13.2) and their particular relationship with input data in the form of digital images. At that time, the performance on a benchmark data set was more or less matched by suitably constructed support vector machine classifiers (compare [55]).

13.1 Deep Neural Networks (DNNs) or Size Matters

Beyond a massive, in particular, commercial investment in R&D of corresponding systems, the success stories of deep neural networks or, more general, deep learning during the last 1.5 decades are based on several prerequisites:

- the availability of massive training datasets with $N > 10^6$ or even much larger;
- the presence of efficient and partly dedicated hardware, e.g., graphic boards with performant GPUs and enough storage were particularly suited for calculating convolution operations in CNNs;
- progress in numerical optimization as well as in parallel computing allowed to train really large networks.

The main reason for the sudden considerable improvements in fields like object recognition from images or language processing was the sheer size, characterized by the number of trainable parameters, of layered learning systems, which now could be handled, and the access to training sets of correspondingly massive size, which were needed for learning all the parameters. Often, progress was achieved by just increasing neural networks a lot by adding **depth**, i.e., number of layers, and also breadth, i.e., number of hidden neurons in the layers. Afterward, a clever engineering of the architecture of such successful networks allowed for cutting down the size a bit without decreasing the performance, but the final systems still were large. A review of deep learning starting from the famous AlexNet [53] for object recognition is given by Alom et al. [3], who, in particular, illustrate the increase in performance of classification systems in relation to their size until they were on the same level as humans.

AlexNet, e.g., is based on a similar but larger architecture than LeNet-5 but involves about 60 million of trainable parameters in a convolutional neural network, discussed in the next section, with 5 convolution and 2 fully connected layers together with few pooling layers. For training the algorithm, specific measures against overfit, in particular *dropout* [75], were also included, which are related in spirit to the principles used in random forest systems (compare Sect. 9.4) and do not lead to much greater computational load which in view of the already long time it takes to train such large systems is important. For the task of recognizing objects from images, AlexNet achieved a significant decrease in the misclassification error rate compared to previous state-of-the-art systems.

Deep learning applications are not restricted to the use of neural networks. Some of them also involve large classification or regression trees or random forests and apply boosting and bagging techniques, i.e., most tools from the history of statistical learning. The distinguishing feature to older machine learning approaches is the size and, here, in particular, the depths in terms of an architecture consisting of many layers where different kinds of operations are combined (compare the next section for an example).

Learning the parameters of a deep neural network involves numerically optimization of the loss function, which, in view of the extremely large number of variables, is a challenging task. Deep learning, therefore, not only relies on high-performance computing (HPC) systems and their efficient use but also on various tricks of the trade. For example, computationally not-so-ideal classical activation functions like (48) are replaced by others, where the currently most popular one is the simple *rectified linear unit* (ReLU)

$$\psi(u) = \max(u, 0) = u \text{ for } u > 0, \text{ and } = 0, \text{ else.}$$

The optimization routines are, of course, also crucial. Classical gradient descent would be too expensive in terms of time and storage as training sets are extremely large. Therefore, often, a similar idea as in simple backpropagation (compare Sect. 8.2) is used, where, in the individual steps of the iteration, the gradient is calculated on the basis of only relative small, randomly chosen batches of training data. This so-called **stochastic gradient descent** method has been successful in many deep learning applications though it is not guaranteed to always provide satisfactory solutions. Fortunately, data analysts have nothing to worry about the details of numerical optimization as stochastic gradient descent software is widely available.

There are other reasons for the success of deep learning beyond the sheer size of training sets and parameter space. On the one hand, they are building upon more than 20 years of experience with statistical learning. For example, awareness of the danger of overfit in view of the many parameters of, e.g., DNNs was the motivation to include some kind of regularization prospectively into the training algorithms, where, besides the already mentioned dropout, also classical methods like weight decay related to ridge regression are used.

On the other hand, it is not only the size but also the many-layered structure that is responsible for the performance of deep learning. In the previous chapter, we have discussed the problem of unnecessarily high dimension of the feature vectors, although due to the dependence between the features, the relevant information is of much lower dimensionality. If the inputs are images, then this effect is pretty obvious. Raw data are the gray-colored values of pixels, the number of which, depending on the resolution, ranges from 10,000 to several millions, but obviously, many neighboring pixels will have similar color values. In a manner similar to the auto-encoder, some layers of DNNs usually are performing a dimension reduction, i.e., they not only solve the forecasting or classification task, but, on the way, they also extract essential features out of the raw data.

13.1.1 Interpretability of Statistical Learning

Hinting at the impressive performance of deep learning at difficult data analytic tasks, which allows to teach artificial intelligence systems to recognize objects or individuals on digital photographs, to follow moving objects on videos, and to translate or even write texts, they are sometimes advertised as the solution for all data problems, in particular by people having a commercial interest. From a more objective viewpoint, it is definitely useful to consider DNNs and other related approaches in situations for which they have been designed. However, if they blindly are applied to problems from industrial production or engineering research, they likely will give some results on training sets, but there is no guarantee that they will perform well in practice afterward or that they cannot be considerably improved. DNNs are black boxes par excellence in contrast to simpler learning methods where we have a better understanding how the system produces a forecast or a classification. Moreover, in many applications, we do only have rather small training sets available, and due to cost constraints, there is no chance to generate an amount of data comparable to those use cases where deep learning excels.

The main problem of generic DNNs and similarly complex deep learning methods is, in particular in applications with high-dimensional feature vectors, the lack of interpretability (compare [69] for an extensive discussion of this topic). They can provide excellent solutions to forecasting or classification problems, but we do not really understand how they arrive at their solutions where the concept of understanding depends of course on the domain of application. In engineering applications, it is a desirable goal that a domain expert has at least a rough idea how the input-output relation produced by an algorithm works. In this respect, simpler learning methods like the LASSO and classification and regression trees or gray-box models, which combine engineering knowledge with statistical learning (compare Sect. 2.6), lead often to interpretable results as the function, represented by them, have a simpler structure.

Interpretability does not necessarily imply a (much) worse performance. Depending on the kind of application, the increase in forecasting error or misclassification rate may be negligible, and, moreover, the simpler approaches in general increase the stability of solutions (compare Sect. 2.2 and [69]).

James et al. [47] (Section 10.6) present a nice discussion that is in line with our experience from consulting. They consider a forecasting problem, where a neural network regression with 1409 parameters was beaten by a careful application of the LASSO (Sect. 3.3) plus crossvalidation resulting in a linear regression model with only five parameters. Such simple forecasting methods are easy to compare with the experience of experts who can judge if the inputs of the system are chosen in a reasonable manner. They also are likely to be more stable against future changes in the underlying data-generating mechanism, e.g., in modifications in a production system. In a similar spirit, for short-term forecasting of time series, the classical decade-old methods, which involve a careful preprocessing of the data followed by fitting a linear autoregression with data-adaptive choice of the hyperparameter (compare Sect. 14.1) or alternatively, the slightly more general Box-Jenkins approach [7] is often already giving satisfactory results.

To summarize, if the amount of data allow for advanced statistical learning at all, one should try it but also apply the corresponding classical methods, which usually are much easier. If the latter show the same or an only slightly worse performance, then they are preferable in view of interpretability and stability. For problems where deep learning has been originally developed, there were, however, no simple known alternatives. For example, classical methods in object recognition required a lot of expertise in image analysis, so they were not simple too, and nevertheless, they were considerably outperformed by deep learning.

Also, in situations without critical implications, interpretability is desirable, but not necessary. If a DNN is used for optimizing a production process and results in a solution that is better than current practice, then we would be satisfied. For safety-critical applications, however, not-interpretable DNNs may be problematic. For, e.g., a self-driving car, we should understand how it makes its decisions. Only then can we trust the algorithm that it would not produce crashes in case of particular data configurations that have not been part of the training.

13.2 Convolutional Neural Networks (CNNs)

Convolutional neural networks have a similar structure as common neural networks but use different data transformations, in particular convolutions. They are very popular in solving learning tasks where the input data are signals (1d-case), images (2d-case), or videos and tomographic images (3d-case). The first great successes have been achieved for image classification problems like automatic reading of handwritten digits and characters or object recognition. Use cases, where the application of CNNs is reasonable, also appear in the context of industrial production too, where, e.g., time series of measurements from sensors are used for the supervision of processes, the quality control of surfaces is based on images or the structure of materials is investigated by looking at three-dimensional tomographic images (compare the chapters *Cracks in Concrete* and *Machine Learning for Inline Surface Inspection Systems* in this book).

Convolutions are operations on functions or, in the digital world, sequences or arrays of real numbers, which have been successfully used in engineering for a long time, in particular in the context of signal analysis. Let us consider a discrete signal $Z_t, t = \ldots - 1, 0, 1, 2, \ldots$, usually modeled as sequence of random variables due to noise. Convolution filters are linear methods for, e.g., filtering the information of interest out of a noisy signal or forecasting a signal. They are characterized by a sequence of filter coefficients $a_k, -\infty < k < \infty$. The convolution of both sequences results in a new sequence $S_t, -\infty < k < \infty$, given by

$$S_t = (a * Z)_t = \sum_{k=-\infty}^{\infty} a_k Z_{t-k}.$$

In practice, usually finite filters are considered, where only a finite number of coefficients do not vanish, and, hence, the sum is finite too.

We already have discussed an example where a time series Z_t is a linear autoregression of order p (12). The subsequent best linear forecasts of Z_t given the last p observations are (compare Sect. 2.5)

$$\widehat{Z}_t = \sum_{k=1}^{p} a_k(p) Z_{t-k} = (a(p) * Z)_t,$$

where $\ldots, 0, a_1(p), \ldots, a_p(p), 0, \ldots$ is the sequence of coefficients of a finite-memory prediction filter.

Convolution is closely related to Fourier transform (compare Sect. 6.1). If a_k resp. b_k denote the Fourier coefficients of two functions $m_a(x), m_b(x)$, then $c_k = (a * b)_k$ are the Fourier coefficients of the product function $m_c(x) = m_a(x)m_b(x)$. This fact and the computational efficiency of the *fast Fourier transform* (FFT) are used in many algorithms, also in machine learning, for a fast calculation of convolutions by applying first the discrete Fourier transform, followed by multiplication and, finally, an inverse discrete Fourier transform.

Convolutional neural networks consist of several layers where each one involves the application of convolution operators to the layer inputs usually alternating with pooling operations to reduce the size of inputs for the subsequent layers. At the end, the output of this first part of a CNN is vectorized and used as input of a common neural network, which, as usual, is trained to solve a classification or forecasting problem. The convolution part of a CNN, thus, can be interpreted as a method to preprocess the original features, represented, e.g., as images, in a similar spirit as the methods discussed in the previous chapter. The convolution and pooling operations transform an image or signal into a final vector of features, which represent the information particularly relevant for solving, e.g., a classification task.

We discuss the detailed structure of CNNs in a simple one-dimensional example where the notation is less cumbersome. The input is a finite part $z = (z_1, \ldots, z_d)$ of a signal or time series, e.g., an audio recording or a sequence of measurements from a sensor. Based on this information, we want to judge the state of the corresponding system, e.g., if it still works reliably, i.e., we have to handle a classification task. First, z is used as the input of a hidden *convolution layer* with H neurons similar to 78. For calculating the output of each such neuron $h = 1, \ldots, H$, we use a linear filter characterized by p coefficients $w_{kh}, k = 1, \ldots, p$. We calculate the convolution of the weight vector $w_{\bullet h} = (w_{1h}, \ldots, w_{ph})$ with the input signal z and additionally add bias weights $w_{0h}, h = 1, \ldots, H$. Finally, we apply a nonlinear activation function ψ, e.g., sigmoid or ReLU. Summarizing, the mapping from the input signal z to the neuron h of the convolution layer is given by

$$s_t^{(h)} = \psi\big(w_{0h} + \sum_{k=1}^{p} w_{kh} z_{t-k}\big) = \psi\big(w_{0h} + (w_{\bullet h} * z)_t\big), \tag{79}$$

$t = p+1, \ldots, d,\ h = 1, \ldots, H$. This results in new signals $s^{(h)} = (s_{p+1}^{(h)}, \ldots, s_d^{(h)})$ of length $d - p$ attributed to each of the H neurons in the convolution layer. All the weights $w_{kh}, k = 0, \ldots, p, h = 1, \ldots, H$ have to be learned by applying suitable learning algorithms to the complete network.

We expect that the filters represented by the coefficient vectors $w_{\bullet h}$ extract features of the signal z, which are particularly informative for classification. If, e.g., the states of the underlying system mainly differ with respect to the power that the signal has in certain frequency bands, then the CNN likely will learn the coefficients of corresponding bandpass filters.

In the subsequent *pooling layer*, we partition each signal $s^{(h)}$ into small disjoint neighborhoods of equal size and pool the signal values in those neighborhoods, e.g., by averaging them or retaining only the maximal value (*average* resp. *maximum pooling*). If we pool, e.g., $q = 2$ neighbors and, for the sake of simplicity, assume d, p to be even numbers, then the H neurons of the pooling layer represent the following signals:

$$r_t^{(h)} = \frac{1}{2}\left(s_{2t-1}^{(h)} + s_{2t}^{(h)}\right) \quad \text{or} \quad = \max\left\{s_{2t-1}^{(h)}, s_{2t}^{(h)}\right\}, \quad t = \frac{p}{2} + 1, \ldots, \frac{d}{2},$$
$$h = 1, \ldots, H. \tag{80}$$

Note that the pooling layer involves no network parameters that have to be learned. It is completely determined by the hyperparameter q, representing the size of the neighborhoods, which appears additional to the hyperparameter H, i.e., the number of neurons.

The resulting signals $r_t^{(h)}$ are of considerably smaller size, here $(d - p)/2$, than the original input z. Thus, the pooling layer reduces the computational load in subsequent parts of the whole CNN. Furthermore, pooling makes the whole procedure less sensitive to small shifts in the time axis. A classification method should depend on the presence of certain effects like large spikes in the signal z, e.g., brief exceptional high values registered by a pressure sensor, but the exact instance of time, where it happens, is usually not important.

Convolutional neural networks consist of several convolution and pooling layers, usually alternating but sometimes also with several convolution layers in a row. Passing the signal through those layers, the length of the signals attributed to each neuron decreases, and the number of neurons in the subsequent layers increases such that the last pooling operation results in many neurons containing only a short signal each. They are fed into *fully connected network* with K hidden neurons in the next step. For sake of simplicity, we proceed with the case of only a single convolution and pooling layer from above. The outputs $r^{(h)}, h = 1, \ldots, H$ of a pooling layer are used as inputs into a usual neural network. However, as they are vectors of length $D = \frac{d-p}{2}$ and not scalar, the weights $u_{kh}, k = 1, \ldots, K, h = 1, \ldots, H$, linking the pooling layer with the fully connected layer are vectors of size D themselves, and multiplication is replaced by the scalar product, i.e., the value attributed to neuron k of the fully connected layer is

$$x_k = \psi\big(u_{k0} + \sum_{h=1}^{H} \langle u_{kh}, r^{(h)} \rangle\big), \quad k = 1, \ldots, K. \tag{81}$$

The x_k, now, are real numbers, and the vector $x = (x_1, \ldots, x_K)$ finally may be used as input vector for a network like (49) or for some common multilayer classification network.

One might ask why the original signal z is not immediately passed through a standard neural network, e.g., (78), for the first step. The main reason is that this kind of network architecture treats the inputs as exchangeable. Each input component z_i gets its own weight $w_{ih}, i = 1, \ldots, d$, and the enumeration of the input components is arbitrary. The information that z has a temporal structure as a signal or a time series is lost. Usually, we model the observed signal as the realization of a random process Z, and then, we typically expect that Z_t shows a much stronger dependence on the immediate predecessor Z_{t-1} than on values $Z_{t-k}, k \gg 1$, in the remote past. It is likely advantageous to exploit this kind of knowledge in constructing a classification algorithm, and CNNs are able to do it. On the first glance, they pay the price of additional computational complexity, but usually the size, p in (79), of convolution filters is much smaller than the size of the signal, d in (79), which keeps the number of network parameters in check.

The majority of the CNN literature considers applications in image analysis. In the simplest case of a grayscale image, the input is after digitalization a two-dimensional array of number, principally real but, in practice, often also with a discrete range of values, e.g., from 0 (black) to a maximum of $2^c - 1$ (white) for some integer c. If X_{ij} denotes the gray value of the pixel at location (i, j), measuring the local brightness of the image and then the $d \times d'$ array $x = \big(x_{ij}\big)_{i=1,\ldots,d, j=1,\ldots,d'}$ is the input available, replacing the one-dimensional signal vector z in (79).

The main structure of convolutional networks does not change whether the input is a vector or an array. We briefly sketch how they look like for grayscale images. The convolution filter is characterized by an array, usually quadratic, of coefficients $w = \big(w_{ij}\big)_{i,j=1,\ldots,p}$ of odd dimension p. Using now the convolution of arrays, the convolution layer, corresponding to (79), is given by transformations like

$$y_{k\ell} = \psi\big(w_0 + (w * x)_{k\ell}\big) = \psi\big(w_0 + \sum_{i=1}^{p}\sum_{j=1}^{p} w_{ij} x_{k-i,\ell-j}\big),$$

$k = p+1, \ldots, d, \ell = p+1, \ldots, d'$. If the convolution layer consists of H hidden neurons, then this transformation is repeated H times with different filter parameters, and the results $y^{(h)} = \big(y_{k\ell}^{(h)}\big)_{k=p+1,\ldots,d,\ell=p+1,\ldots,d'}$ are images themselves of somewhat smaller size. Corresponding to (80), pooling is now done over small disjoint subarrays of size $q \times q$. For, e.g., $q = 2$ and max pooling, we would have

$$r_{11} = \max\{y_{11}, y_{12}, y_{21}, y_{22}\}, \ r_{12} = \max\{y_{13}, y_{14}, y_{23}, y_{24}\} \text{ etc.},$$

resulting in H images $y^{(h)} = \left(r_{k\ell}^{(h)}\right)$ of smaller size $\frac{d-p}{2} \times \frac{d'-p}{2}$.

The final step of moving to a fully connected network is identical to (81). The spatial structure of the images from the last pooling layer is no longer taken into account, but the arrays are vectorized in the usual manner, i.e., by putting all rows or columns one behind the other, and long vectors are generated, which then serve as inputs to (81).

Color images are split up into their respective color channels and passed separately through a CNN. In the case of a RGB image, we get three arrays representing the contributions of red, green, and blue to the brightness of the pixels. Up to a threefold number of parameters, the algorithms work similarly as for grayscale images.

13.3 Recurrent Neural Networks (RNNs)

Recursive neural networks consist of a sequence of applications of the same simple neural network to changing inputs from a sequence $X_1, \ldots, X_D$ of input vectors. The recursiveness results from the fact that they not only use the recent X_ℓ as input but additionally the state $S^{(\ell)}$ of the hidden neurons from the previous activation of the network. In this respect, they resemble the state space representation of linear systems, which form, e.g., the basis of Kalman filtering.

We briefly sketch how RNNs work with a simple classification example with two states labeled 0 and 1. Note that the whole sequence $X_\ell, \ell = 1, \ldots, D$ acts as long feature vector from which we have to decide to which of the two classes the object under investigation belongs. We assume that all inputs $X_\ell = (X_{\ell 1}, \ldots, X_{\ell d})$ are d-dimensional vectors.

We consider a single-layer neural network with H neurons and output given by (50). We initialize the recursion by presenting the first vector X_1 of the sequence to the network and get

$$S_h^{(1)} = \psi\big(w_{0h} + \sum_{i=1}^{d} w_{ih} X_{1i}\big), \ h = 1, \ldots, H.$$

The H-dimensional vector $S^{(1)}$ consists of the values attributed to the neurons in the hidden layer and represents the state of the hidden layer after this initialization.

In the following, the vectors of the feature sequence are fed one after the other into the same network, and, additionally, the previous state of the system is used as additional input, i.e., the vector $S^{(\ell)}$ attributed to the neurons after entering X_ℓ into the network is given by

$$S_h^{(\ell)} = \psi\big(w_{0h} + \sum_{i=1}^{d} w_{ih} X_{\ell i} + \sum_{k=1}^{H} \tilde{w}_{kh} S_k^{(\ell-1)}\big),\ h = 1, \dots, H,\ \ell = 2, \dots, D.$$

In each step, we can also calculate a 0–1 output from the states of the neurons as in (50)

$$R_\ell = 1_{[0,\infty)}\big(v_0 + \sum_{h=1}^{H} v_h S_h^{(\ell)}\big),\ \ell = 1, \dots, D,$$

though for the final classification, only the last output R_D, which through the recurrent scheme depends on the whole sequence $Z_1, \dots, Z_D$, is used.

The RNN has $w_{ih}, i = 0, \dots, d, \tilde{w}_{kh}, k = 1, \dots, H, v_h, h = 1, \dots, H$ and v_0 as trainable parameters. To learn these parameters, we need a training set $Y_j, X_1(j), \dots, X_D(j), j = 1, \dots, N$, consisting of many sequences and of the corresponding classes $Y_j \in \{0, 1\}$. As the output $R_D(j)$ of the RNN with the jth sequence as input also assumes only values 0 and 1, minimization of the empirical quadratic loss

$$\frac{1}{N} \sum_{j=1}^{N} \big(Y_j - R_D(j)\big)^2$$

is equivalent to minimizing the number of misclassifications on the training set. For more than two classes, we would proceed as in Sect. 2.4 and use the RNN to learn about class probabilities as functions of the input sequence and, then, apply the Bayes classifier.

As RNNs expect vectors of real numbers as input, the handling of sequences of qualitative values, like words in the analysis of text documents, has to include an initial preprocessing (compare Sect. 12.1). If this step does not use publicly available generic coding, e.g., for words by vectors of real numbers, then this might be the computationally most expensive part of the whole RNN application. Even then, straightforward applications of RNNs might not lead to really satisfactory results. Successful implementations often require additional measures, like in text classification, the use of certain structural properties of language (compare [47], Section 10.5.1 for details).

RNNs can handle a wide range of sequences of features. Due to their structures, where network weights are shared between steps of the recursion, it is possible to work with really long sequences, in general even of varying length. If inputs are however sequences of short real-valued vectors, then they compete with convolutional networks as described in the previous section. It depends on the type of data and the problem to be solved which learning approach leads to better results. If, e.g., sequences used as inputs consist of real-valued measurements from technical devices, then after some preprocessing, they often can be modeled in the normal state of the system as stationary time series with a rather short memory. For such data, classical filtering methods are tried-and-tested tools, and it is to be expected that CNNs will work well.

14 Time Series in Statistical Learning

As already briefly discussed in Sect. 2.5, many data in industrial production or in engineering research have the form of time series, i.e., of sequences of consecutive observations with a time stamp attached to each. For further analysis by statistical learning methods, even whole observed functions are digitalized and transformed into sequences. We consider in the following only sequences of observations Z_t, which are equidistant in time, for the sake of simplicity labeled by integers t. We also assume that the single data are real valued. Integer-valued time series or time series of counts have similar properties but differ in some details (see [22] for a review).

We first give a brief review of classical linear time series models that are related to linear regression (compare Sect. 3.1). Correspondingly, in statistical learning, methods developed for regression models are also applied to time series. For certain purposes, this is unproblematic, as we are only interested in estimates or forecasts, but we have to be aware of some important differences arising through the dependence of the data. These mainly have an impact on inference and on hyperparameter selection and are discussed in the second section.

In the third section, we briefly present the main ideas of spectral analysis of time series. These methods, which are tried-and-tested tools of signal analysis in engineering, are also quite helpful in statistical learning, as they can transform time series data into meaningful and almost independent features as potential inputs in learning algorithms. The fourth section is dedicated to preprocessing of time series data before feeding them into learning algorithms. We briefly discuss the removal of instationarities like trends or seasonal effects, the supplementation of missing data, and pre-whitening techniques.

In the last section, we address a kind of data-analytic problem that often arises in the context of time series data. It is of interest if the data-generating process, i.e., the structure of the time series, changes as this may indicate that the system moves from one state into another or that some event of interest, e.g., a defect, has occurred. Depending on the background of these events, they can be detected with change-point analysis or by fitting switching time series models.

Before looking at specific models, let us briefly discuss the concept of **stationarity** of time series. In Sect. 2.5, we have addressed the problem of forecasting the value Z_t of a time series from the last p observations $Z_{t-1}, \ldots, Z_{t-p}$, i.e., we are looking for a predictor $m(Z_{t-1}, \ldots, Z_{t-p})$ for Z_t. This is closely related to a regression problem, where the output Y_j should be well approximated by a function $m(X_{j1}, \ldots, X_{jd})$ of the inputs. Assume that our information about the time series consists of a long sequence of observations $Z_1, \ldots, Z_T$. Then, the time series forecasting problem can formally be written as learning a regression function m from the training set $Y_t, X_t, t = p+1, \ldots, T$, with $d = p, N = T - p$, outputs $Y_t = Z_t$ and corresponding inputs $X_t = (Z_{t-1}, \ldots, Z_{t-p}))$. We can use all the tools of statistical learning in regression, which we have discussed in the previous chapters. However, we have tacitly made the crucial assumption that the stochastic structure does not change over time, i.e., the distributions of random vectors $(Z_{t-1}, \ldots, Z_{t-p})$, do not depend on t. In this case, we know that we may use the same function m for forecasting Z_t from the last p observations for all t.

Time series in applications are frequently not stationary and should be transformed to a stationary process in advance, which then is used as input for learning algorithms (compare Sect. 2.5 for an example). Sometimes, that will not be so easy as the observed system changes with time in a more complicated manner. The methods and models discussed in this chapter are nevertheless applicable in the frequent case where the changes of the data structure are slow compared to the flow of observations. Appropriate modifications of stationary time series analysis are available under the keyword of *local stationarity* (see [16] for a review). A simple example for such a process would be an autoregressive process (12) of order $p = 1$ with slowly changing coefficient: $Z_t = a_1(t)Z_{t-1} + \varepsilon_t$.

A quite different form of nonstationarity is observed in piecewise stationary time series. At a few instants in time, the data-generating process changes, but between them, it is stationary. This kind of instationarity either can be handled by change-point analysis (compare Sect. 14.5) or in terms of *Markov switching* processes, where the latter is in particular appropriate for systems that alternate between a finite number states and the changes depend on the values of hidden state variables (see [25] for a review).

Figure 46 shows on the left $T = 240$ observations from a stationary autoregressive process of order $p = 3$, whereas the data on the right are from a piecewise-stationary process. Before $t = 81$ and after $t = 160$, the data-generating process coincides with the example on the left, whereas, in between, the system changes to an autoregressive scheme of order $p = 1$ and, also, to innovations with smaller variance. With some experience, the instationarity can be seen from the plot

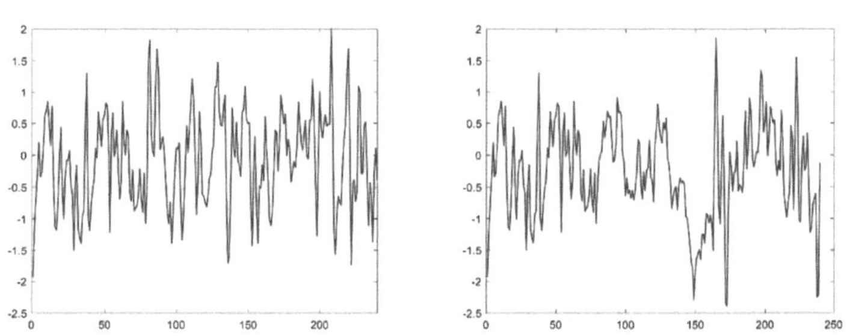

Fig. 46 $T = 240$ observations from a stationary AR(3)-process (left) and from a only piecewise stationary time series (right)

as the local amplitudes are smaller in the central part. For an objective decision, there are algorithms for detecting such situations like those in Sect. 14.5.

14.1 Linear Time Series

In the following, we assume that the data are part of a time series that is stationary and has a finite variance. For linear time series models, the mean and the covariances of the data are essential. Due to stationarity, the mean is the same for all Z_t, and the covariance as a measure of dependence of Z_t, Z_s is a function of the time lag $t - s$ between the two measurements only:

$$\mathrm{E}\, Z_t = \mu, \quad \mathrm{cov}\,(Z_t, Z_s) = r_{t-s} \quad \text{for all } t, s. \tag{82}$$

In particular, $r_0 = \mathrm{var}\, Z_t$ is the common variance of the observations. $r_t, -\infty < t < \infty$, are called the **autocovariances** of the time series.

A very special stationary time series that serves as an important building block for more complex models is the so-called *white noise* $\varepsilon_t, -\infty < t < \infty$,, which consists of random variables with identical mean μ_ε and variance σ_ε^2, and for which the autocovariances vanish for all lags greater than 0, i.e., $\mathrm{cov}\,(\varepsilon_t, \varepsilon_s) = 0$ for $t \neq 0$. Often, it is even assumed that the ε_t are independent and identically distributed. Then, they are called *strict white noise*.

Many time series models assume that the observation X_t at time t is partly determined by the past up to time $t - 1$ and some **new** random variable ε_t, the **innovation** at time t, which is generated at time t and which does not depend on the past at all. An example is the rather flexible *causal linear time series*, which allows for a representation

$$Z_t = \sum_{k=0}^{\infty} c_k \varepsilon_{t-k} \quad \text{with} \sum_{k=0}^{\infty} c_k^2 < \infty. \tag{83}$$

where ε_t are strict white noise. Stationary linear autoregressions of order p or AR(p)-processes (12) are a special case of this class of time series models. If we do not assume that their mean is 0, they are given by

$$Z_t = \mu + \sum_{k=1}^{p} a_k (Z_{t-k} - \mu) + \varepsilon_t, \tag{84}$$

where the ε_t are strict white noise with mean $\mathrm{E}\, \varepsilon_t = 0$.

Autoregressive processes are widely applied as models for time series, as they allow for simple forecasts, but they are not flexible enough to approximate all simple time series well enough without requiring too many parameters. A more general class of models consists of the so-called *mixed autoregressive-moving average processes* or ARMA(p, q)-processes

$$Z_t = \mu + \sum_{k=1}^{p} a_k(Z_{t-k} - \mu) + \sum_{k=1}^{q} b_k \varepsilon_{t-k} + \varepsilon_t \tag{85}$$

where again the ε_t are zero mean white noise. Here, Z_t is allowed to depend explicitly on past innovations, too. Obviously, an ARMA$(p, 0)$-process is just an AR(p)-process. The other extreme case ARMA$(0,q)$ is called *moving average* of order q or MA(q)-process. ARMA-models, which in the electrical engineering literature are sometimes also called pole-zero models, are the most widely used class of linear parametric time series models.

AR- and ARMA-models are not necessarily stationary, but there are well-known conditions on the parameters that guarantee it: the zeroes of the *generating polynomial* or *z-transform* of the autoregressive parameters $A(z) = 1 - a_1 z - \ldots - a_p z^p$ have to lie outside of the closed unit circle $\{z;\ |z| \leq 1\}$ in the plane of complex numbers. If there would be a zero inside the unit circle, then the $|Z_t|$ would explode, i.e., grow exponentially fast in time. For a zero on the unit circle, the process would also be unstable but behave more like a random walk with linearly increasing variance over time.

Note that ARMA-processes are in a rather general sense equivalent to stochastic processes, which have a linear *state space* or *Markovian representation*

$$X_t = \mathbf{F} X_{t-1} + e_t\ , \quad e_t = (\varepsilon_t, 0 \ldots, 0),$$

where X_t is a column state vector of appropriate dimension and $\mathbf{F}$ is the system matrix (compare section 10.4.4 of [64]). For an AR(2)-process, we would, e.g., have $X_t = (Z_t - \mu, Z_{t-1} - \mu)^T$ and $e_t = (\varepsilon_t, 0)^T$, and the system matrix has the entries $F_{11} = a_1,\ F_{12} = a_2,\ F_{21} = 1,\ F_{22} = 0$.

ARMA-processes allow for an explicit decomposition $Z_t = \pi_t + \varepsilon_t$ where the innovation ε_t is independent of the past and π_t is completely determined by the past of the process, i.e., by $Z_s, s < t$. Hence, π_t is the best forecast of Z_t based on all the information available up to time $t - 1$. In case of an autoregression,

$$\pi_t = \mu + \sum_{k=1}^{p} a_k(Z_{t-k} - \mu),$$

i.e., it depends only on finitely many of the past observations. This is no longer true for general ARMA-processes, as here, π_t also depends on past innovations, which are not observed. Given $Z_1, \ldots, Z_{t-1}$ for large enough t, the forecast π_t can however be well approximated by $\hat{\pi}_t$ derived from the following recursion, which is motivated by (85) and $\varepsilon_s = Z_s - \pi_s$. As initialization, we set $\hat{\varepsilon}_\ell = 0, \ell \le \max(p, q)$, and, then, proceed for $s = \max(p, q) + 1, \ldots, t$:

$$\hat{\pi}_s = \mu + \sum_{k=1}^{p} a_k(Z_{s-k} - \mu) + \sum_{k=1}^{q} b_k \hat{\varepsilon}_{s-k}, \quad \text{and } \hat{\varepsilon}_s = Z_s - \hat{\pi}_s. \tag{86}$$

Up to know, we have pretended that the parameters of (85) are known. In practice, of course, we have to estimate them from a sample $Z_1, \ldots, Z_T$. For the mean μ and the autocovariances r_t, we can apply the *law of large numbers*, which also holds for stationary time series under weak conditions, in particular for linear processes, and get

$$\hat{\mu} = \frac{1}{T} \sum_{t=1}^{T} Z_t \quad \text{and} \quad \hat{r}_t = \frac{1}{T} \sum_{k=1}^{T-t} (Z_{t+k} - \hat{\mu})(Z_k - \hat{\mu}) \tag{87}$$

as estimates converging to μ resp. r_t for $T \to \infty$. For an autoregressive process, estimates $\hat{a}_k$ for the autoregressive parameters are derived from a least-squares approach, i.e., we minimize w.r.t. a_k

$$\frac{1}{T} \sum_{t=p+1}^{T} \left(Z_t - \hat{\mu} - \sum_{k=1}^{p} a_k(Z_{s-k} - \hat{\mu})\right)^2.$$

The minimum achieved is an estimate $\hat{\sigma_\varepsilon}^2$ of the innovation variance σ_ε^2. Least-squares estimates for autoregressive parameters are practically identical to the so-called *Yule-Walker estimates*, up to for larger T negligible terms, which are usually provided by statistical software. From the viewpoint of statistical learning, they have one important advantage: the so-called *Levinson-Durbin algorithm* (compare [10], Section 5.2) provides a fast recursive method for the simultaneous calculation of autoregressive parameter estimates $\hat{a}_k(p), k = 1, \ldots, p$, fitted to the time series data for a whole range of AR(p)-models. Therefore, choosing the hyperparameter p based on the training data is greatly facilitated. For this purpose, usually information criteria like AIC or BIC, discussed in Sect. 3.5, are minimized. For autoregressions, for which it has originally been developed, AIC is, up to constants that are not relevant for minimization, given by

$$\text{AIC}_p = T \log \hat{\sigma_\varepsilon}^2(p) + 2p,$$

and in BIC, the factor 2 on the right-hand side is replaced by $\log T$.

As an example, Fig. 47 shows AIC_p and BIC_p, $p = 1, \ldots, 20$, for a sample of $T = 120$ from a stationary AR(3)-process. Both criteria become minimal at the correct order $p = 3$. Note that BIC_p increases faster beyond the true order, whereas

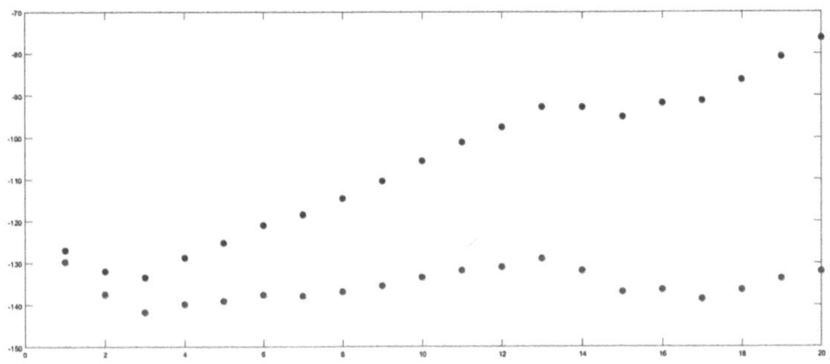

Fig. 47 AIC_p (red) and BIC_p (blue) for $N = 120$ data from AR(3)-process

AIC_p flatters around its minimum. Therefore, AIC_p is, in general, more prone to become minimal at somewhat larger orders than necessary by chance.

Estimating parameters in the general ARMA-model by least squares is slightly more involved as the innovations ε_s are not observed. However, we can set up a similar recursion as in (86) to calculate approximations $\hat{\varepsilon}_s$ simultaneously with the parameter estimate $\hat{a}_k, \hat{b}_k$. The minimum values achieved is again an estimate of σ_ε^2. It is also possible to calculate the parameter estimates of ARMA(p, q)-models simultaneously for a whole range of p, q by adopting a similar approach as in the Levinson-Durbin recursion (see [24]). AIC and BIC are identical to the pure autoregressive case except that the factor p on the right-hand side is replaced by $p + q$.

Up to now, we have only considered one-dimensional time series. In practice, frequently several times series are observed simultaneously, which are modeled as multivariate stochastic processes. There are also AR- and ARMA-models for higher-dimensional time series, which require a somewhat involved notation such that we do not go into details here and refer to the literature, e.g., [58]. However, we like to mention a problem important for practice. Multivariate ARMA-models suffer from non-identifiability, i.e., there are various different models and parameter sets describing the same stochastic process (see [42]). Without particular precautions, this may cause trouble for estimation and, then, for forecasting.

14.2 Nonlinear Time Series in Statistical Learning

Many of the statistical learning methods presented in the previous chapters can also be applied to time series analysis, in particular to forecasting problems. In the following, we focus on the case of a univariate time series Z_t observed at $t = 1, \ldots, T$. We assume that it is part of a stationary time series that is weakly dependent in the sense that Z_t, Z_{t-s} become almost independent pretty fast for large enough $s > 0$. Stationary linear AR- and ARMA-processes have this property, and, e.g., $|r_s| = |\text{cov}\,(Z_t, Z_{t-s})|$ decreases exponentially fast with increasing s. Under certain conditions on the function m, this also holds for *nonlinear autoregressions* of order p satisfying

$$Z_t = m(Z_{t-1}, \ldots, Z_{t-p}) + \varepsilon_t, \tag{88}$$

where ε_t are as usual zero-mean innovations, which are independent and identically distributed. Formally, this resembles the general regression model (2) with input vectors $X_t = (Z_{t-1}, \ldots, Z_{t-p})$.

Weakly dependent stationary time series are in many respects behaving similar as independent data in regression. We have already mentioned that laws of large numbers are still valid, which are responsible for many common estimates being consistent in the sense that, for increasing sample size, they become more and more close to the quantity of interest. For example, we may estimate the forecasting function m by a kernel smoother as in (63), e.g., for $p = 1$

$$\widehat{m}(x, h) = \frac{\frac{1}{N}\sum_{t=1}^{T-1} K_h(x - Z_t)\, Z_{t+1}}{\hat{p}(x, h)} \qquad \text{with } \hat{p}(x, h) = \frac{1}{N}\sum_{t=1}^{T} K_h(x - Z_t)$$

or, for general order p, by a single-layer neural network output function (45),

$$f_H(x; \hat{\vartheta}) = \hat{v}_0 + \sum_{h=1}^{H} \hat{v}_h \psi\big(\hat{w}_{0h} + \sum_{i=1}^{p} \hat{w}_{ih} x_i\big),$$

where the network weights combined in the parameter vector $\hat{\vartheta}$ minimize the empirical squared forecasting error

$$\sum_{t=p+1}^{T} \big(Z_t - f_H(Z_{t-1}, \ldots, Z_{t-p}; \vartheta)\big)^2$$

w.r.t. ϑ.

If, however, we look at inference, e.g., getting an estimate for the forecasting error and hence an idea about the reliability of our forecast, or if we try to apply resampling methods for data-adaptive selection of hyperparameters, e.g., of p, H in the case of the neural networks-based predictor, then the dependence of the data Z_t and hence the input vectors $(Z_{t-1}, \ldots, Z_{t-p})$ becomes relevant and is the cause of differences to the regression case.

Even in the most simple example of estimating the mean of a linear autoregressive process $Z_t = \mu + a(Z_{t-1} - \mu) + \varepsilon_t$ of order 1, the influence of dependence on inference is apparent. We know that the sample mean $\hat{\mu}$ of the previous section is a reasonable estimate of μ. The width of confidence intervals for gauging the reliability of the estimate will be proportional to the standard deviation, i.e., the square root of the variance of $\hat{\mu}$, where the latter is, for large T, given by

$$\text{var}\,\hat{\mu} \approx \frac{1}{T}\frac{\sigma_\varepsilon^2}{(1-a)^2}$$

for all $-1 < a < 1$, which is the stationarity condition for an AR(1)-process. $a = 0$ corresponds to the case of independent inputs as in regression. We see that for strong positive dependence between neighboring observations from time series, i.e., for $0 \ll a < 1$, $\hat{\mu}$ will have a much larger variance and be a less reliable estimate of μ than in the independent case, and this situation is, unfortunately, typical for most engineering time series being discrete observations of smooth signals. For $a < 0$, the quality of the estimate will be even improved, but that is a rare case in engineering applications. This general tendency of still reasonable estimates, which however are not so good as we are used from regression situations, holds in most cases where statistical learning meant for independent data is applied to time series.

On the other hand, there are also situations where statistical learning profits from the dependence of time series with considerable practical implications. For example, [20] (Chap. 8.1.2) considered how large a dike has to be built if it should withstand all floods of the next 100 years with probability 95%. Applying methods of risk analysis, it turned out that the dike can be built considerably lower if the flood levels are modeled as a time series instead of a sequence of independent data. Similar problems also show up in industry if a guarantee for the reliability of components has to be given.

The serial dependence of time series data may also impact the choice of hyperparameters of statistical learning algorithms considerably. For single-layer neural networks (45), e.g., we know that the number H of neurons in the hidden layer has to increase with the size of the training set to get consistency, i.e., convergence of $f_H(.;\vartheta)$ to $m(.)$. Franke and Diagne [26] showed that the rate of increase of H is qualitatively much slower in the time series case than for independent input data. This can be interpreted as an effective reduction of sample size due to the serial dependence of the data, as, e.g., in the extreme case of a strong positive correlation and small innovations, $Z_t \approx Z_{t+1}$ for all t.

These differences between time series and independent input data is particularly important for the application of data-adaptive hyperparameter selection procedures. Hart [43] showed, e.g., that for positively dependent data, plain **crossvalidation** may not work at all. More precisely, he considered a regression model with equidistant design and the corresponding Priestley-Chao estimate (60), where the hyperparameter is the bandwidth h. However, he allowed the residuals ε_j to be generated by a time series instead of being independent identically distributed. In case of positive dependence like for an AR(1)-process with $a > 0$, crossvalidation chose much too small bandwidths resulting in a considerably overfitted rough function estimates. The intuition for this behavior does not rely on the particular local smoothing setting but likely holds for all kinds of statistical learning procedures.

Crossvalidation is essentially a resampling method like the bootstrap (compare Sect. 9.1). For forecasting, it numerically calculates the mean-squared prediction error of the estimate by Monte Carlo simulation and, then, minimizes it w.r.t. the hyperparameter (compare Sect. 3.4). It relies on the fact that the pseudo samples, which in the leave-one-out case are just the training set without one observation, have at least approximately the same structure as the original data. Removing, e.g., one observation from a set of independent identically observations only lowers the sample size by 1, but has no impact at all on the stochastic structure. If we remove, however, one observation, say Z_s, from a time series of length T, then the resulting data vector $Z'_t = Z_t, t < s, Z'_t = Z_{t+1}, s \leq t \leq T-1$, is no longer generated by a stationary time series, as, e.g.

$$\text{cov}\,(Z'_{s-1}, Z'_s) = \text{cov}\,(Z_{s-1}, Z_{s+1}) \neq \text{cov}\,(Z_{t-1}, Z_t) = \text{cov}\,(Z'_{t-1}, Z'_t) \quad \text{for } t \neq s.$$

The pseudo-samples are not representative, and we cannot take it for granted that tried-and-tested hyperparameter selection methods developed for independent data also work with time series. Already [43] has modified simple crossvalidation to take into account dependence of the data, which led to better results, but if one relies on statistical learning software, then one has to check if the hyperparameter selection methods included are also able to handle time series data. For example, [66] discuss an estimation problem, where crossvalidation works for a stochastic process of dependent data.

In contrast to crossvalidation, **validation** (compare Sect. 3.4) can be applied to time series data without qualm. We only have to take care that the input data are generated by a stationary, short-memory process, but that is anyhow advisable as it is a prerequisite for applying many statistical learning procedures at all. In Sect. 14.4, we discuss how to transform original time series, which do not satisfy these conditions, in a preprocessing step. In validation, we use $Z_1, \ldots, Z_T$ as training set for estimating, e.g., a forecasting function based on (88), and $Z_{T+1}, \ldots, Z_{T+M}$ as validation set for approximating the mean-squared prediction error. Both sets are not strictly independent as in regression, but, due to the short memory condition, which implies exponentially fast convergence of covariances to 0, only the last few values of the training set and the first few values of the validation set will show a significant correlation which may be neglected.

Bootstrap methods can be applied to time series in classical settings with parametric models or in statistical learning. However, which type of bootstrap is useful depends on the kind of statistics under investigation. Here, we only consider briefly the autoregressive models (88) and the residual-based bootstrap (compare Sect. 9.1.2). For an extensive review, we refer to [52].

To generate pseudo-samples that have similar properties as the original time series and may be used for Monte Carlo approximations, we first need an initial estimate $\widetilde{m}$ of the autoregressive function m. Then, we calculate the *sample innovations*

$$\tilde{\varepsilon}_t = Z_t - \widetilde{m}(Z_{t-1}, \ldots, Z_{t-p}), \; t = p+1, \ldots, T,$$

from which we, as in Sect. 9.1.2, subtract their average to get $\hat{\varepsilon}_t$. We resample, i.e., draw with replacement, the bootstrap pseudo innovations ε_t^b from the $\hat{\varepsilon}_{p+1}, \ldots, \hat{\varepsilon}_T$:

$$\mathrm{pr}^*\{\varepsilon_t^b = \hat{\varepsilon}_k\} = \frac{1}{T-p}, \; t, k = p+1, \ldots, T, b = 1, \ldots, B. \tag{89}$$

Then, we construct the bootstrap pseudo time series from (88), replacing the unknown m by its initial estimate

$$Z_t^b = \widetilde{m}(Z_{t-1}^b, \ldots, Z_{t-p}^b) + \varepsilon_t^b, \; t = p+1, \ldots, T, b = 1, \ldots, B,$$

where we initialize the recursion by setting $Z_t^b = Z_t, t = 1, \ldots, p$.

Now we have B bootstrap pseudo time series $Z_t^b, t = 1, \ldots, T,$, and from each, we calculate the estimate $\widehat{m}^b(.), b = 1, \ldots, B$. We may use these estimates for approximating the whole distribution or certain measures of variability of the forecasting function estimate $\widehat{m}(.)$ by Monte Carlo simulation.

As an example of this procedure, [27] have investigated kernel estimates $\widehat{m}(.; h)$ like (63) for m and shown that the residual-based bootstrap and some alternatives like the *wild bootstrap* work. A crucial assumption was that the initial estimate $\widetilde{m}(.)$ should be smoother than the estimate $\widehat{m}(.)$, which is investigated by bootstrap methods. More precisely, if $\widetilde{m}(.) = \widehat{m}(.; \tilde{h})$ is also chosen as a kernel estimate, then $\tilde{h} \to 0$ with a slower rate than the bandwidth h of $\widehat{m}(.) = \widehat{m}(.; h)$, i.e., in practice $\tilde{h} > h$. This prerequisite of the bootstrap in statistical learning can be explained observing that good, in the sense of leading to small estimation or forecasting error, estimates $\widehat{m}(.)$ are usually a bit rougher than the true underlying function $m(.)$. To mimic the behavior of original data, we therefore have to use a relatively smooth autoregressive function $\widetilde{m}(.)$ in generating the pseudo samples. This feature of bootstrapping in statistical learning is not peculiar to time series but also holds in regression models with independent inputs (compare Sect. 9.1.2, where we briefly mentioned it for spline smoothers).

14.3 Spectral Analysis of Time Series

The dependence of data from a stationary time series is mainly characterized by its autocovariances r_t. In the short memory case, we already mentioned that $|r_t| \to 0$ exponentially fast, but for other time series, we usually have at least $\sum_{t=-\infty}^{\infty} |r_t| < \infty$. In that case, we have pointwise convergence of the Fourier series (compare Sect. 6.1)

$$f(\omega) = \sum_{t=-\infty}^{\infty} r_t e^{-it\omega}, \qquad \text{where also } r_t = \frac{1}{2\pi} \int_{-\pi}^{\pi} e^{it\omega} f(\omega) d\omega. \tag{90}$$

The function $f(\omega)$, $-\pi \le \omega \le \pi$, is called the (power) **spectral density** of the time series. Here and in the following, for brevity, we call ω *frequency* though it really is the circular frequency proportional to the usual frequency in the physical sense. Note that the spectral density of *white noise* is a constant $f_\varepsilon(\omega) = r_0 = \sigma_\varepsilon^2$, i.e., all frequencies contribute equally to the total power of the signal as in white light, which was the motivation for the name of this special kind of time series.

In case of a linear process (83), the spectral density also can be represented as

$$f(\omega) = \sigma_\varepsilon^2 \left| \sum_{k=0}^{\infty} c_k e^{-ik\omega} \right|^2.$$

The spectral density provides a characterization of the dependence structure of the time series that is equivalent to the sequence of autocovariances. Before we look at its interpretation, we briefly discuss estimation. From observing $Z_1, \ldots, Z_T$ from a stationary time series, we get estimates $\hat{r}_t, 0 \le t \le T-1$, from (87) and, due to symmetry, also $\hat{r}_{-t} = \hat{r}_t, 1 \le t \le T-1$. Replacing the unknown autocovariances in (90) by their estimates as far as available, we get the so-called *periodogram*

$$I_T(\omega) = \sum_{t=-T+1}^{T-1} \hat{r}_t e^{-it\omega}$$

as a seemingly reasonable estimate of $f(\omega)$. Let us for the moment assume that $\mathrm{E}\, Z_t = \mu = 0$. In view of the efficiency of the fast Fourier transform (FFT) algorithm, the periodogram is usually calculated by its equivalent representation as

$$I_T(\omega) = \left| D_T(\omega) \right|^2, \qquad \text{where } D_T(\omega) = \frac{1}{\sqrt{T}} \sum_{t=1}^{T} Z_t e^{-it\omega}$$

is the *discrete Fourier transform* (DFT) of the sample $Z_1, \ldots, Z_T$. Usually, those functions are calculated on a grid $\omega_k = \frac{2\pi k}{T}, -\frac{T}{2} < t \le \frac{T}{2}$, i.e., at the so-called *Fourier frequencies.*

The periodogram indeed is for large enough T close to the spectral density on the average, i.e., $\mathrm{E}\, I_T(\omega) \approx f(\omega)$, but it nevertheless is not an estimate at all, as its variance does not become smaller with increasing sample size T. The reason for this behavior is the fact that the DFT is a one-to-one transformation of the random sample $Z_1, \ldots, Z_T$ of size T to another random sample $D(\omega_k), -\frac{T}{2} < t \le \frac{T}{2}$, of again size T. Note that the $D_T(\omega_k)$ are complex-valued data, but as $D(-\omega)$ is the complex conjugate of $D(\omega)$, they are determined by T real numbers. From the latter,

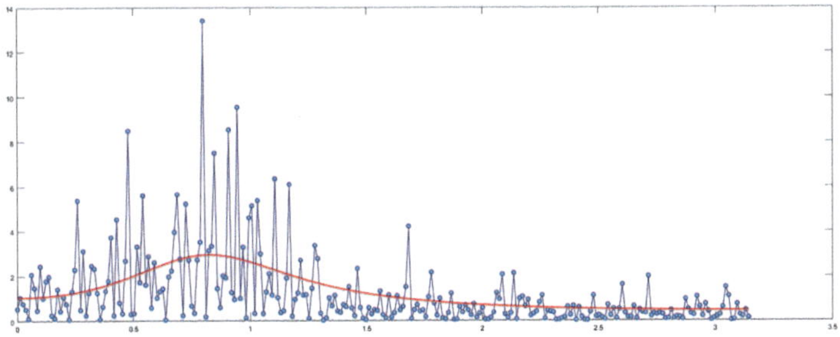

Fig. 48 Spectral density (red) and periodogram (blue) based on a sample of size $N = 512$ from an ARMA(2,1)-process

we also can reconstruct the original data by the inverse discrete Fourier transform

$$Z_t = \frac{1}{\sqrt{T}} \sum_{-\frac{T}{2}<k\leq\frac{T}{2}} D_T(\omega_k) e^{it\omega_k}$$

$$= \frac{1}{\sqrt{T}} \sum_{-\frac{T}{2}<k\leq\frac{T}{2}} A_k \cos(t\omega_k + \Phi_k), \quad t = 1, \ldots, T, \tag{91}$$

where for the second representation, we have written the complex DFT-values as $D_T(\omega_k) = A_k e^{i\Phi_k}$, with $A_k \geq 0$, and used that the data Z_t are real valued. Therefore, the $D_T(\omega_k)$ and, hence, the periodogram values $I_T(\omega_k)$, $-\frac{T}{2} < t \leq \frac{T}{2}$, are as random as the original data and show a quite erratic behavior. We cannot expect them to converge to the values of the usually smooth function $f(\omega)$. If the mean of the data Z_t is not 0, then above arguments still hold if we replace Z_t by the centered $Z_t^0 = Z_t - \hat{\mu}$ everywhere.

Figure 48 shows as an example the spectral density $f(\omega)$ of an ARMA(2,1)-process together with the periodogram based on a sample of size $T = 512$. $I_T(\omega)$ fluctuates around $f(\omega)$ in an erratic manner where the variability is in particular large where the spectral density assumes large values.

The periodogram is not a reasonable estimate of the spectral density, but if we smoothen it, we get good estimates similar to the Priestley-Chao kernel smoothers in regression, depending on a bandwidth h as hyperparameter too:

$$\widehat{f}(\omega, h) = \frac{2\pi}{T} \sum_{-\frac{T}{2}<k\leq\frac{T}{2}} I_T(\omega_k) K_h(\omega - \omega_k),$$

where K_h is a scaled kernel function as in Sect. 10.1.

Figure 49 shows on the right the true spectral density (red) and a kernel spectral estimate (blue) using the same ARMA data as for the previous figure and a

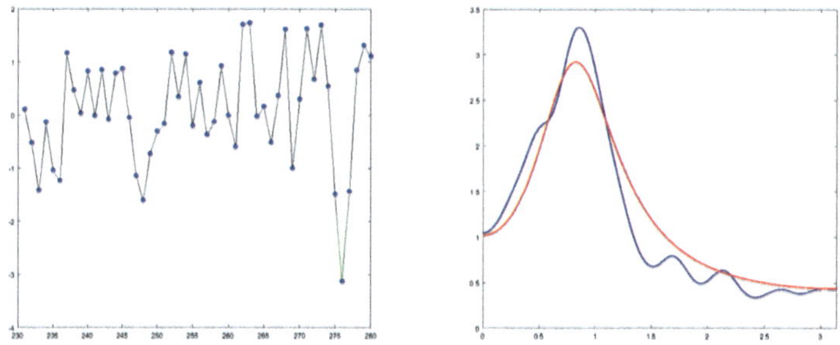

Fig. 49 Part of an ARMA process on the left and spectral density (red) and spectral density estimate (blue) on the right

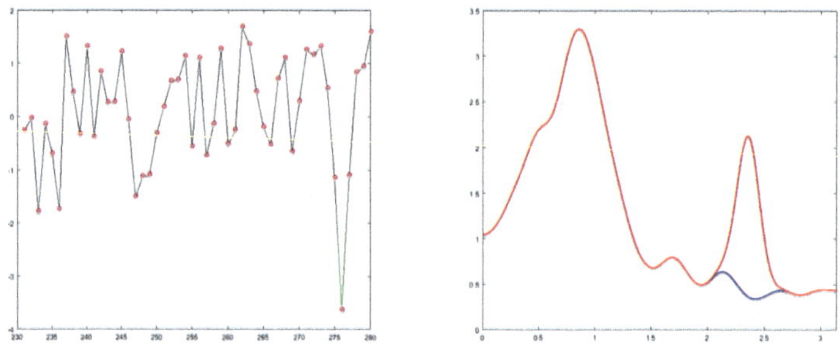

Fig. 50 Part of a disturbed ARMA process on the left and spectral density estimates of the original data (blue) and the contaminated data (red) on the right

bandwidth $h = 0.1$. On the left, a subsample of size 50 from the central part of the data is plotted for better visibility of the details.

From the theory of spectral analysis, we know that the DFT values $D_T(\omega)$, $D_T(\omega')$ at different frequencies $0 \leq \omega < \omega' \leq \pi$ for $T \to \infty$ become more and more uncorrelated and, in the case of Gaussian processes, even independent. Hence, (91) is a decomposition of the dependent time series data Z_t into weakly dependent cosine components with frequencies ω_k and amplitudes A_k (compare Chapters 4 and 10 of [11] for the underlying theory). As $A_k^2 = |D_T(\omega_k)|^2 = I_T(\omega_k)$ and E $I_T(\omega) \approx f(\omega)$, the spectral density measures the average contribution of the spectral component with frequency ω to the time series. For example, large spikes in the spectral density show which frequency components of the signal Z_t have particularly large power.

Structural properties of a time series are often much better seen from the periodogram $I_T(\omega)$ or a spectral density estimate $\widehat{f}(\omega, h)$ than from the raw data Z_t or from estimates of the autocovariances.

As an example, Fig. 50 shows part of a sample of size $N = 512$, which coincides with the ARMA data used for the last two figures. However, now, a small periodic disturbance has been added, represented by a cosine wave with frequency $\omega_d = 0.75 * \pi$. This is typical for the surveillance of machinery, which sometimes starts to vibrate due to external influences or internal defects. Comparing the subsamples on the respective left-hand sides of Figs. 49 and 50, there is hard to see any difference by just visual inspection. Comparing the spectral density estimates on the right, the difference between both signals is immediately visible as the spectrum of the disturbed data has a prominent additional peak around ω_d.

Looking at (91), the DFT provides a decomposition of the signal $Z_1, \ldots, Z_T$ into weakly correlated and in the Gaussian case weakly dependent terms, where A_k measures the amount contributed by component k to the signal. In this respect, it resembles the principal component analysis, which we discussed in Sect. 12.2, as a tool for extracting impactful features from the originally observed input vectors. In particular, in classification problems based on time series observations, using $I_T(\delta_k), 0 \leq k \leq K$, as inputs for statistical learning algorithms often results in a much better performance than working with the strongly dependent Z_t. Here, for some $K \ll T$, $\delta_k = \frac{2\pi k}{K}, 1 \leq k \leq \frac{K}{2}$ form a grid of Fourier frequencies on the interval $[0, \pi]$ coarser than the grid given by the original ω_k to reduce the dimension of the input vector and, additionally, make the considered periodogram values even less dependent. As the number of principal components used as features, K is a hyperparameter of the learning method. If not only amplitudes but also the phase shifts of the spectral components of the signal Z_t are important, one could also use A_k, Φ_k derived from the DFT $D_T(\delta_k)$ as inputs.

14.4 Preprocessing of Time Series

Statistical learning methods for classification and forecasting, which have been developed for training sets of independent observations, work also reasonably well for time series if the latter are stationary and have a reasonably short memory. This does not always hold in practice, but there are various transformations, which allow to modify time series data such that they become suitable inputs for learning algorithms.

We already have discussed de-seasonalization as an example for such preprocessing in Sect. 2.5. Besides such *seasonal effects*, time series often show *trends*, i.e., the mean $\mathrm{E}\, Z_t$ is not constant, as required by stationarity, but, e.g., increases with time. A simple way to cope with this kind of instationarity is transforming the original data, e.g., by applying repeatedly, say ℓ times, the difference operator $\Delta Z_t = Z_t - Z_{t-1}$. $\ell = 1$ ($\ell = 2$) would, e.g., suffice to remove a linear resp. quadratic trend. If the transformed time series can be modeled as an ARMA(p, q)-process, then the original process is called an integrated ARMA-process or **ARIMA**(p, ℓ, q)-process. Analogously a time series showing a seasonal effect with period s may be de-seasonalized by applying the s-step difference operator $\Delta_s Z_t = Z_t - Z_{t-s}$. Usually, one application of this transformation suffices, but, in principle, we could repeat it say L times. Combining both transformations, Box and Jenkins [7] have developed an extensive methodology of analyzing and forecasting time series based on seasonal ARIMA or **SARIMA**$(p, \ell, q) \times (P, L, Q)$ models for which $Z_t^o =$

$\Delta^\ell \Delta_s^L Z_t$ is a special type of ARMA-process allowing for a suitable, parsimonious parametrization. Then, the Z_t^o are better suited for further processing in a statistical learning framework.

In practice, in time series as in other kinds of data, some values may be **missing**, e.g., due to a passing defect of a sensor or a brief stop of production due to maintenance. Most statistical learning algorithms expect, however, complete data, such that the gaps have to be filled somehow. Here, the dependence of time series observations Z_t in time is helpful. We only discuss isolated missing data and how to interpolate them, but similar ideas may also be applied to handling larger gaps.

If Z_t is missing, then we know that usually it will be rather strongly depend on the observed neighbors Z_{t-1}, Z_{t+1}. A quick and dirty interpolation methods would be to replace the missing values just by the plain average $\widehat{Z}_t = \frac{1}{2}(Z_{t-1} + Z_{t+1})$. This is correct in the mean, but not in the scaling. A more refined method of *linear interpolation* is based on fitting a linear model to the available data, e.g., by using an EM-algorithm [11, section 8.7] and then using it for filling the gaps. Brockwell and Davis [11, Section 8.6] discuss this for linear state space models and the equivalent ARMA-models. In the simplest case of an AR(1)-model, this results in an interpolation $\widehat{Z}_t = \frac{a_1}{1+a_1^2}(Z_{t-1} + Z_{t+1})$ where the autoregressive parameter a_1 has of course to be replaced by an estimate.

Interpolation of missing values is the same kind of problem as forecasting of future values. Hence, we also may use statistical learning approaches to solve this kind of problem. Given times series data, we try to supplement missing values Z_t by a function $m(Z_{t-q}, \dots, Z_{t-1}, Z_{t+1}, \dots, Z_{t+q})$ of the $q \geq 1$ left and right neighbors. We can approximate m by any of the statistical learning approaches presented in the previous chapters, e.g., by a single-layer neural network output function $f_H(x; \vartheta)$ with $(2q)$-dimensional input as in (45). Let S_q be the set of all time instants $q + 1 \leq s \leq T - q$, for which the blocks $Z_{s-q}, \dots, Z_{s+q}$ of length $2q + 1$ are completely observed. Assume that there are N such blocks, which are used for the training set. Then, we learn the network weights ϑ from minimizing the empirical squared loss

$$\sum_{s \in S_q} \left(Z_s - f_H(Z_{t-q}, \dots, Z_{t-1}, Z_{t+1}, \dots, Z_{t+q}; \vartheta)\right)^2 = \min_{\vartheta}!$$

Then, $\widehat{Z}_t = f_H(Z_{t-q}, \dots, Z_{t-1}, Z_{t+1}, \dots, Z_{t+q}; \hat{\vartheta}_N)$ is used for replacing missing observations Z_t. This approach also can be generalized to filling larger gaps in the data provided that there are enough complete pieces of the observed time series.

In classical time series analysis, it has been observed that rather weakly correlated Z_t work better for estimating relationships between different processes. In statistical learning, time series are not only used for forecasting their own future values but also as regressors for other data, e.g., for the output Y_t of a system like

$$Y_t = m(X_t, Z_{t-1}, \dots, Z_{t-p}) + \varepsilon_t,$$

where the X_t are independent identically distributed feature vectors and the last p values of the time series act as additional feature components. In such a situation, due to the same reasoning as for using principal component analysis (compare Sect. 12.2), it is helpful to apply *prewhitening* to $Z_1, \ldots, Z_T$ during the preprocessing phase, which results in transformed time series data $Z_1^{(w)}, \ldots, Z_T^{(w)}$, which are less correlated among themselves. There are various techniques for finding a suitable transformation. We could, e.g., pretend that Z_t is part of a linear AR(p)-process, estimate the autoregressive parameters, and set

$$Z_t^{(w)} = Z_t - \sum_{k=1}^{p} \hat{a}_k Z_{t-k}, \quad t = p+1, \ldots, T.$$

If Z_t is really an AR(p)-process, the prewhitened data $Z_t^{(w)} \approx \varepsilon_t$ will almost be white noise for large enough T, but even if the autoregressive prewhitening filter is misspecified, it will decrease the dependency between the time series data. Brockwell and Davis [11] (Section 10.1) discuss prewhitening based on ARMA-models, but other procedure being part of the statistical software of choice could be used.

14.5 Changepoint Detection

In practice, it may be of interest, e.g., for the purpose of predictive maintenance, if an observed time series changes from a certain point, say τ, onward. This may be a sudden change, e.g., caused by the failure of a component in production, or a gradual change starting at τ, e.g., due to wear and tear. Often, such events are not observed directly, but we can infer them only indirectly, e.g., from sensor measurements, from energy consumption, or from the quantity or quality of the output. In advanced settings, one is not only interested in deciding from (usually multivariate) time series observations if and when a change has happened, but one also likes to learn which type of defect has happened in which component. Here, we only consider the problem of detecting so-called *changepoints* in a univariate time series. For illustrating how changepoint detection works, we mainly focus on the *offline* problem of deciding retrospectively if a change has happened in an already observed sample $Z_1, \ldots, Z_T$. For applications in production or process engineering, *sequential* or *online* procedures, which are designed to detect the occurrence of a change as soon as possible from a consecutive stream of data $Z_1, Z_2, \ldots$, are of even greater interest. They use the same kind of methodology but are slightly more complicated, such that we refer to the literature (compare [4, 13] for extensive reviews).

Common changepoints tests are intuitively easy to understand, though in the background some advanced stochastic process theory is needed for calculating the critical bounds. We restrict our discussion to so-called CUSUM (*cumulative sums*) procedures and start with a simple setting where the observed time series is just strict white noise, i.e., the Z_t are independent and identically distributed with finite variance σ^2. We consider a possible change of the mean, i.e., given the data $Z_1, \ldots, Z_T$, we want to know if the mean μ is stable over the whole observation period or if there is a changepoint $1 \leq \tau < T$ such that $\mathrm{E}\, Z_t = \mu, t \leq \tau$ and $\mathrm{E}\, Z_t = \mu' \neq \mu, t > \tau$.

We formulate the changepoint detection as a testing problem where we have to decide if the hypothesis "no change" holds or if, alternatively, a change in mean has happened at some time point $\tau < T$. As usual in statistical testing, we calculate a test statistic, and we reject the hypothesis if this value lies above a critical bound, depending on our chosen level of confidence. In our changepoint problem, we consider all $s < T$ and compare the sample means of the data before and after s:

$$D_s = \left| \frac{1}{s} \sum_{t=1}^{s} Z_t - \frac{1}{T-s} \sum_{t=s+1}^{T} Z_t \right|.$$

By the law of large number, both sample means should be roughly equal ($\approx \mu$) and, hence, D_s should be small if the mean is not changing. If, on the other hand, there is a change at $s = \tau$, all data of the first mean have mean μ, and all data of the second have mean μ' such that the difference will be particularly large if there is a change. This intuitive argument motivates the weighted CUSUM test statistic

$$M_T = \max_{1 \leq s < T} \sqrt{\frac{s(T-s)}{T}}\, D_s = \max_{1 \leq s < T} \sqrt{\frac{T}{s(T-s)}} \left| \sum_{t=1}^{s} (Z_t - \overline{Z}_T) \right| = \max_{1 \leq s < T} C_s,$$

where $\overline{Z}_T = \frac{1}{T} \sum_{t=1}^{T} Z_t$ denotes the mean of all the data. The hypothesis is rejected in favor of the presence of a changepoint if $M_T > c_{\alpha,T}$. The scaling of the difference between the two sample means is chosen such that, based on theory (compare [15]), it is possible to calculate the critical bound $c_{\alpha,T}$ such that in case of rejection of the "no change" hypothesis, this decision is correct up to the small error probability $\approx \alpha$ for large enough T:

$$c_{\alpha,T} = \frac{\hat{\sigma}_T}{c_{1,T}} \left(-\log \frac{-\log(1-\alpha)}{2} + c_{2,T} \right), \tag{92}$$

where $c_{1,T}, c_{2,T}$ are universal constants depending only on T and $\hat{\sigma}_T$ is any reasonable estimate of the variance, e.g., the sample variance, which converges to σ^2 for increasing T under the assumption of the hypothesis.

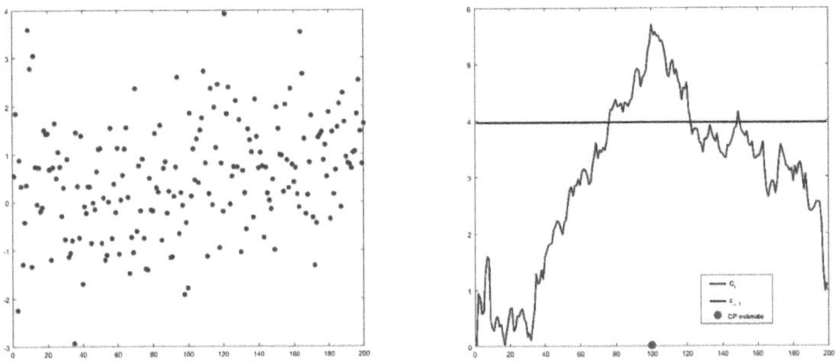

Fig. 51 Data (left) and rescaled CUSUM statistics including changepoint (CP) estimate

If the test detects a changepoint, we immediately have also an estimate $\hat{\tau}$ for the location of the changepoint τ as the time for which the rescaled sample mean difference D_s becomes maximal:

$$\sqrt{\frac{\hat{\tau}(T-\hat{\tau})}{T}} D_{\hat{\tau}} = \max_{1 \le s < T} \sqrt{\frac{s(T-s)}{T}} D_s .$$

The test can be visualized by plotting the rescaled cumulative sum statistics C_s, $1 \le s < T$, and the critical bound $c_{\alpha,T}$. If the former exceeds the bound, a changepoint is detected, and its estimate is the location of the maximum of the C_s-plot. Figure 51 shows a sample of size $T = 200$ of independent Gaussian random variables with variance 1. At $\tau = 100$, the mean changes from 0 to 1, which is not easy to detect visually. On the right-hand side, the rescaled CUSUM statistics C_s are plotted. For $\alpha = 0.05$, they exceed the critical bound $c_{\alpha,T}$ shown as a black horizontal line such that the changepoint is detected. By chance, the changepoint estimate $\hat{\tau}$ coincides exactly with the true instant $\tau = 100$ of the change.

Note that the plot C_s, $1 \le s < T$, also may provide some clues (not present in the example) if there is more than one changepoint. In that case, there will be pronounced local maxima of the plot at different locations. A quick-and-dirty approach to handle such a situation would be to split the sample at the detected changepoint and to repeat the test now for both subsamples.

CUSUM tests can be generalized to time series provided that they coincide with stationary, short memory processes before and after the changepoint τ. We consider as an example a nonlinear but parametric autoregressive process of order 1 where

$$Z_t = g(Z_{t-1}; b) + \varepsilon_t ,$$

where g is a function known up to a q-dimensional parameter vector b and ε_t is white noise with variance σ_ε^2 as usual. This includes, e.g., the linear AR(1) model (84) where $b = (\mu, a_1)$. We want to check if there is a changepoint $2 \le \tau < T$, such that the model parameter is b up to time t and is $b' \ne b$ for $t > \tau$.

In spite of the dependence of the time series data Z_t, we can use the same idea of looking at CUSUMs, but no longer of the data but of the sample residuals $\hat{\varepsilon}_t$. To get those approximations of the unobservable innovations ε_t, we first calculate an estimate $\hat{b}$ of the parameter by least squares, pretending that there is no change, i.e., we minimize the empirical loss

$$Q_T(b) = \frac{1}{T}\sum_{t=2}^{T}\left(Z_t - g(Z_{t-1}; b)\right)^2$$

w.r.t. b, and then, we have $\hat{\varepsilon}_t = Z_t - g(Z_{t-1}; \hat{b}), t = 2, \ldots, T$. Note that by construction, these sample residuals have average 0, such that the scaled CUSUM statistics here is

$$M_T = \max_{2 \le s < T-1} \sqrt{\frac{T-1}{s(T-1-s)}}\left|\sum_{t=2}^{s}\hat{\varepsilon}_t\right|.$$

The asymptotic (for $T \to \infty$) distribution of this test statistic is the same as in the independent case such that we reject the hypothesis and detect a changepoint if $M_T > c_{\alpha,T}$, where the critical value is given by (92) with an estimate $\hat{\sigma_\varepsilon}$ of σ_ε replacing $\hat{\sigma}_T$. In that case, we again get an estimate $\hat{\tau}$ of the location of the change as that value s for which the maximum in the definition of M_T is achieved. For the estimate of the residual variance, we could use the usual sample variance satisfying $\hat{\sigma_\varepsilon}^2 = Q_T(\hat{b})$. Kirch et al. [51] discuss some alternatives that in particular are meant to increase the power of the test, i.e., the probability to detect a change if there is one.

The procedure is easily generalized to nonlinear autoregressions of arbitrary order $p \ge 1$, where $g(.; b)$ depends on $Z_{t-1}, \ldots, Z_{t-p}$. Also, we may use the same kind of changepoint test for general autoregressive models (11). Here, we are interested if the forecasting function m is stable over the whole observation period $1 \le t \le T$ or if it changes to $m' \ne m$ somewhere in between at $p + 1 \le \tau < T$. There is one important assumption to be aware of which, however, is usually fulfilled in practice.

For data generated by (11), we have first to estimate m by $\widehat{m}$, using one of the statistical learning methods described in the previous chapters, to get hold of the sample residuals $\hat{\varepsilon}_t = Z_t - \widehat{m}(Z_{t-1}, \ldots, Z_{t-p})$ as ingredients of our CUSUM statistic M_T. For example, if we use neural network output functions as in Sect. 14.2 for estimating m, then $\widehat{m}(.) = f_H(.; \hat{\vartheta})$. Note that we usually are in a misspecified situation, i.e., $m(.) \ne f_H(.; \vartheta_{\text{opt}})$, where ϑ_{opt} is the parameter value for which we get the best approximation of m by a neural network output function (compare Sect. 8.4). The CUSUM test nevertheless works as shown by Kirch and Tadjuidje Kamgaing [51] under the crucial assumption that the changes of interest in the autoregression function also impact the best neural network approximations. The change from $m(.)$ to $m'(.)$ at τ must imply $f_H(.; \vartheta_{\text{opt}}) \ne f_H(.; \vartheta'_{\text{opt}})$ and, hence,

$\vartheta_{\text{opt}} \neq \vartheta'_{\text{opt}}$, where the latter is the parameter that results in the best neural network approximation of m'.

As an example, we follow [51] and consider two so-called threshold autoregressive process of order 1 with autoregressive functions

$$m(x) = (0.1{+}0.2\,\text{sgn}(x))\,x \;\text{ resp. }\; m'(x) = (0.1{+}0.4\,\text{sgn}(x))\,x{+}0.25(1{+}\text{sgn}(x)),$$

where $\text{sgn}(x)$ denotes the sign of x. The sample size is $T = 500$, and the changepoint is located at $\tau = 250$, i.e., up to $t = \tau$, the data are generated by $Z_t = m(Z_{t-1}) + \varepsilon_t$ and afterward by $Z_t = m'(Z_{t-1}) + \varepsilon_t$, where the innovations ε_t are independent identically distributed standard normal random variables.

Figure 52 shows the data on the left-hand side. We fit a feedforward neural network with one hidden layer of $H = 3$ neurons to the data and applied the changepoint test with level 5%. On the right-hand side, the rescaled CUSUM statistics C_s are plotted. A changepoint is detected at $\hat{\tau} = 251$, which is almost correct. The black horizontal line marks the critical bound $c_{\alpha,T}$ for $\alpha = 0.05$. Here, we estimated the innovation variance σ_ε^2 by a combination of the sample innovation variances before and after the estimated changepoint $\hat{\tau}$ (compare Section 3 of [51]).

To illustrate that the example satisfies the crucial assumption for changepoint detection algorithms, which are using misspecified models for learning, Fig. 53 shows the two autoregressive functions $m(x) \neq m'(x)$ before and after the changepoint on the left-hand side. On the right-hand side, the optimal neural network approximations $f_3(.;\vartheta_{\text{opt}})$ resp. $f_3(.;\vartheta'_{\text{opt}})$ are plotted. As the latter are also different, we may hope to detect a change even in the case of misspecification.

CUSUM tests are likely applicable together with other sieve estimates instead of neural networks as well. The applied literature contains some examples (e.g., [71], [83]), but to our knowledge, there has not been a detailed theoretical investigation if this practice is justified. However, the arguments used by Kirch and Tadjuidje Kamgaing [51] for neural networks are of a kind that presumably apply also to regression trees, etc.

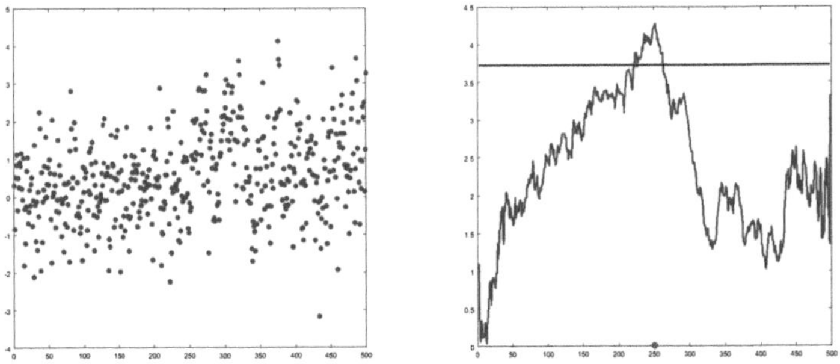

Fig. 52 Nonlinear autoregressive data of order 1 (left) and rescaled CUSUM statistics based on a neural network fit including changepoint (CP) estimate (right)

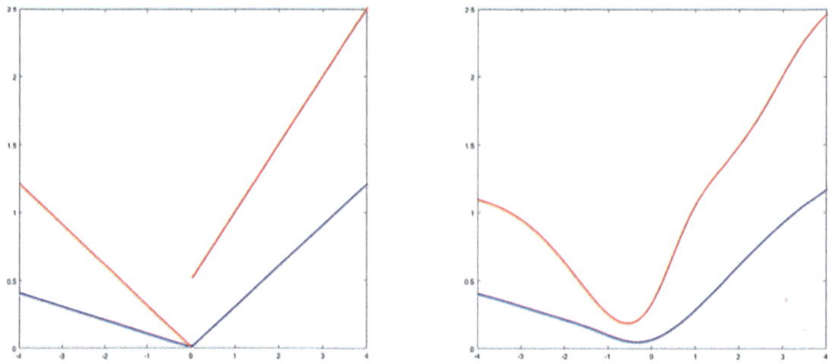

Fig. 53 On the left-hand side, threshold autoregressive functions $m(x)$ (blue) and $m'(x)$ data (red); on the right-hand side, corresponding optimal approximations by a neural network with $H = 3$ neurons

Let us conclude with two extensions that are practically relevant. The CUSUM test, used for Fig. 51, has been constructed for detecting a sudden instantaneous change, here the mean of independent identically distributed data from μ to μ'. There are also similar tests being tuned for **gradual changes**, starting at some τ (compare [84]). For example, for a mean change, we would have $\mathrm{E}\, X_t = \mu, t \leq \tau$, and $\mathrm{E}\, X_t = \mu + (t - \tau)\delta, t > \tau$, for some small $\delta \neq 0$. Due to, e.g., wear and tear, a system would slowly move out of the stable state from τ onward, and we want to detect that from the data as soon as possible.

An even more important situation is the problem of data observed sequentially or **online**. Usually, we start from some historical data $Z_1, \ldots, Z_r$ for which we are certain that the data-generating process is stable. Afterward, we observe one observation Z_t after the other, $t > r$, and we repeatedly apply a CUSUM like test until a change is detected. The details of the procedure for neural network-based estimates of an autoregressive function are given by Kirch and Tadjuidje Kamgaing [50].

References

1. Akaike, H. (1974). A new look at the statistical model identification. *IEEE Transactions on Automatic Control, 19*, 716–723.
2. Akaike, H. (1977). On entropy maximization principle. In P. R. Krishnaiah (Ed.), *Applications of statistics* (pp. 27–41). North Holland.
3. Alom, M. Z., Taha, T. M., Yakopcic, C., Westberg, S., Sidike, P., Nasrin, M. S., van Essen, B. C., Awwal, A. A. S., & Asari, V. K. (2018). The history began from AlexNet: A comprehensive survey on deep learning approaches. arXiv:1803.01164. https://arxiv.org/pdf/1803.01164.pdf. Cited 15 June 2023
4. Aminikhanghahi, S., & Cook, D. J. (2017). A survey of methods for time series change point detection. *Knowledge and Information Systems, 51*, 339–367.
5. Bishop, C. M. (2006). *Pattern recognition and machine learning*. Springer.

6. Box, G. E., & Cox, D. R. (1964). An analysis of transformations. *Journal of Royal Statistical Society, Series B, 26*, 211–243.
7. Box, G. E. P., & Jenkins, G. M. (1976). *Time series analysis, forecasting and control*. Holden-Day.
8. Breiman, L. (1996). Bagging predictors. *Machine Learning, 24*, 123–140.
9. Breiman, L., Friedman, J., Olshen, R. A., & Stone, C. J. (1984). *Classification and regression trees*. Chapman & Hall.
10. Brockwell, P. J., & Davis, R. A. (1987). *Time series: Theory and methods* (2nd ed.). Springer.
11. Brockwell, P. J., & Davis, R. A. (1996). *An introduction to time series and forecasting*. Springer.
12. Charnes, A., Frome, E. L., & Yu, P. L. (1976). The equivalence of generalized least squares and maximum likelihood estimates in the exponential family. *Journal of the American Statistical Association, 71*, 169–171.
13. Cho, H., & Kirch, C. (2021). Data segmentation algorithms: Univariate mean change and beyond. *Econometrics and Statistics*. https://doi.org/10.1016/j.ecosta.2021.10.008
14. Cortes, C., & Vapnik, V. (1995). Support-vector networks. *Machine Learning, 20*, 273–297.
15. Csörgö, M., Csörgö, M., & Horváth, L. (1997). *Limit theorems in changepoint analysis*. Wiley.
16. Dahlhaus, R. (2012). Local stationary processes. In T. Subba Rao, S. Subba Rao, & C. R. Rao (Eds.), *Handbook of Statistics* (Vol. 30, pp. 351–413). Elsevier.
17. Daubechies, I. (1992). *Ten Lectures on Wavelets*. SIAM.
18. DiCiccio, T. J., & Efron, B. (1996). Bootstrap confidence intervals. *Statistical Science, 11*, 189–228.
19. Donoho, D. L., & Johnstone, I. M. (1994). Ideal spatial adaptation by wavelet shrinkage. *Biometrika, 81*, 425–455.
20. Embrechts, P., Klüppelberg, C., & Mikosch, T. (1997). *Modelling extremal events for insurance and finance*. Springer.
21. Engle, R. F. (1984). Wald, likelihood ratio, and Lagrange multiplier tests in econometrics. In Z. Griliches, & M. D. Intriligator (Eds.), *Handbook of econometrics* (Vol. II, pp. 775–826). North-Holland.
22. Fokianos, K. (2012). Count time series models. In T. Subba Rao, S. Subba Rao, C. R. Rao (Eds.), *Handbook of statistics* (Vol. 30, pp. 315–347). Elsevier.
23. France, S. L., & Carroll, J. D. (2010). Two-way multidimensional scaling: A review. *IEEE Transactions on Systems, Man, and Cybernetics, Part C (Applications and Reviews), 41*, 644–661.
24. Franke, J. (1985). A Levinson-Durbin recursion for autoregressive-moving average processes. *Biometrika, 72*, 573–581.
25. Franke, J. (2012). Markov-switching time series models. In T. Subba Rao, S. Subba Rao, & C. R. Rao (Eds.), *Handbook of statistics* (Vol. 30, pp. 99–122). Elsevier.
26. Franke, J., & Diagne, M. (2006). Estimating market risk with neural networks. *Statistics & Decisions, 24*, 233–253.
27. Franke, J., Kreiss, J. P., & Mammen, E. (2002). Bootstrap of kernel smoothing in nonlinear time series. *Bernoulli, 8*, 1–37.
28. Franke, J., & Neumann, M. (2000). Bootstrapping neural networks. *Neural Computation, 12*, 1929–1949.
29. Freedman, D. A. (1981). Bootstrapping regression models. *Annals of Statistics, 9*, 1218–1228.
30. Freund, Y., & Schapire, R. E. (1997). A decision-theoretic generalization of on-line learning and an application to boosting. *Journal of Computer and System Sciences, 55*, 119–139.
31. Friedman, J. H., & Stuetzle, W. (1981). Projection pursuit regression. *Journal of the American Statistical Association, 76*, 817–823.
32. Furnival, G. M., & Wilson, R. W. (1974). Regressions by leaps and bounds. *Technometrics, 16*, 499–511.
33. Golub, G. H., Heath, M., & Wahba, G. (1979). Generalized cross-validation as a method for choosing a good ridge parameter. *Technometrics, 21*, 215–223.
34. Goodfellow, I., Bengio, Y., & Courville, A. (2016). *Deep learning*. MIT Press.

35. Grenander, U. (1981). *Abstract inference*. Academic Press.
36. Györfy, L., Kohler, M., Krzyzak, A., & Walk, H. (2002). A distribution-free theory of nonparametric regression. Springer.
37. Härdle, W. (1990). *Applied nonparametric regression*. Cambridge University Press.
38. Hald, A. (1960). *Statistical theory with engineering applications*. Wiley.
39. Hall, P. (1992). *The bootstrap and Edgeworth expansion*. Springer.
40. Hampel, F. R. (2001). Robust statistics: A brief introduction and overview. In *Research Report/Seminar für Statistik* (Vol. 94). Eidgenössische Technische Hochschule (ETH).
41. Hannan, E. J. (1980). The estimation of the order of an ARMA process. *Annals of Statistics, 8*, 1071–1081.
42. Hannan, E. J., & Deistler, M. (1988). *The statistical theory of linear systems*. Wiley.
43. Hart, J. D. (1991). Kernel regression estimation with time series errors. *Journal of Royal Statistical Society, Series B, 53*, 173–178.
44. Hastie, T., Tibshirani, R., & Friedman, J. (2001). *The elements of statistical learning: Data mining, inference and prediction*. Springer.
45. Heidenreich, N. B., Schindler, A., & Sperlich, S. (2013). Testing for parameter stability in nonlinear autoregressive models. *AStA Advances in Statistical Analysis, 97*, 403–433.
46. Hornik, K., Stinchcombe, M., & White, H. (1989). Multilayer feedforward networks are universal approximators. *Neural Networks, 2*, 359–366.
47. James, G., Witten, D., Hastie, T., & Tibshirani, R. (2021). *An introduction to statistical learning: With applications in R*. Springer.
48. James, G., Witten, D., Hastie, T., Tibshirani, R., & Taylor, J. (2023). *An introduction to statistical learning: With applications in Python*. Springer
49. Jeffers, J. N. (1967). Two case studies in the application of principal component analysis. *Journal of Royal Statistical Society, Series C, 16*, 225–236.
50. Kirch, C., & Tadjuidje Kamgaing, J. (2011). An online approach to detecting changes in nonlinear autoregressive models. Preprint (2011). https://kluedo.ub.rptu.de/frontdoor/deliver/index/docId/2772/file/pp_online.pdf. Cited 25 Mai 2023
51. Kirch, C., & Tadjuidje Kamgaing, J. (2012). Testing for parameter stability in nonlinear autoregressive models. *Journal of Time Series Analysis, 33*, 365–385.
52. Kreiss, J. P., & Paparoditis, E. (2011). Bootstrap methods for dependent data: A review. *Journal of the Korean Statistical Society, 40*, 357–378.
53. Krizhevsky, A., Sutskever, I., & Hinton, G. E. (2012). ImageNet classification with deep convolutional neural networks. *Algorithms, 14*(12), 345.
54. Kurnatowski, M. v., Schmid, J., Link, P., Zache, R., Morand, L., Kraft, T., Schmidt, I., Schwientek, J., & Stoll, S. (2021). Compensating data shortages in manufacturing with monotonicity knowledge. *Algorithms, 14*. https://doi.org/10.3390/a14120345
55. LeCun, Y., Bottou, L., Bengio, Y., & Haffner, P. (1998). Gradient-based learning applied to document recognition. *Proceedings of the IEEE, 86*, 2278–2324.
56. Lin, L., & Dunson, D. B. (2014). Bayesian monotone regression using Gaussian process projection. *Biometrika, 101*, 303–317.
57. Link, P., Poursanidis, M., Schmid, J., Zache, R., von Kurnatowski, M., Teicher, U., & Ihlenfeldt, S. (2022). Capturing and incorporating expert knowledge into machine learning models for quality prediction in manufacturing. *Journal of the Intelligent Manufacturing, 33*, 2129–2142.
58. Lütkepohl, H. (2005). *New introduction to multiple time series analysis*. Springer.
59. Morgan, K. L., & Rubin, D. B. (2012). Rerandomization to improve covariate balance in experiments. *Annals of Statistics, 40*, 1263–1282.
60. Morgan, K. L., & Rubin, D. B. (2015). Rerandomization to balance tiers of covariates. *Journal of the American Statistical Association, 110*, 1412–1421.
61. Murata, N., Yoshizawa, S., & Amari, S. I. (1994). Network information criterion-determining the number of hidden units for an artificial neural network model. *IEEE Transactions on Neural Networks, 5*, 865–872.

62. Parzen, E., Tanabe, K., & Kitagawa, G. (Eds.). (2012). *Selected papers of Hirotugu Akaike*. Springer.
63. Patil, P., Wei, Y., Rinaldo, A., & Tibshirani, R. (2021). Uniform consistency of cross-validation estimators for high-dimensional ridge regression. In *International Conference on Artificial Intelligence and Statistics* (pp. 3178–3186). PMLR.
64. Priestley, M. B. (1981). *Spectral analysis and time series* (Vol. 2). Academic Press.
65. Rasmussen, C. E., & Williams, C. K. I. (2006). *Gaussian processes for machine learning*. MIT Press.
66. Richter, S., & Dahlhaus, R. (2019). Cross validation for locally stationary processes. *Annals of Statistics, 47*, 2145–2173.
67. Ripley, B. D. (1996). *Pattern recognition and neural networks*. Cambridge University Press.
68. Rosenblatt, F. (1958). The perceptron: A probabilistic model for information storage and organization in the brain. *Psychological Review, 65*, 386.
69. Rudin, C., Chen, C., Chen, Z., Huang, H., Semenova, L., & Zhong, C. (2022). Interpretable machine learning: Fundamental principles and 10 grand challenges. *Statistic Surveys, 16*, 1–85.
70. Schorning, K., Dette, H., Kettelhake, K., Wong, W. K., & Bretz, F. (2017). Optimal designs for active controlled dose-finding trials with efficacy-toxicity outcome. *Biometrika, 104*, 1003–1010.
71. Severo, M., & Gama, J. (2010). Change detection with kalman filter and cusum. In *Ubiquitous knowledge discovery: Challenges, techniques, applications* (pp. 148–162). Springer.
72. Shao, J. (1997). An asymptotic theory for linear model selection. *Statistica Sinica, 7*, 221–242.
73. Shibata, R. (1980). Asymptotically efficient selection of the order of the model for estimating parameters of a linear process. *Annals of Statistics, 8*, 147–164.
74. Sigillito, V., Wing, S., Hutton, L., & Baker, K.: Ionosphere. In *UCI machine learning repository* (1989). https://doi.org/10.24432/C5W01B
75. Srivastava, N., Hinton, G., Krizhevsky, A., Sutskever, I., & Salakhutdinov, R. (2014). Dropout: A simple way to prevent neural networks from overfitting. *Journal of Machine Learning Research, 15*, 1929–1958.
76. Stone, C. J. (1982). Optimal global rates of convergence for nonparametric regression. *Annals of Statistics, 10*, 1040–1053.
77. Stone, M. (1977). An asymptotic equivalence of choice of model by cross-validation and Akaike's criterion. *Journal of Royal Statistical Society, Series B, 39*, 44–47.
78. Vanhatalo, J., Riihimäki, J., Kartikainen, J., Jylänki, P., Tolvanen, V., & Vehtari, A. (2013). GPstuff: Bayesian modeling with Gaussian processes. *Journal of Machine Learning Research, 14*, 1175–1179.
79. Vapnik, V. N. (1995). *The nature of statistical learning theory*. Springer.
80. Vapnik, V. N. (1998). *Statistical learning theory*. Wiley.
81. White, H. (1989). Some asymptotic results for learning in single hidden-layer feedforward network models. *Journal of the American Statistical Association, 84*, 1003–1013.
82. Woods, H., Steinour, H. H., & Starke, H. R. (1932). Effect of composition of Portland cement on heat evolved during hardening. *Industrial Engineering and Chemistry, 24*, 1207–1214.
83. Xu, Q., Lu, S., Zhai, Z., & Jiang, C. (2020). Adaptive fault detection in wind turbine via RF and CUSUM. *IET Renewable Power Generation, 14*, 1789–1796.
84. Yi, F., & Qiu, P. (2022). An adaptive CUSUM chart for drift detection. *Quality and Reliability Engineering International, 38*, 887–894.
85. Yu, B. (2013). Stability. *Bernoulli, 19*, 1484–1500.
86. Zhang, H., Yin, G., & Rubin, D. B. (2021). PCA Rerandomization. arXiv:2102.12262. https://arxiv.org/pdf/2102.12262.pdf. Cited 30 Mai 2023.
87. Zhou, Q., Ernst, P. A., Morgan, K. L., Rubin, D. B., & Zhang, A. (2018). Sequential rerandomization. *Biometrika, 105*, 745–752.
88. Zobel, N., Herzog, N., Lehwald, A., & Kolomiichuk, S. (2017). Predictive maintenance in process manufacturing. In *European Conference on Process Safety and Big Data*. https://doi.org/10.24406/publica-fhg-402024

Machine Learning for Inline Surface Inspection Systems: Challenges, Approaches, and Application Example

Henrike Stephani, Thomas Weibel, Ronald Rösch, and Ali Moghiseh

Abstract Using deep learning in complex online surface inspection systems is challenging due to different framework conditions. First, time restrictions in production are usually fixed in terms of clock rate and response time. Furthermore, these methods need a lot of data, while typically, the data situation is thin in the beginning as well as continuously unbalanced: defects occur rarely, thereby providing few example data for learning, while the desired detection rate is 100%. Another important issue is that although defect catalogs exist, they often change, especially when automatic inspection is applied for the first time. This is due to imaging systems usually being able to detect more defects than visual-manual inspection, therefore production, management, and quality assurance usually reiterate their prior defect catalogs. However, data-driven methods depend heavily on consistent annotation. Therefore, respective parties must be made aware of this issue on the one hand; on the other hand, annotation and reannotation must be easy and useable by non-experts. Related is the issue of parametrization and traceability. Both are not inherent to neural networks but must be provided to some level to help build trust in machine learning methods. In this chapter, we present a quality inspection system that uses deep neural networks for defect detection under real production conditions in wood manufacturing. We will address how we systematically deal with the above issues both in terms of process and algorithm.

1 Introduction

In many industrial production sites, manual inspection is still the method of choice when it comes to surface control. Among the reasons for this are the flexibility of manual laborers and their ability to make decisions under unexpected circumstances.

H. Stephani (✉) · T. Weibel · R. Rösch · A. Moghiseh
Fraunhofer ITWM, Kaiserslautern, Germany
e-mail: henrike.stephani@itwm.fraunhofer.de; thomas.weibel@itwm.fraunhofer.de; ronald.roesch@itwm.fraunhofer.de; ali.moghiseh@itwm.fraunhofer.de

J. Franke, A. Schöbel (eds.), *Statistical Machine Learning for Engineering with Applications*, Lecture Notes in Statistics 227,
https://doi.org/10.1007/978-3-031-66253-9_2

However, there are several reasons why computer vision (CV)-based inspection systems are constantly on the rise [11]. The main reason is that computers mostly outperform humans in repetitive calculations, in speed as well as consistency. Furthermore, long-running systems are more cost-efficient than manual inspection [9].

Up until today, CV-based inline inspection algorithms are often done with so-called "classical" CV algorithms. That means experts build algorithm chains. These chains consist of filtering algorithms, statistical methods, adaptive thresholding, modeled feature extraction, and usually some classification step in the end [8, 13]. This approach has various advantages:

- With a few example images, first approaches can be developed. This is especially relevant as most inline inspection systems start with hardware ordering and integration phases that can take several months. Customers (i.e., production) expect prototypical solutions quickly after hardware integration.
- Lean solutions are possible. Simple (sub-)tasks can be solved with very efficient algorithms. With high production speed, this can be very relevant.
- Single steps of the solution are understandable, explainable, and parametrizable. Especially with new defects and product variants, the parameters can be adapted, and the algorithm does not have to be redesigned.

However, there are also several disadvantages: There is a plethora of parameters to be manually set. Design and maintenance of these algorithm chains is an expert task. The image quality should be high, i.e., defects should have high contrast compared to the rest of image. But most importantly, their performance is sometimes poor and limited. Some problems, such as the inspection of natural materials, are almost impossible.

In the last 15 years, these "classical" algorithms have been replaced by data-driven deep learning approaches [8]. Several problems, such as the detection of cars in images or the identification of faces, are exclusively solved with these methods and often outperform human recognition [5, 10]. General advantages are:

- Data-driven solutions are usually generic and depend mostly on good data.
- There are many openly available architectures and datasets that can be adapted to the respective application domain.
- They find defects also in poor-quality images.
- They outperform classical image processing/machine learning methods if sufficiently many data are available.

Because of these advantages, various deep learning CV system have been developed for defect detection. In some application areas, good results have been shown. However, the applicability of these methods highly depends on the precise definition of the application domain (e.g., steel [7, 15]) and the defect types and example datasets (e.g., cracks [1, 4, 6]. Providing this expert knowledge as well as the necessary database presents several difficulties:

- The data is very unbalanced. Defects occur rarely (in comparison to non-defective parts) and vary in importance.
- Data collection must be done in accordance with the production schedule—as not every variety is produced at the same time. Production is usually not changed for the purpose of data collection
- Data annotation is necessary and needs precise and consistent definition of defect classes

In this chapter, we will present a process on how to implement a data-driven solution for image-based surface inspection under industrial conditions. This process will especially focus on the issue of data collection and annotation. We will discuss challenges and present an application example.

2 Methods

In this section, we want to show how the process of developing a surface inspection system must change depending on the used methodology. We will first explain the typical process with classical image processing and then explain why deep learning-based methods require a different access to the problem. Finally, we will explain our proposed hybrid approach.

2.1 Classical Image Processing for Inline Surface Inspection

A classical image processing pipeline usually consists of filtering the image, adaptive thresholding to identify candidates for defects, and finally classification of these defects [2]. A typical process of developing image processing solutions for inline surface inspection involves expert knowledge from domain experts but only at specific points. In Fig. 1, the process steps displayed in blue represent their input. Steps in yellow represent CV expert work. Initially, defect catalogs are shared, and production/manual inspection expert interviews are conducted. However, after

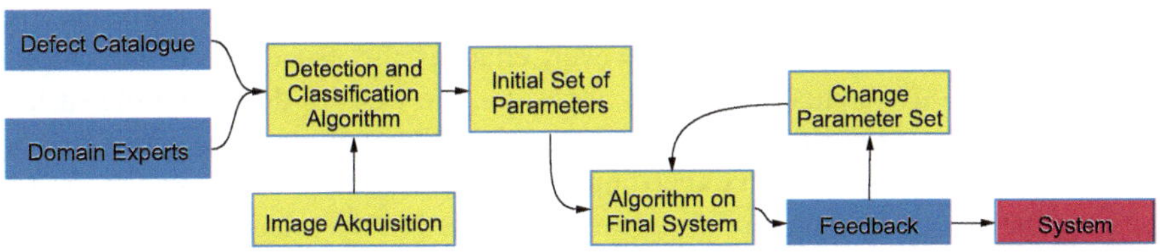

Fig. 1 Typical development steps of a defect detection algorithm with classical image processing. Domain experts, blue; image processing experts, yellow. Copyright: Fraunhofer ITWM (self generated)

this, even with only very limited image material, a first algorithm chain can be developed. Image processing experts combine parameterizable algorithms. A first iteration is then run on the system; at this point again, production experts check the outcome of the automatic analysis. Typically, too many defects are found on the one hand (false positives), and some might also be missed (false negatives). This leads to a reparameterization of the developed algorithms. If the system is well designed, unknown variants of the product/new kinds of defects can be included into the system by adapting only a few parameters. However, challenges of this expert modeled image processing are that the advantage of being able to parameterize each step also is a disadvantage: parameterization has to be done manually and usually with image processing expert knowledge.

2.2 Deep Learning for Inline Surface Inspection

We will focus on deep learning in this work as this is the dominant form of data-driven approaches. Within deep learning approaches, convolutional neural networks (CNN) are the most relevant architectures. Images are not a random accumulation of pixel values but rather contain spatial information. Features should therefore represent this spatial connection. CNNs are designed to encode this information on several resolution layers. They consist first of an image encoding through so-called convolutional layers, in which image information is encoded, and max pooling layers, where most relevant features are selected, and then, depending on the desired output, for example, of a classification step done by a so-called fully connected layer, where the network decides how to sort the images into classes [14], or of a decoder step, where not only classification but also segmentation can be performed [3].

In both cases, thousands to millions of weights must be learned. Not only efficient algorithms and computational power are needed for this purpose but also a lot of data. Data acquisition, however, is only a small part of the task. For the networks to learn their weights properly, they must know "how" to learn them. In case of classification on an image level, they need to be annotated. In case of detection, annotation is not performed on an image level, but the various classes contained in an image must also be localized.

The most important task for data-driven methods is to build a representative and large-enough database. This, however, changes the procedure of developing surface inspection systems. In a classical system, the work of designing the surface inspection system is mostly done by image processing experts who "translate" the defect catalog and the domain expert input into algorithms—see the yellow part of the pipeline in Fig. 1. This modeling can be avoided in deep learning solutions, as it is done by the data itself. Data must therefore be reliable and well categorized. Therefore, domain experts must be involved to a higher degree in building the database as illustrated by the green parts in Fig. 2. This can be very time-consuming and difficult. Another issue is that in the traditional approach, interpretability and parametrization are inherent to the solution. In contrast to that, in data-driven

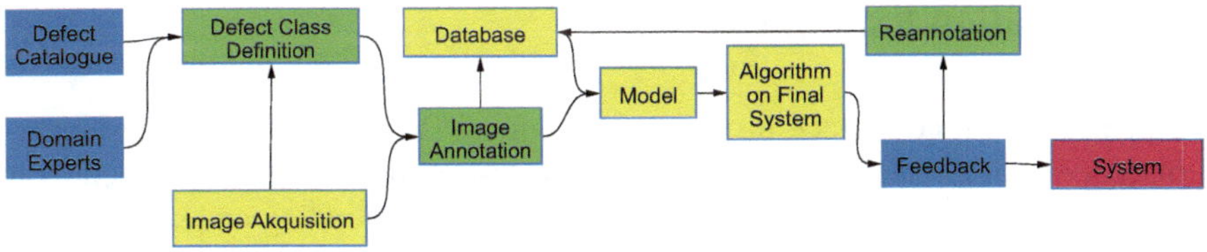

Fig. 2 Process of building a defect detection based on deep learning algorithms. Domain experts, blue; image processing experts, yellow; both, green. Copyright: Fraunhofer ITWM (self generated)

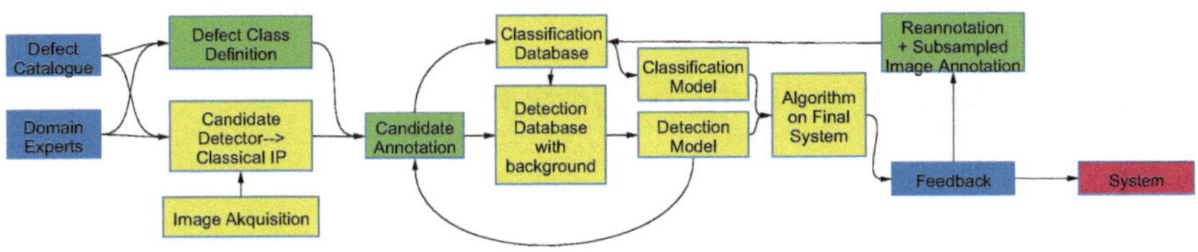

Fig. 3 Process of building a defect detection based on deep learning algorithms. Domain experts, blue; image processing experts, yellow; both, green. Copyright: Fraunhofer ITWM (self generated)

methods, most of this is "hidden" in the annotation. This annotation must be consistent over the whole dataset. Future changes of requirements and variants that might influence current annotation must therefore be anticipated.

2.3 *Hybrid Setup for Inline Surface Inspection*

The most problematic part of deep learning systems is the collection and annotation of data. Therefore, instead of exclusively developing a data-driven system, we propose to use a hybrid solution as displayed in Fig. 3. A candidate detector is developed by means of classical image processing. A focus should be in detecting rather too many defects than too little and on building a first solution. The candidate annotation is then usually much easier and faster than the original image annotation. The discussion of how to annotate defects is similar in the full deep learning solution. However, first, defect candidates can be used as a basis for discussion. This is significant mainly for two reasons: First, to get an impression of the expression of defects in the images—as they are usually different from the human-visual impression. Second, the number of detected defects can be estimated—often automatic inspection systems detect more defects than is desirable for production management. The database that is build can either be used for classification or can be supplemented with "background" images for additional localization as described above. Even if some "raw image annotation" has to be done too (marked here as "subsampled image annotation"), the bulk of the necessary data is done by candidate annotation.

3 Application Example

We present an inline-surface inspection system for shelve panels. The system will replace a visual manual inspection. There are several fixed parameters such as the production pace (about 50 m/min) and response time (depending on length of panel (7–12 s). Requirements to the system are that it meets the production times and be as good as the manual inspection (or better). The reasons for exchanging the current inspection are the higher reliability and consistency of an automatic system.

3.1 Hybrid Setup for Inline Surface Inspection

3.2 Image Acquisition

Product varieties consist of unicolored panels as well as wood pattern. Wood pattern is not just applied by coloring the panels but also by imprinting wooden structures. Panels have different sizes, ranging from about 50 cm up to 2 m in length and 20 cm to 50 cm width. The defect catalog consists of typical defects such as scratches and cavities and several specific defects such as discoloring, excess of glue or lines. Higher adaptability and irregular pattern of surfaces are the reason this system was developed with a deep learning algorithm. Independently of the algorithmic approach, the first task is to define an imaging setup that ensures that defects are covered. We decided to use three different kinds of images: a brightfield setup for scratches and cavities, a darkfield setup for even surface defects, and a full-reflection setup that is specifically integrated for defects in the imprinted wooden structure. The setup can be seen in Fig. 4.

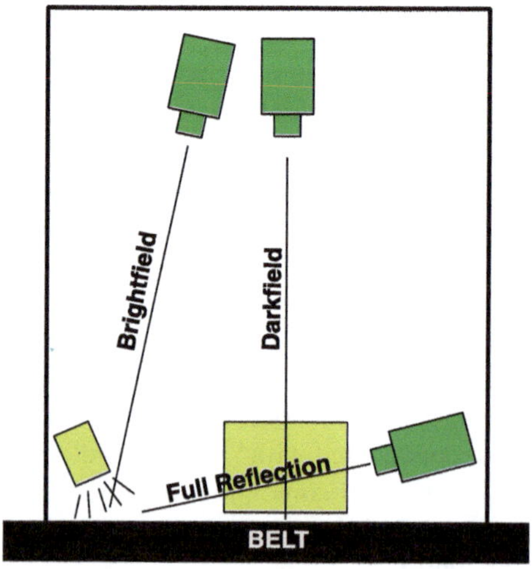

Fig. 4 Imaging setup: light sources are in yellow (left one spotlight, right one incident light), cameras in green. Copyright: Fraunhofer ITWM (self generated)

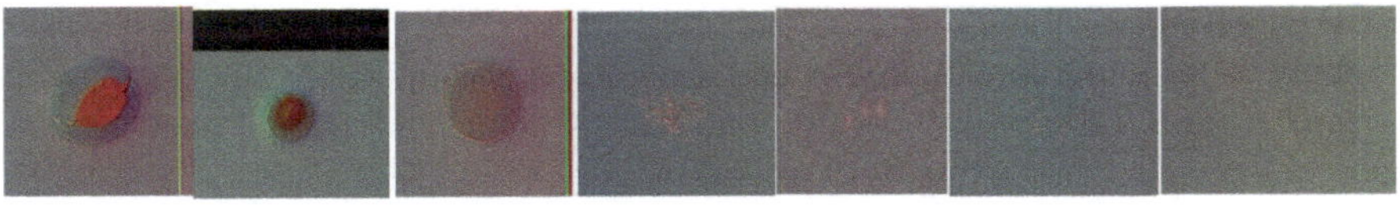

Fig. 5 Customer-defined defect type "sinkhole" with different intensities, strong (left) to weak (right). Copyright: Fraunhofer ITWM (self generated)

3.3 *Defect Definition*

Defect definition must be done in discussion between production/domain experts and CV experts. The former usually have a focus on the genesis of the defect as well as the possible consequences (e.g., fixable vs. unfixable). Image-processing experts must consider how to define classes that comply with the mathematical logic of the deep learning algorithm. In Fig. 5, one can see an example of the defect class "sinkhole." The expression of these defects is very different: on the left-hand side, strong sinkholes can be seen, which are characterized by round, clear structures, while weak sinkholes appear through a cluster of dots. Sinkholes are an example of defects that have a varying appearance, and to ensure a detection by a data-driven method, the full variety must be covered with multiple examples.

Therefore, in this system, we decided to build a classifier that is based on the image expression of defects as much as possible while still fulfilling customers' requirements. We have five main categories that are of most interest to the customer: two general categories of critical (positive) and uncritical defects (negative) and three further defect types that are of high importance (Staub=dust, not critical; Nest = big local defects that are critical, e.g., sinkhole would be one of them; Rillispur = line defects, also critical).

3.4 *Candidate Detection*

Tasks that can be solved by classical image processing in a stable manner should be solved because it reduces the amount of data needed. In many production cases, the segmentation of object and belt is such a pre-processing step. This step is also executed for the three images from this system. After this, we have three exact segmentations of the same object. One could treat them as separate grayscale images, process them separately, and join the results in the end. Instead of this, we decided to register them onto each other, to directly deal with one multidimensional image instead of three one-dimensional ones. Furthermore, as we have exactly three channels, it is very useful to treat them as the standard format of three-dimensional color images (e.g., RGB as in red-green-blue). Libraries for image processing and image display are generally available for RGB images and can directly be used on this "mixed" image. Additionally, pretraining databases are more often available for

RGB than for general multidimensional images. The CNN might therefore be easier to apply. After foreground detection and registration, the algorithm pipeline consists of typical steps, edge preserving filtering, adaptive thresholding, and mathematical morphology, all done on this RGB image.

3.5 Candidate Annotation

A panel of 2 meters consists of about 9000 pixels in length. Defects down to 10 pixels need to be detected (without our RGB image registration on three images per panel). This is tedious at best, not doable at worst. We built a custom annotation tool as displayed in Fig. 6. A visualization of the detailed defect itself (three different images, original and normalized for clearer visibility) is displayed alongside—as classification decision depends on size and position on the panel—the RGB image of the whole panel. The annotation expert should have the definition of the defect classes at hand. Annotation itself can then be done without further image processing/AI expertise with one or two clicks per defect.

3.6 Deep Learning

Our goal is to build a database that enables the use of different deep learning techniques. The model used for classification is the VGG16 [12] architecture,

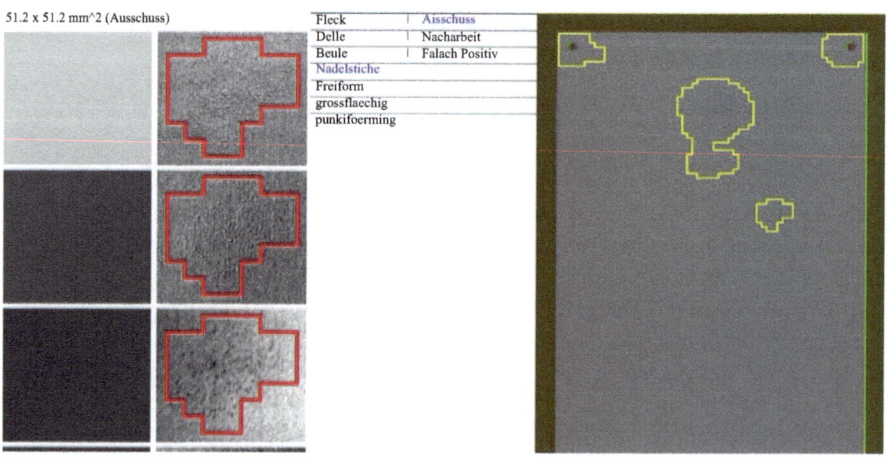

Fig. 6 Custom annotation tool. Left: three different images of candidate (brightfield, darkfield, full reflection) original and normalized. Right side: position of candidate in panel. Middle: annotation choices. Copyright: Fraunhofer ITWM (self generated)

Table 1 Precision as an indicator of the number of items that are correctly identified, recall as an indicator of the number of items that were missed, and support, aka the total number of available examples. All for CV-expert defined categories

Category	Precision	Recall	Support
Nest	0.970	0.780	1072
Rillispur	0.972	0.946	2013
Staub	0.812	0.623	334
Negative	0.830	0.967	2095
Positive	0.728	0.725	630
Overall accuracy	0.890	0.884	6144

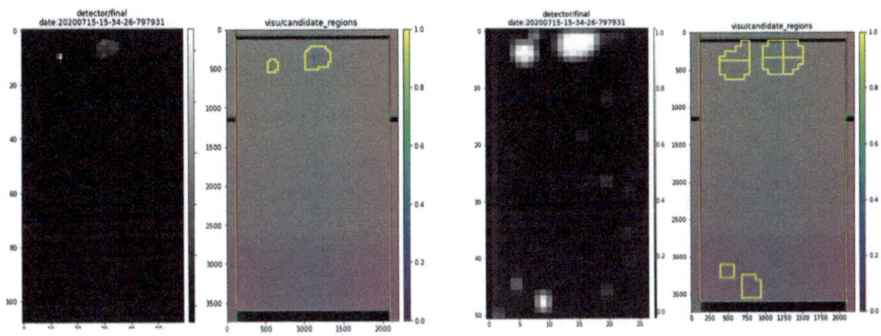

Fig. 7 Defect detection with classical image processing (left) and sliding window NN on the right. Copyright: Fraunhofer ITWM (self generated)

pretrained on the ImageNet [10] database. In Table 1, recall and precision of the CV-expert defined annotation is shown. While the overall accuracy of the domain-expert defined classes was 58%, the overall accuracy of the CV-expert defined classes is above 88%. Careful labeling with understanding of the underlying mathematical algorithms can therefore make a big difference. Additionally, some confusion are in between the classes of "Staub" and "negative," which do not count as defects. They can therefore be neglected. It must be said that the database for CV-expert labeling was slightly bigger and not as biased. This might also have influenced the results. For detection, we use the same model with an additional background class and combine all the labeled foreground classes. On the right-hand side of Fig. 7, one can see the resulting response of the classifier of this foreground class after applying a sliding window with overlap to detect areas of interest.

4 Discussion

The used methods enable a fast annotation of the database. However, in contrast to classical approaches, it remains difficult to change annotations iteratively. Therefore, the definition of defect classes is one of the most crucial points in designing such a system. In our approach, it became clear that previously defined defect classes are not necessarily suitable for a data-based defect detection algorithm. Especially the dataset bias and the need for sufficient representation of all classes lead to poor results in the beginning. In addition to well-defined defect classes, we therefore advise a two-step approach, where a first coarse classification is performed. It is also our experience that the importance of defects changes over time, with changing production facilities and changing supplier/customers of the production. This is also a reason why a graded defect annotation is sensible. This way, reannotation can also be minimized by only targeting subgroups of the whole dataset, if necessary. The collection of additional data without any candidate detection first also has to be performed. In our case, this was done by monitoring a graphical display of the system by image processing experts. In the future, we will additionally implement an automatic feedback system that enables production to do feedback sessions and thereby improve the system.

5 Conclusion

We have shown that using deep learning for inline defect detection systems is possible but at the same time has various challenges in comparison to a traditional inspection automatization approach. Production experts and image processing experts must communicate constantly to ensure that defect definition is correct. However, if a correct database is produced, constant advances in deep learning can easily be applied on this database. Even tasks that might not be solvable, for example, due to time restrictions might be solvable in the future by a new architecture/hardware setup based on the existing data set. Careful dataset generation can therefore be seen as a solid investment into the future of quality inspection as well as a means to realize a current solution.

References

1. Chen, F.-C., & Jahanshahi, M. R. (2017). Nb-cnn: Deep learning-based crack detection using convolutional neural network and naïve Bayes data fusion. *IEEE Transactions on Industrial Electronics, 65*(5), 4392–4400.
2. Gonzalez, R. C., & Woods, R. E. (2018). *Digital image processing* (Fourth edition, Global edition). Pearson.
3. Jing, J., Wang, Z., Rätsch, M., & Zhang, H. (2022). Mobile-UNet: An efficient convolutional neural network for fabric defect detection. *Textile Research Journal, 92*(1–2), 30–42.
4. Kim, B., Yuvaraj, N., Sri Preethaa, K. R., & Pandian, R. A. (2021). Surface crack detection using deep learning with shallow CNN architecture for enhanced computation. *Neural Computing and Applications, 33*(15), 9289–9305.
5. Krizhevsky, A., Sutskever, I., & Hinton, G. E. (2017). ImageNet classification with deep convolutional neural networks. *Communications of the ACM, 60*(6), 84–90.

6. Müller, O., Moghiseh, A., Stephani, H., Rottmayer, N., & Huang, F. (2018). Application of deep learning for crack segmentation on concrete surface. In *Forum Bildverarbeitung 2018* (p. 209). KIT Scientific Publishing.
7. Neogi, N., Mohanta, D. K., & Dutta, P. K. (2014). Review of vision-based steel surface inspection systems. *EURASIP Journal on Image and Video Processing, 2014*(1), 1–19.
8. O'Mahony, N., Campbell, S., Carvalho, A., Harapanahalli, S., Hernandez, G. V., Krpalkova, L., Riordan, D., & Walsh, J. (2019). Deep learning vs. traditional computer vision. In *Science and information conference* (pp. 128–144). Springer.
9. Rauhut, M., & Stephani, H. (2018). Konzeption und Aufbau eines Online-Oberflächeninspektionssystems. In M. Sackewitz (Ed.), *Handbuch zur industriellen Bildverarbeitung. Qualitätssicherung in der Praxis*. Fraunhofer Verlag.
10. Russakovsky, O., Deng, J., Su, H., Krause, J., Satheesh, S., Ma, S., Huang, Z., Karpathy, A., Khosla, A., Bernstein, M., et al. (2015). ImageNet large scale visual recognition challenge. *International Journal of Computer Vision, 115*(3), 211–252.
11. Shirmohammadi, S., & Ferrero, A. (2014). Camera as the instrument: The rising trend of vision based measurement. *IEEE Instrumentation & Measurement Magazine, 17*(3), 41–47.
12. Simonyan, K., & Zisserman, A. (2014). Very deep convolutional networks for large-scale image recognition. *arXiv preprint arXiv:1409.1556.*
13. Sun, X., Gu, J., Tang, S., & Li, J. (2018). Research progress of visual inspection technology of steel products—a review. *Applied Sciences, 8*(11), 2195.
14. Wang, T., Chen, Y., Qiao, M., & Snoussi, H. (2018). A fast and robust convolutional neural network-based defect detection model in product quality control. *The International Journal of Advanced Manufacturing Technology, 94*(9), 3465–3471.
15. Wang, Y., Liu, M., Zheng, P., Yang, H., & Zou, J. (2020). A smart surface inspection system using faster R-CNN in cloud-edge computing environment. *Advanced Engineering Informatics, 43*, 101037.

Gaussian Process Regression for the Prediction of Cable Bundle Characteristics

Lilli Burger, Vanessa Dörlich, and Michael Burger

Abstract In modern cars, a huge number of different cables can be found. They are typically combined in hoses and bundles in various different ways. For virtual product development and simulation-based design, it is necessary to know the characteristic physical parameters, like effective bending or torsion stiffness, of these cable systems. In early stages of the development process as well as for highly customized individual cable configurations, measuring effective stiffness properties is, however, often very challenging. In this contribution, we show results from our current research activities aiming at data-based modeling and estimating effective stiffness parameters for cable bundles. On the basis of an available dataset consisting of measured stiffness values for varying cable bundles, the overall goal is to identify a model out of this data that predicts bundle stiffness values for unknown bundles. We outline our approach to solve this nonlinear identification task with Gaussian Process Regression. Besides a short introduction to the industrial application area, we demonstrate and illustrate the applicability and prediction quality of Gaussian Process Regression for this task.

1 Introduction

The amount of cables used in modern cars sums up to a total length of several kilometers and is still increasing due to a rise in the use of electronic devices such as sensors or entertainment systems. Different cable types from 50 to 100 are bundled in up to 900 different combinations in the wiring harness of a car specifically tailored to the customers' demand.

L. Burger · V. Dörlich · M. Burger (✉)
Mathematics for Vehicle Engineering, Fraunhofer Institute for Industrial Mathematics ITWM, Kaiserslautern, Germany
e-mail: lilli.burger@itwm.fraunhofer.de; vanessa.doerlich@itwm.fraunhofer.de; michael.burger@itwm.fraunhofer.de

J. Franke, A. Schöbel (eds.), *Statistical Machine Learning for Engineering with Applications*, Lecture Notes in Statistics 227,
https://doi.org/10.1007/978-3-031-66253-9_3

For virtual product development and simulation-based design, it is necessary to know the characteristic physical parameters, mainly the effective bending or torsion stiffness, of these cable systems. In principle, measuring effective stiffness values for all cables and assembled cable bundles is possible. However, especially in early stages of the development process, those measurements are challenging, and they might be impractical (due to time and costs) or merely impossible, for instance, due to the lack of physical prototypes.

In this contribution, we show results from our current research activities aiming at data-based modeling and estimation of effective stiffness parameters for cable bundles. On the basis of an available dataset consisting of measured stiffness values for varying cable bundles and characteristic parameters of the single cables, the overall goal is to identify a model out of this data. Furthermore, this model is supposed to be used to estimate stiffness values for new bundles with inputs that can be specified efficiently without complex measurement efforts. An overview of the dataset is given in Sect. 1.2.

We outline our approach to solve this nonlinear identification task with Gaussian Process Regression (GPR), which is introduced in Sect. 2 briefly. In Sect. 3, the applicability of GPR for this task is illustrated, and the performance of this approach in terms of prediction results for different cable bundle types is discussed.

1.1 Simulation of Cable Bundles

A high number of highly flexible and slender structures such as cables and cable bundles has to be assembled in the narrow installation space of modern cars, e.g., in the engine compartment. Therefore, the wiring harness has to be taken into account in early design phases. Optimization of cable lengths and mounting positions must ensure that cables do not conflict with other components for static routing as well as during the vehicle's lifetime. Additionally, mountability of the wiring harness is essential in assembly planning. During assembly, large bending curvatures may occur, while large twists are typically avoided as they cause damage. The time-consuming and cost-intensive construction of wiring harness hardware prototypes can be avoided by performing a physically accurate simulation of the cable routing in early design phases. The software IPS Cable Simulation is well suited for this purpose, since it enables real-time assembly simulations and analysis of flexible slender structures based on a geometrically nonlinear beam model taking into account the CAD data of the vehicle [12]. Examples of applications from automotive industry are given in Chapter 6 of [2].

A brief overview of the used model is given in [11] and a more detailed description in [4]. The necessary model parameters are the length density ρA (mass per length) of the cable, its effective bending stiffness $[EI]$, torsion $[GJ]$, and tensile stiffness $[EA]$. In general, stiffnesses are composed of a material parameter (modulus of elasticity E, shear modulus G) and a geometric quantity (cross-sectional area A, axial moment of inertia I, and polar moment of inertia

J). However, considering the complex composite structure of cables and cable bundles, they are only considered as composite parameters and not separated into a material and geometrical property. In order to achieve realistic simulation results, model parameters that are as accurate as possible are required. Accordingly, a user must determine the parameters of the cable or harness segment to be simulated by measurement or at least estimate them using appropriate methods. Bending stiffness $[EI]$ and torsion stiffness $[GJ]$ are the main focus, while length density $[\rho A]$ is comparatively easy to determine (by measuring the length and mass), and a rough estimate is usually sufficient for the tensile stiffness $[EA]$.

The measurement system MeSOMICS® ("Measurement System for the Optically Monitored Identification of Cable Stiffnesses") was specifically designed for the measurement of bending and torsion stiffness of flexible cables [6]. It allows a robust and reproducible measurement of the mechanical parameters in the deformation ranges relevant for typical applications of cables.

Executing the measurement procedure for up to 100 individual cable types occurring in a passenger car represents a reasonable cost and time investment. The measurement of the parameters of several 100 bundle variations occurring in one car model is, however, on the one hand too expensive and on the other hand often not possible in early design phases, since hardware prototypes of the wiring harnesses are not available yet. In such situations, bundle parameters have to be estimated. IPS Cable Simulation provides the Advanced Bundle Editor module for this purpose, which can be used to assemble a virtual bundle and estimate its mechanical parameters (see Fig. 1). An estimate of the bundle tensile stiffness can easily be calculated since the cables are typically approximately parallel to

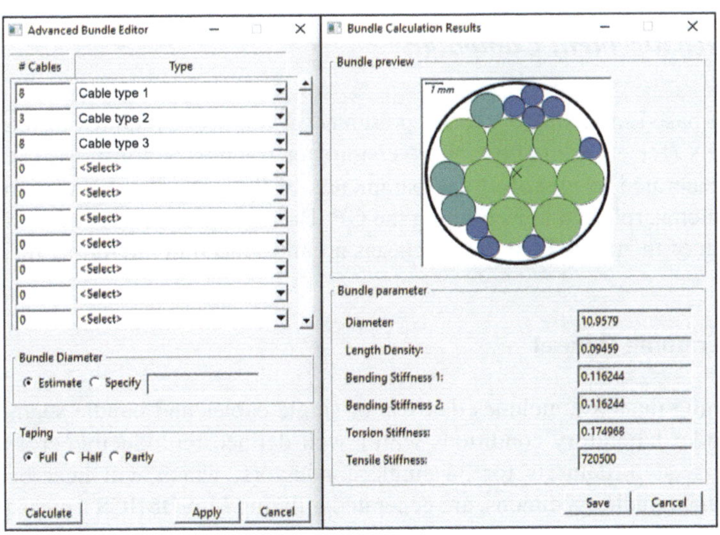

Fig. 1 Creating a virtual cable bundle segment using the Advanced Bundle Editor (left) and results of the parameter estimation including the visualization of a possible bundle cross section (right)

each other in a cable bundle, i.e., arranged in a parallel circuit. Thus, the bundle's tensile stiffness $[EA]_B$ equals in good approximation the sum of the cables' bending stiffnesses $[EA]_c$,

$$[EA]_B = \sum_i [EA]_{c_i} . \tag{1}$$

The bundle's bending and torsion stiffness are not only determined by the single cable stiffnesses but additionally by effects such as contact and friction between the constituents. Cables are furthermore usually held together using textile tape, which has a non-negligible influence on the bundle bending and torsion stiffnesses. The mechanical parameters of the textile tape are hardly measurable separately. One possibility to treat these mesoscopic effects is the inclusion of contact and friction in physical models. In our approach, we treat these various complex effects using a data-based approach from machine learning for the estimation of bundle bending and torsion stiffnesses. We aim at an estimation accuracy of at least $\pm 50\%$ relative to measured bundle stiffnesses, which originates basically from two reasons: On the one hand, the way cable bundles are build can vary highly (e.g., one bundle is taped more tightly than another bundle, but both consist of exactly the same components and tape), which leads consequently to a comparatively high measurement scattering even under ideal experimental conditions. On the other hand, the three-dimensional shape of a cable bundle does typically not deviate within the range of $\pm 50\%$. It has to be noted that higher accuracy of the bundle bending and torsion stiffnesses may be required, if forces or moments on mounting elements are relevant.

1.2 Measurement Campaign

The data-based modeling approach presented in this work requires different types of datasets (for each bundle taping version) for training and testing. The training data is generated to identify hyperparameters of the considered Gaussian Process (GP), which is referred to as training the GP. The test database is used afterward for validation of the trained GP. Both datasets are introduced in the following sections.

1.2.1 Training Dataset

The training database includes datasets of single cables and bundle segments produced under laboratory conditions with a well-defined reproducible cross section. In the first step, datasets for 34 single conductors, which will later be used to produce the bundle specimens, are generated utilizing MeSOMICS® measurements. Each dataset of a single cable contains cable parameters radius r_c, length density $[\rho A]_c$, bending stiffness $[EA]_c$, and torsion stiffness $[GJ]_c$. These 34 cables were combined in different variations to create bundles of varying complexities. For all

Fig. 2 Examples of bundle cross sections with different taping patterns: partly, half, and full, from left to right. Cables in training bundles are ordered in concentric layers (three bundles on left); cables in test bundles are ordered randomly (bundle on the right)

bundle types, we aimed at a regular well-defined cross section with approximately closest packing. We achieved this by ordering cables in cross-section concentric layers (see Fig. 2, left). Training bundles consist of up to three different cable types, which contain up to 80 single cables, and their diameters range from 1.2 mm to 10.3 mm. Inspired by wiring harnesses from automotive applications, the bundles are taped in three patterns. The first pattern represents applications where the cables are held together at discrete points by tape or a cable tie ("partly"). For the second pattern, the tape is applied in a spiral, leaving a gap between the tape ("half"). The last pattern is a "full" wrap of the cables, achieved by applying the tape in a spiral with a minimal overlap. For each cross-section composition, all three patterns are used and treated as separate bundle types. This yields a training database that contains 500 bundles, where the data of each bundle includes the bundle parameters radius r_B, length density $[\rho A]_B$, bending stiffness $[EA]_B$, and torsion stiffness $[GJ]_B$. Additionally, information of the composition of the bundle is stored, i.e., the amount of cables per base cable type n_c and the taping pattern type.

1.2.2 Test Dataset

In addition to the bundles with a well-defined structure, ten bundles with a random order of cables in the cross section have been manufactured (see Fig. 2, right). The bundles contain up to ten different base cable types summing up to a maximum of 50 cables per bundle. Three taping patterns are applied to these ten randomly ordered bundle types, and MeSOMICS® measurements have been performed on these specimens to generate validation datasets. Datasets include the same information as the training data and will serve for the initial validation of the trained GP. It is crucially required that the model generalized these random bundles, since this is the situation in practical process application. However, this generalization can be expected because the GP model trained here can be seen as a surrogate for physical mechanism.

2 Gaussian Process Regression

In this application, we consider Gaussian Process Regression as an underlying technique. In principle, also other methods could be used here; the choice taken here is due to the fact that GPs are known to perform well also on datasets of moderate size and, moreover, they are able to deliver a predictive distribution rather than just a point estimate. In the sequel, we briefly summarize the basics of GP Regression.

Let $D = \left\{\left(\vec{X}, y\right)\right\}$ denote the considered database with inputs $\vec{X} = (x_1, \ldots, x_N)^T \in \mathbb{R}^{N \times d}$ consisting of d different input features; $x_i = \left(x_i^{(1)}, \ldots, x_i^{(d)}\right)^T$, at N measurement points; and corresponding outputs $y = (y_1, \ldots, y_N)^T$.

We assume that there is a mapping f between inputs and outputs to be approximated. Moreover, we assume independent and normally distributed uncertainty ε with standard deviation σ_ε^2 in the observed data (e.g., due to measurement errors):

$$y_i = f(x_i) + \varepsilon, \quad \varepsilon \sim \mathcal{N}(0, \sigma_\varepsilon^2). \tag{2}$$

The central task is to choose a model (family) for the mapping f, train a specific model f based on the observed data D, and finally use that model for predicting an output y^* for a suitably chosen (new) input x^*, such that $y^* = f(x^*)$. In our approach, we use a stochastic framework, in order to solve this model identification task, and in particular, we choose a *Gaussian Process* with mean function $m : \mathbb{R}^d \to \mathbb{R}$ and covariance function $k : \mathbb{R}^d \times \mathbb{R}^d \to \mathbb{R}$ as model structure for the mapping f to be identified,

$$f \sim GP(m, k). \tag{3}$$

A GP is a collection of random variables, for which any finite subset is assumed to have a joint (multivariate) Gaussian distribution. It is characterized by its *mean function* m and its *covariance function* k; the latter is also referred to as *kernel function* [8]. A more detailed introduction and discussion of GPRs, especially in the context of machine learning applications, can be found in [7].

Very often, it is assumed that the GP has a mean function identically equal to zero. In our application, however, it is preferable to choose a non-zero mean function (see Sect. 3.1). The covariance function describes *similarity* or *nearness* between data points on input level. One of the most common covariance functions is the *squared exponential* covariance function

$$k(x, x') = \exp(-\frac{1}{2}|x - x'|^2). \tag{4}$$

This example illustrates how *similarity* can be interpreted in this case, as $k(x, x')$ is almost equal to one for inputs x and x' very near to each other and decreases with growing distance.

An overview about the mean and covariance functions as well as examples of commonly used functions is given in [7] as well.

After specifying the GP by choosing the mean and covariance function, the key step to understand how the GP is used for modeling f and for prediction is to consider the joint distribution of the observed target values y and a requested function value y^* at an input x^*, which is assumed to be Gaussian as well:

$$\begin{pmatrix} y \\ y_* \end{pmatrix} \sim \mathcal{N}\left(\begin{pmatrix} \vec{M}_{\vec{X}} \\ m(x_*) \end{pmatrix}, \begin{pmatrix} \vec{K} + \sigma_n^2 I & \vec{K}_* \\ \vec{K}_*^T & \vec{K}_{**} \end{pmatrix}\right), \tag{5}$$

with mean vector $\vec{M}_{\vec{x}} = [m(x_1), \ldots, m(x_N)]^T$ and a covariance matrix consisting of

$$\vec{K} = \begin{pmatrix} k(x_1, x_1) & \ldots & k(x_1, x_N) \\ \vdots & \ddots & \vdots \\ k(x_N, x_1) & \ldots & k(x_N, x_N) \end{pmatrix}, \tag{6}$$
$$\vec{K}_{**} = k(x_*, x_*), \;\; \vec{K}_* = [k(x_*, x_1), \ldots, k(x_*, x_N)]^T .$$

Recall that the quantity of interest is y^* as response to x^* (prediction), in the described stochastic framework with a GP, $f \sim GP(m, k)$, as model structure. We consider, consequently, the conditional density $p(y^*|x^*, D)$ to derive information about y^*, given x^* and the observed database. According to Eq. (5) and using basic properties of Gaussian distributions, this conditional distribution is Gaussian too, and it can be calculated as

$$p(y^*|x^*, D) = \mathcal{N}(m(x_*) + \vec{K}_*(\vec{K} + \sigma_\varepsilon^2 I)^{-1}(y - \vec{M}_{\vec{X}}), \vec{K}_{**} - \vec{K}_*(\vec{K} + \sigma_\varepsilon^2 I)^{-1}\vec{K}_*^T). \tag{7}$$

Thus, using GPs in this way allows us to derive a *predictive distribution* rather than just a point estimate. That is, in addition to a point estimate,

$$y^* = m(x_*) + \vec{K}_*(\vec{K} + \sigma_\varepsilon^2 I)^{-1}(y - \vec{M}_{\vec{X}}), \tag{8}$$

a GP prediction automatically provides an uncertainty quantification for a specific estimation in terms of the underlying assumed variance of the predictive distribution,

$$\sigma_*^2 = \vec{K}_{**} - \vec{K}_*(\vec{K} + \sigma_\varepsilon^2 I)^{-1}\vec{K}_*^T . \tag{9}$$

After specifying the form of the mean and covariance function, a set of hyperparameters θ, containing among others the parameter of the (non-zero) mean function and the parameters of the covariance function, have to be derived. In a Bayesian framework, this can be done by maximizing an appropriate posterior distribution for the given data set D. With the additional assumption that the prior distribution of the hyperparameters is approximately constant, i.e., $p(\theta) \approx const$, in fact, no prior information is considered, one can obtain a suitable θ by maximizing the log-likelihood $p(y|\vec{X}, \theta)$ (e.g., with a standard gradient ascent method):

$$p(\theta|\vec{X}, y) \propto p(y|\vec{X}, \theta) \cdot p(\theta) \rightarrow \theta^* = \text{argmax}_\theta\, p(\theta|\vec{X}, y) \approx \text{argmax}_\theta\, p(y|\vec{X}, \theta). \tag{10}$$

Again, due to the fact that only Gaussian distributions are involved, that log-likelihood can be derived analytically (cf. [10]):

$$p(y|\vec{X}, \theta) = -\frac{1}{2}(y - m)^T \left(K + \sigma_n^2 I\right)^{-1} (y - m) - \frac{1}{2} log|K + \sigma_\varepsilon^2 I| - \frac{n}{2} log 2\pi, \tag{11}$$

where the vector θ again subsumes all hyperparameters of the mean function m and the covariance matrix K.

3 Prediction of Cable Bundle Stiffnesses Using Gaussian Process Regression

3.1 *Gaussian Process Setup*

We consider one GP per taping pattern and effective bundle stiffness, which results in six GPs in total, which have to be trained.

Depending on which stiffness value is considered, the effective bending, or torsional stiffness of the bundle, we choose appropriate features. Due to different physical behavior and considerations, the feature selection differs for both effective stiffness, as is described in Sect. 3.1.1. All features are characteristics of the base cables, which the bundles consist of, or describe the bundle composition. Also, they have in common that they already exist for the simulation application and are easy to measure or evaluate respectively.

3.1.1 Selected Features

For the estimation of the effective bending stiffness of a bundle, we choose parameters for the single base cables and the bundle itself as well as a nonlinear combination of some of these parameters and combine them in the input vector

$$\vec{x} = \left[\sum_i [EI]_{c_i}, \sum_i [\rho A]_{c_i}, \sum_i r_{c_i}, r_b, \sum_i [\rho A]_{c_i} \cdot \sum_i r_{c_i}/n_c \right], \quad (12)$$

with

- $[EI]_c$: base cable effective bending stiffness,
- $[\rho A]_c$: base cable length density,
- r_c: base cable radius,
- n_c: number of cables in the bundle,
- r_b: bundle radius.

The number of base cables combined to a bundle differs for each bundle. To guarantee a constant number of features per bundle, we sum up some of the base cable parameters or use the maximum, respectively, as is done for the base cable torsional stiffness (cf. the input vector in (13)). The model output is the effective bending stiffness of the bundle itself, $\vec{y} = [EI]_b$.

The input vector $\vec{x}$ for the estimation of the effective torsional stiffness of a bundle is

$$\vec{x} = \left[\sum_i [EI]_{c_i}, \max_i([GJ]_{c_i}) \sum_i [\rho A]_{c_i}, \sum_i r_{c_i}, r_b \right], \quad (13)$$

whereby $[GJ]_c$ is the base cable effective torsional stiffness. The model output is the effective torsional stiffness of the bundle, $\vec{y} = [GJ]_b$.

3.1.2 Selected Mean and Covariance Function

For our application, the *rational quadratic* covariance function

$$k(x_i, x_j) = \sigma_f^2 \left(1 + \frac{1}{2\alpha} \sum_{m=1}^{d} \frac{(x_i - x_j)^2}{l_m^2} \right)^{-\alpha} \quad (14)$$

has proven to be a good choice for our database and use case. It is defined by a length-scale $\alpha \in \mathbb{R}$ and the process variance $\sigma_f^2 \in \mathbb{R}$, which controls the magnitude. Moreover, we choose a kernel function with characteristic length-scale l_m controlling the smoothness in each input feature direction, because GPs with a single length-scale parameter l are not suited to efficiently characterize regression surfaces in which the mean of the response changes more rapidly in certain input directions than others [10].

Furthermore, we choose the quadratic mean function

$$m(x) = \vec{h}(x)^T \beta, \vec{h}(x) := \left[1, x^{(1)}, \ldots, x^{(d)}, \left(x^{(1)}\right)^2, \ldots, \left(x^{(d)}\right)^2\right]^T, \quad (15)$$

with parameter vector $\beta = [\beta_1, \ldots, \beta_{2d+1}]^T$.

3.2 Estimation Results

The GP has been trained with the input and output data of the training bundles described in Sect. 1.2.1. With the hyperparameters determined during the training process, we can easily calculate the covariance matrix $\vec{K}$ according to definition (6) and covariance function described in Eq. (14) as well as the mean vector $\vec{M}_{\vec{X}}$ according to Eq. (15).

Furthermore, the covariance matrices $\vec{K}_*$ and $\vec{K}_{**}$ (Eq. (6)) can be calculated with the known input parameters of the new bundle of interest. So these linear algebra computations remain for the calculation of the point estimate Eq. (8) and the estimation variance Eq. (9), which makes the estimation using GPs very efficient (for an amount of data not too high as it is the case in this application).

A process overview for the estimation of effective bundle stiffnesses is summarized in Fig. 3.

For the validation of the described effective bundle stiffness estimation, we use our test dataset, consisting of irregular composed bundles introduced in Sect. 1.2.2.

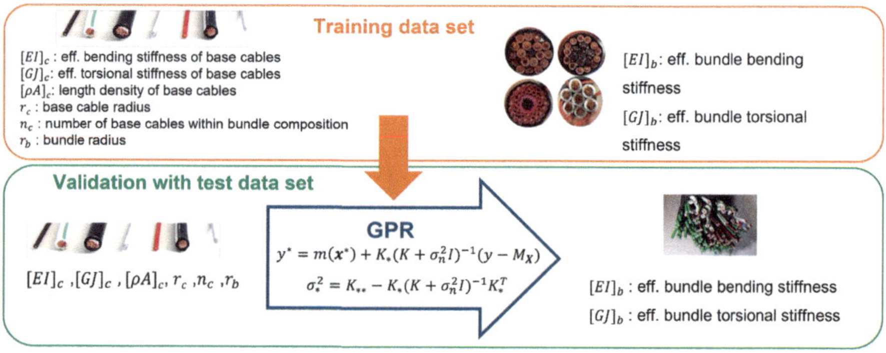

Fig. 3 Estimation process overview

Table 1 Estimation errors for $[EI_b]$ for three bundle taping options

Taping	$mean(e_{abs})$ [Nm^2]	$max(e_{abs})$ [Nm^2]	$mean(e_{rel})$	$max(e_{rel})$	$r_{\|e_{rel}\|\leq 0.5}$
partly	0.0078	0.0166	0.2153	0.7298	0.90
half	0.0213	0.0836	0.2106	0.4721	1.00
fully	0.0110	0.0281	0.1344	0.2556	1.00

Table 2 Estimation errors for $[GJ_b]$ for three bundle taping options

Taping	$mean(e_{abs})$ [Nm^2]	$max(e_{abs})$ [Nm^2]	$mean(e_{rel})$	$max(e_{rel})$	$r_{\|e_{rel}\|\leq 0.5}$
partly	0.0126	0.0307	0.3448	0.6603	0.80
half	0.0094	0.0353	0.2400	0.5742	0.89
fully	0.0204	0.1043	0.3218	0.9439	0.80

We consider the absolute and relative prediction errors,

$$
\begin{aligned}
e_{abs} &= |y^* - y^{meas}|, \\
e_{rel} &= \frac{e_{abs}}{y^{meas}},
\end{aligned}
\tag{16}
$$

where y^* is the estimated effective bending or torsional bundle stiffness and y^{meas} is the stiffness we measured for validation purpose.

We know from cable simulation studies that an effective stiffness deviation of ±50% still leads to sufficient results concerning many applications, e.g., geometry layout. Thus, to assess the prediction quality of a GP, we also consider the ratio of estimations with a relative error below or equal to ±50% compared to all estimations done, $r_{|e_{rel}|\leq 0.5}$, in our validation.

In Tables 1 and 2, the results for both effective bundle stiffness estimations of the test dataset and all taping options are summarized.

Ratios $r_{|e_{rel}|\leq 0.5}$ in Tables 1 and 2 show that most of all estimations are of a quality, which makes them useful for further bundle simulations. The estimation of the effective torsional stiffness is less accurate than the effective bending stiffness simulation, nevertheless, with a mean of the relative estimation error of maximal 34.48%, quite successful.

As an example, in Fig. 4, the result for the effective bending stiffness estimation is illustrated for partly taped bundles. On the left side, the absolute measured effective bundle stiffness and the estimated bending stiffnesses ± standard deviation are presented. The diagram on the right illustrates the relative estimation error (cf. Eq. (16)) with a relative representation of the standard deviation of these estimations. Additionally, the ±50% error bound is presented by red dashed lines.

The estimation for bundle no. 4 is the only effective bending estimation with a relative error outside our desired error boundaries, as illustrated in Fig. 5, even if the standard deviation range is considered.

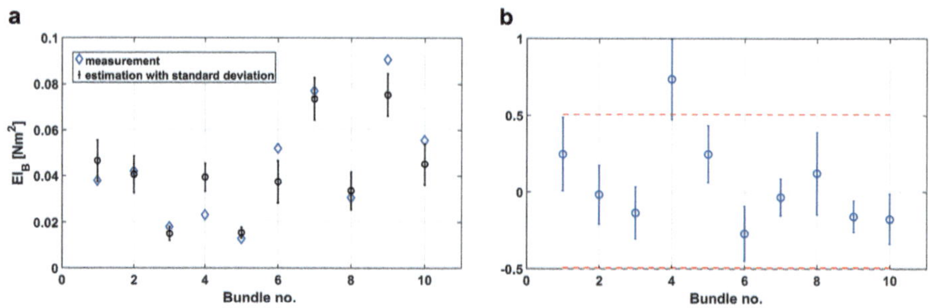

Fig. 4 (**a**) Comparison between estimated and measured effective bending stiffnesses for partly taped bundles. (**b**) Relative estimation error with ±50% error bound for partly taped bundles

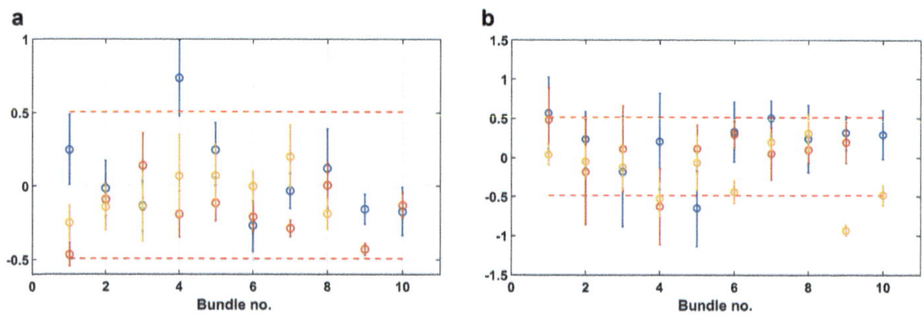

Fig. 5 (**a**) Relative estimation error with ±50% error bound for all effective bending stiffness estimations. (**b**) Relative estimation error with ±50% error bound for all effective torsional stiffness estimations

Table 3 Mean value of the relative standard deviation for all estimations

Taping	$[EI]_b\ mean(\sqrt{\sigma_*^2}/y^{meas})$	$[GJ]_b\ mean(\sqrt{\sigma_*^2}/y^{meas})$
partly	0.1872	0.4165
half	0.1124	0.3726
fully	0.1792	0.1880

In Fig. 5, for all estimations done for the validation bundles, the relative error together with the relative standard deviation $\sqrt{\sigma_*^2}/y^{meas}$ is illustrated, and in Table 3, mean values of these relative standard deviations for each effective bundle stiffness and taping variant are summarized.

As mentioned in Sect. 2, we do not only have a point estimate but a prediction distribution using GPs; thus, we also know the uncertainty of our prediction in terms of its variance σ_*^2, cf. Eq. (9), or its standard deviation.

Relative standard deviations, summarized in Table 3, show that the estimation results for the effective torsional stiffness are more uncertain than the effective bending stiffness results, which also fits to the more accurate estimation results of the effective bending stiffnesses (see Fig. 5).

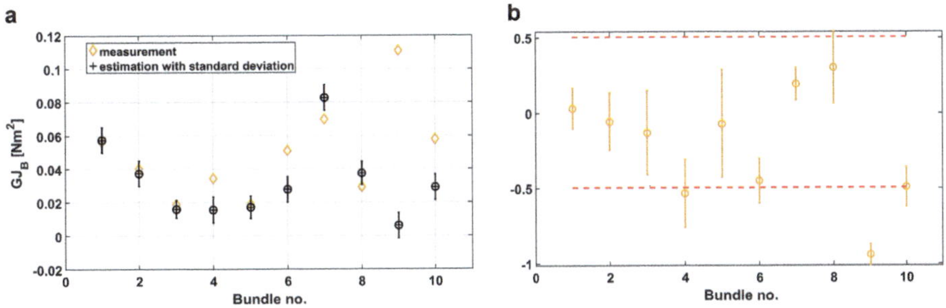

Fig. 6 (**a**) Comparison between estimated and measured effective torsional stiffnesses for half-taped bundles. (**b**) Relative estimation error with ±50% error bound for half-taped bundles

In Fig. 6, estimation results for the effective torsional stiffness are exemplary illustrated for fully taped bundles. The maximum relative error is with a value of 94.39%, the biggest error for all test bundles, and can be assigned to the estimation result of bundle no. 9.

The variance of a GP increases with the distance between two data points increasing. That means the more uncertain the estimation of our test bundles becomes, the more the test bundle (its feature values, respectively) differs from the training bundles, because the matrix values for $\vec{K}_*$ decrease (cf. Eq. (9)).

Comparing the Euclidean norm between the test bundle input and the training bundle input, it is bundle no. 9, which has the biggest distance to the training dataset. Furthermore, comparing the distance between training and test data of each input feature, we can see that the feature $[\rho A]_c$ for bundle no. 9 is even outside the training dataset (see Fig. 7). Here, in addition to the test data input and the minima and maxima of training data, which are illustrated as bars, the distribution of the training data feature is plotted to get an impression of where the test data is located compared to the training dataset.

A test data point with a higher distance to the training dataset has a higher probability to be estimated incorrectly. Thus, it makes sense to compare the input features of the test data to those of the training data to detect possible uncertainties and to decide how to enhance the training dataset. Consequently, estimation results can be improved if the GP is trained with training data enhanced in regions with a lower data distribution, e.g., the range between 0.4 and 0.9 [kg/m] for the feature $[\rho A]_c$ (cf. Fig. 7).

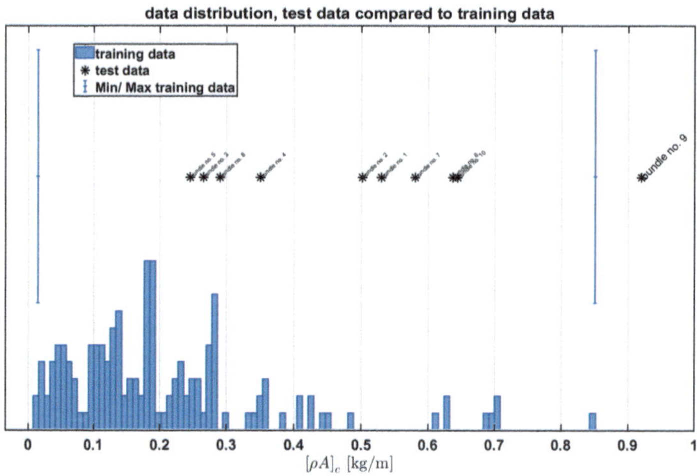

Fig. 7 Feature $[\rho A]_c$ in training and test dataset

4 Summary

This chapter presents an approach to determine unknown mechanical model parameters for the simulation of cable harness segments using GPR. For this purpose, first, a database with measurements of cable bundles as well as individual base cables is required in order to train a GP once. If the training is successful, the GP and easily measured cable parameters can be used to estimate the effective stiffnesses of a bundle. In addition to the effective bundle stiffness, we derive the variance for each estimation, which allows us to simulate a tolerance envelope by using the standard deviation.

A validation with bundles not included in the training database was successful and shows the great potential of this data-based methodology for automotive applications.

References

1. Antman, S. S. (2005). *Nonlinear problems of elasticity*. Springer.
2. Linn, J. (2020). Discrete Cosserat Rod Kinematics Constructed on the Basis of the Difference Geometry of Framed Curves—Part I: Discrete Cosserat Curves on a Staggered Grid. *Journal of Elasticity, 139*, 177–236.
3. Linn, J., & Dressler, K. (2017). Discrete cosserat rod models based on the difference geometry of framed curves for interactive simulation of flexible cables. In L. Ghezzi, D. Hömberg, & C. Landry (Eds.), *Math for the digital factory*. Mathematics in Industry (Vol. 27). Springer.
4. Linn, J., Hermansson, T., Andersson, F., & Schneider, F. (2017). Kinetic aspects of discrete Cosserat rods based on the difference geometry of framed curves. In *ECCOMAS Thematic Conference on Multibody Dynamics, Prague, Czech Republic*.

5. Dörlich, V., Linn, J., & Diebels, S. (2018). Flexible beam-like structures—experimental investigation and modeling of cables. In *Advances in mechanics of materials and structural analysis* (pp. 27–46). Springer.
6. www.mesomics.eu. Accessed 20 March 2021.
7. Rasmussen, C. E., & Williams, C. K. I. (2006). *Gaussian processes for machine learning*. The MIT Press. www.GaussianProcess.org/gpml
8. Barber, D. (2012). *Bayesian reasoning and machine learning*. Cambridge University Press.
9. Burger, L., Doerlich, V., Burger, M., Linn, J., & Schneider, F. (2022). Estimation of cable bundle stiffness based on Gaussian Process Regression. In *MATHINDUSTRY, Volume 39 Progress in Industrial Mathematics at ECMI 2021*. Springer.
10. Gelman, A. , Carlin, J. B., Stern, H. S., Dunson, D. B., Vehtari, A., & Rubin, D. B. (2020). *Bayesian data analysis third edition*. http://www.stat.columbia.edu/~gelman/book/BDA3.pdf
11. Schneider, F., Linn, J., & Andersson, F. (2021). Digital validation of cable performance for vehicles in operation. In K. Berns, K. Dressler, R. Kalmar, N. Stephan, R. Teutsch, & M. Thul (Eds.). *Proceedings of the Commercial Vehicle Technology 2020/2021*. Springer Vieweg.
12. IPS Cable Simulation website. www.flexstructures.de

Machine Learning for Predictive Maintenance in Production Environments

Alex Sarishvili, Fabian Menzel, Benjamin Adrian, and Julia Burr

Abstract System condition-based predictive maintenance is an important factor for the availability and performance of complex machines and offers cost-saving potential due to shorter and better planned outages. For this purpose, complex models are commonly used to detect possible faults and degradations in an early stage. These models need information about the faults and all operating points as a training set. Often these data sets are not available or sufficient, so that alternatively training data can be generated artificially. For well-researched rotating machines, scientific literature, which provides information about the modeling of different kinds of damage conditions, can be used to achieve this. We develop a demonstrator grid and a system for detecting degrading and sudden faults on rotating systems. This requires data to be recorded from the sensors (acceleration, noises, etc.) under variation of the speed on the demonstrator. Afterwards the training set for the diagnosis models has to be generated using a simulation of a set of possible faults. For this, the identified system and modeling of the possible faults is needed. The models have to be implemented, trained on this data set, and evaluated on real but manually caused damage data at the demonstrator. The aim is to investigate the extent to which models trained with simulated data can be used for diagnosis in reality. Also, different kind of models will be compared regarding their performance.

1 Introduction

In the manufacturing industry, the failure-free operation of machines is the basic prerequisite for production planning. Ensuring this is the task of the maintenance department, which largely carries out maintenance work at noncritical times for production using preventive maintenance windows at intervals of operating hours

A. Sarishvili · F. Menzel · B. Adrian (✉) · J. Burr
Fraunhofer ITWM, Kaiserslautern, Germany
e-mail: alex.sarishvili@itwm.fraunhofer.de; fabian.menzel@itwm.fraunhofer.de; benjamin.adrian@itwm.fraunhofer.de; julia.burr@itwm.fraunhofer.de

J. Franke, A. Schöbel (eds.), *Statistical Machine Learning for Engineering with Applications*, Lecture Notes in Statistics 227,
https://doi.org/10.1007/978-3-031-66253-9_4

based on experience. The advantage of this approach is the avoidance of undesired failures. However, in order to be able to react quickly in the event of unexpected malfunctions or even failures, the maintenance department places employees on standby duty in shift operation and keeps appropriate spare parts in stock. This causes running costs, which have an average effect on the product price.

Condition-based maintenance is based on data that record the operating status of the machines at suitable time intervals. Based on condition monitoring, maintenance windows can be planned more demand-oriented to allow a reduced frequency of maintenance windows without increasing downtime risks. It also allows to reduce the cost-driving shift operation of maintenance technicians.

Predictive maintenance analyzes the dynamics or drift of the measured condition away from a reference point that is considered good and predicts a future point in time when the condition is likely to be degraded to such an extent that continued operation would mean too great a risk of failure. This enables targeted planning of maintenance windows to inform both production and spare part procurement within a reasonable time. Individual machine manufacturers are already working towards establishing leasing models based on predictive maintenance, long-term guarantees, and leasing models that are tailored to actual measured wear. This is being implemented using a data-based approach based on machine learning methods. However, the main challenge for data analysts is that, due to the preventive maintenance strategy, required data on failure and malfunction events is largely missing. This prevents the use of supervised machine learning techniques for trend detection or diagnosis. Furthermore in both scenarios (supervised and unsupervised learning), data quality, measured with methods based on Shannon's information theory and data derivation variability (domain adaptation), plays a crucial role in order to successfully deploy AI in the industry.

In the first part of this chapter, we show an application of ML methods on rolling bearing damage diagnosis based on simulated data. Since there is already a lot of domain knowledge available, we focused on feature engineering and classical machine learning methods. In the second part of the chapter, we show the methods for measuring data quality and try to answer the question about how and why deep learning is prone to overperform shallow learning machines in multi-domain scenario. In particular, we use and manipulate real experimental data from a plastic extrusion process to show the mentioned superiority on an example of SAE (Stacked Auto-Encoder) neural network classification machine.

2 Rolling Bearing Diagnosis

In this study, we show a proven approach for training suitable diagnostic and prognostic models using the example of rolling bearings in rotating drives. Unexpected failures of components of rotating machines lead to unplanned downtimes and thus to significant losses for the operator. Therefore, it is advantageous to constantly monitor machines, to diagnose damage at an early stage, and to forecast degrading trends as accurately as possible.

Depending on the damage type, various operating data, such as pressures and temperatures, are recorded and monitored by means of sensors. In the case of rotating machinery and related damage, vibration data are particularly valuable, as they indicate damage at a very early stage. In the following, the automated diagnosis of rolling bearings will be discussed in more detail.

Rolling bearings are among the most important elements in technical systems. They are used to transfer forces between static and dynamic machine components. Depending on the application, they are produced and used as axial and radial bearings. While wear is caused by insufficient lubrication, water ingress, and corrosion or dirt particles, local damage occurs due to the Hertzian pressure. The alternating load caused by the contact with the balls leads to stress changes in the material. Cracks that result can grow and lead to material breakouts [8]. While insufficient lubrication can be monitored via oil pressure or bearing temperature, the appropriate way to detect local damage is vibration monitoring. The time of replacement when damage is diagnosed depends on the load on the bearing and the type of damage. For example, an unlubricated bearing would overheat and fail in a shorter period of time. In general, bearing damage usually progresses slowly at first and more rapidly later, along with exponential degradation. To avoid recurring damage, it is important to perform a root cause analysis after removing the bearing that has been diagnosed as being damaged [9]. In the following, the focus is set on the vibration diagnosis of local damage.

Local defects and pittings in the inner ring, the outer ring, and the rolling elements produce multiple single peaks in the frequency spectrum depending on the geometry. In the case of several bearings with different geometries, the vibration patterns can be used to localize the damage. Under the assumption of inner ring speed of f_{rot} and no outer ring speed, the following formulas can be derived for the kinematic frequencies of the respective damages, which are explained in the following in more detail. z is the number of rolling elements, d the rolling element diameter, D_m the bearing pitch diameter, and α the contact angle [11]:

$$f_{bpfo} = \frac{1}{2} z f_{rot} \left(1 - \frac{d}{D_m} \cos(\alpha)\right) \tag{1}$$

$$f_{bpfi} = \frac{1}{2} z f_{rot} \left(1 + \frac{d}{D_m} \cos(\alpha)\right) \tag{2}$$

$$f_{bsf} = \frac{1}{2} f_{rot} \frac{D_m}{d} \left(1 - \left(\frac{d}{D_m} \cos(\alpha)\right)^2\right) \tag{3}$$

$$f_{ftf} = \frac{1}{2} f_{rot} \left(1 - \frac{d}{D_m} \cos(\alpha)\right) \tag{4}$$

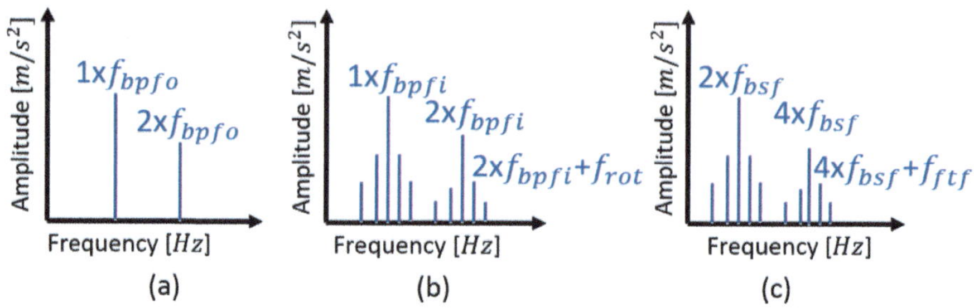

Fig. 1 Local rolling bearing defects on (**a**) outer ring, (**b**) inner ring, and (**c**) rolling element

With a rotating inner ring and a stationary outer ring, local outer ring damage is excited by each rolling element. In the envelope spectrum (spectrum of Hilbert transformation of a signal), shown in Fig. 1, the ball pass frequency outer f_{bpfo} and its multiples can be seen. In contrast, damage of the inner ring is excited more or less strongly depending on its current position due to the influence of the rotor weight. This leads to a modulation of the inner ball pass frequency f_{bpfi} with the speed f_{rot} in the envelope spectrum. The situation is similar for damage of the rolling element. Although the exact rolling path is unpredictable, the rolling element damage is ideally excited twice per rolling element rotation (inner ring and outer ring contact). The envelope spectrum will show twice the ball spin frequency f_{bsf} and its multiples, as well as side bands with a distance of the fundamental train frequency f_{ftf}. This is because, similar to the inner ring faults, defects at the lower half of rotation will be excited with greater load due to the weight of the rotor [6].

Thus, local damages are theoretically explained and should therefore also be easily identified. However, the difficulty lies in the often very different degrees of harmonics, the superposition of different vibrations, and also the differentiation from other frequencies of the technical system, which are in similar frequency ranges. Also, in the case of minor damage, the measurement noise can cover the damage frequencies. For this reason, proper signal preprocessing is very important. Nevertheless, this theoretical knowledge provides a basis for automated classification using machine learning algorithms.

2.1 *Simulation of Bearing Damages*

Since local bearing damage is relatively rare in a technical system and the good condition is predominant, these classification problems are based on highly unbalanced data sets. In order to generate enough data to sufficiently train classification models that can also generalize well, damage data can be simulated. A variety of approaches have been developed for this purpose in recent years. Alternatively, data can also be recorded by means of a test bench or demonstrator and used for training. The simulation can be performed in different ways. On the one hand, complex physical and analytical models can be generated, mostly based on mass-spring-damper models, numerical FEM models, or even simplified mathematical functions, which model the respective excitations as sequences of decaying impulses [5]. In the following, two types (an analytical model and a simple model based on mathematical functions) are explained in more details.

The first simulation model for local bearing damage is the conventional model of McFadden and Smith. It models the excitations as unit impulses $\delta(t)$, which are one for t = 0 and zero otherwise, occurring at the corresponding kinematic damage frequency. The following formula shows this impulse sequence $d(t)$, scaled with the factor d_0 and with an impulse distance of T_d, which is the inverse of the damage frequency. These decay as exponential functions $e(t)$. For inner ring damage, the periodic load function $q(t)$ is included and amplitude modulated with the impulse sequence. A periodic transfer function $a(t)$ is also included, which models the distance to the vibration sensor. Therefore, the modeled local defect can be described by $(d(t)q(t)a(t)) \cdot e(t)$ with amplitudes directly proportional to the load [15]:

$$d(t) = d_0 \sum_{k=-\infty}^{\infty} \delta(t - kT_d) \tag{5}$$

Ho and Randall developed the model further by modeling the decay as free vibration and by varying the time interval between two excitations stochastically [13].

An example of an analytical model is a 5-DoF model (five degrees of freedom model), which provides two degrees of freedom each for the inner ring and outer ring in the radial direction (vertical and horizontal) and one degree of freedom in the vertical direction for the resonator. The model allows to simulate bearing damages with different defect position, shape, and size [18]. Since assumptions must be made for the inner ring and outer ring with regard to stiffness, damping, etc., the modeling effort is significantly greater than for the mathematical model already presented. In general, more complex modeling leads to simulation results that are closer to reality, but also require more assumptions and more system parameters to be determined (Fig. 2).

Figure 3 shows the simulation of an outer ring damage using the traditional model. In the time domain, the impulses, decaying with the resonance frequency, can be seen. In the envelope spectrum, the characteristic frequencies can be observed.

2.2 Feature Extraction and Diagnosis

As mentioned, local damage of rolling bearings is theoretically well described. Therefore, it is reasonable to use the theoretical knowledge in automated diagnosis

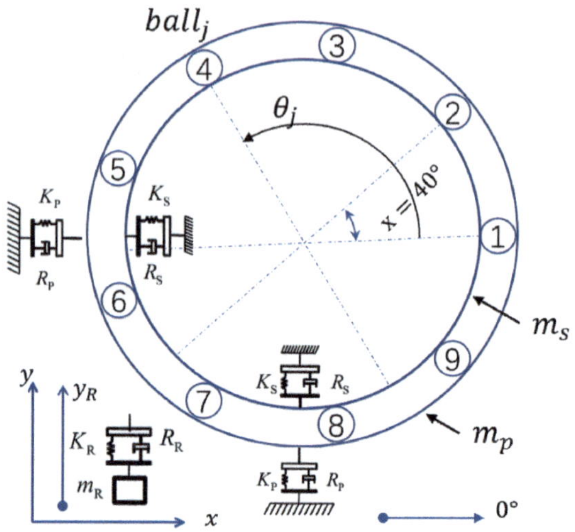

Fig. 2 Analytical model for rolling bearing damage simulation

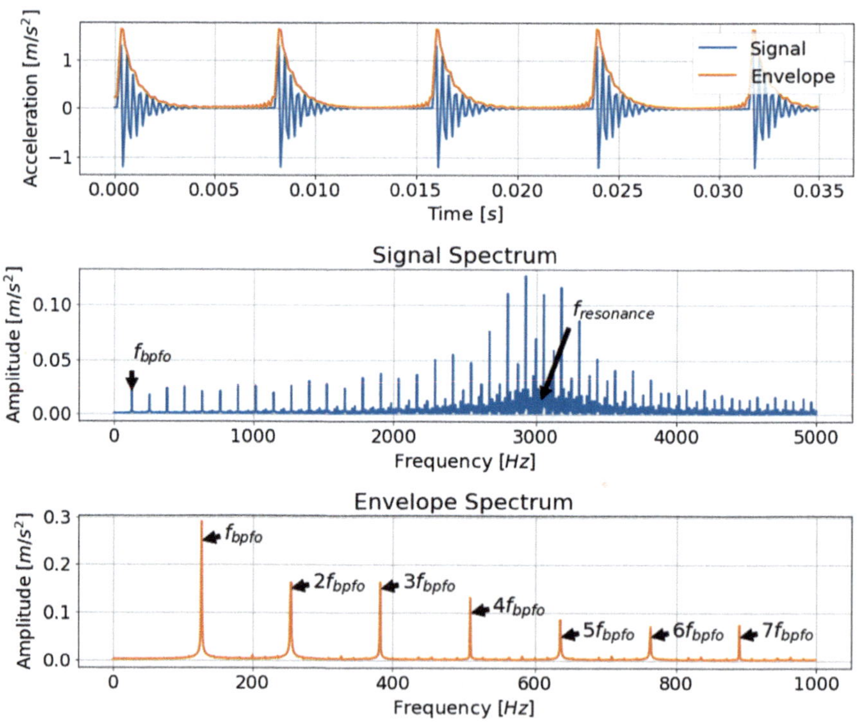

Fig. 3 Simulation of an outer ring damage using the traditional model

using machine learning. In contrast to deep learning with automated feature extraction and selection, the feature extraction in this case can be performed manually in order to incorporate the knowledge into the models. Deep learning models often require a large amount of training data and are more difficult to interpret. For this reason, focusing on feature engineering in combination with classical classification models seems to be of advantage.

In feature extraction from vibration signals, there are several approaches that can be used. For example, statistical quantities can be extracted from the time signal. These have the advantage that they are very easy and fast to calculate, and they can reflect especially the impulsive damage behavior well. Disadvantages are the strong noise dependency and the non-suitability for nonstationary signals. Nevertheless, they are often used in rolling bearing diagnostics. Examples are the crest factor, but also kurtosis. Another approach can be found in the frequency domain. Via the Fourier transform, the signal can be decomposed into harmonic components consisting of frequency and amplitude. Thus, in the frequency spectrum, the frequencies of the signal can be observed. Since excitations caused by rolling element bearing damage occur in the form of impulses, and the excited components therefore vibrate at their resonance, demodulation by means of an envelope spectrum is suitable. In this case, the excitation frequencies are clearly visible. A similar approach is used with the cepstrum [16], where periodicities in the spectrum are determined. Although the envelope spectrum also has the disadvantage of being sensitive to noise, it contains all the frequency information about the damage excitations and is therefore preferred in rolling bearing diagnosis. In order to form features from the various spectra, maximum values can be extracted manually in self-defined intervals (intervals around expected damage frequencies). There are also automatic procedures with subsequent selection, which will not be discussed in detail here. A disadvantage of these automatic procedures is the loss of explainability. There are also procedures for feature extraction from the time-frequency domain. These have a great advantage over the previously mentioned methods in that they can also be used for nonstationary signals. Discrete and continuous wavelet transforms are particularly worth mentioning here, but also the Hilbert-Huang transform [14]. The disadvantages of these methods are the comparatively larger computational effort and the transformation into univariate features. However, these methods are also partly suitable for rolling bearing diagnosis.

After extracting the features, they can be used to train machine learning classification models. When standardizing the features, it is important to store the standard deviations and mean values in order to apply them to the real data. Model selection becomes less important due to the strong focus on the features. Thus, good results could be obtained with different classical models (kernel-based (SVM), ensemble learning (random forest), and neural network (ANN)). Here, the models have exactly as many inputs as features and either one output to predict the class (damage type or good-condition) or as many outputs as classes with respective classification probabilities per class. While in neural networks softmax layers can be used for this purpose, in ensemble learning or, more specifically, in the random forest model, a probability is calculated using the number of trees and their votes. The classification models trained using supervised learning can then be used to diagnose real damage. For this purpose, the same features have to be extracted from the raw data of the plant and, if necessary, have to be standardized. When using simulated data in training, a transfer learning approach is suitable. This means that the model parameters are retrained with features from real damage data. Since, as already described, these damages rarely occur, the models can be iteratively retrained again and again by means of active learning and adapted to the technical system.

2.3 *Example of Application: Demonstrator*

In the following, an example for such classification procedure will be presented. We trained a random forest classifier with simulated damage data. Three damage types were simulated: inner ring damage, outer ring damage, and rotor unbalance. Different feature extraction algorithms, as explained before, were used, to generate features. The trained classifier was then used to classify artificially caused damages at a demonstrator. At the demonstrator multiple computer fans with several rolling bearing damages and rotor unbalance were equipped with acceleration sensors. Bearing damages were manually caused by dissembling, damaging, and reassembling the bearing. Rotor unbalance was created by attaching glue dots to the rotor blades. For different rotor speeds, acceleration and speed measurements were conducted. Especially with statistical quantities of the raw signal as features (mean, standard deviation, skew, kurtosis, crest, median peak) as well as features from the frequency domain (with the help of the discrete wavelet transform), the first four decomposition levels of the simulated signal were computed. For each of the four detail signals and the approximation signal of the last level, the statistical quantities listed above here were calculated. A classification performance of about 70% could be achieved. The confusion matrix of the classification based on statistical quantities can be seen in Fig. 4. It shows that misclassifications mainly occur between both kinds of bearing damage and between healthy and unbalanced condition.

This way, we were able to show that even a model that has never seen the actual damages before during the training can already detect these damages reasonably well.

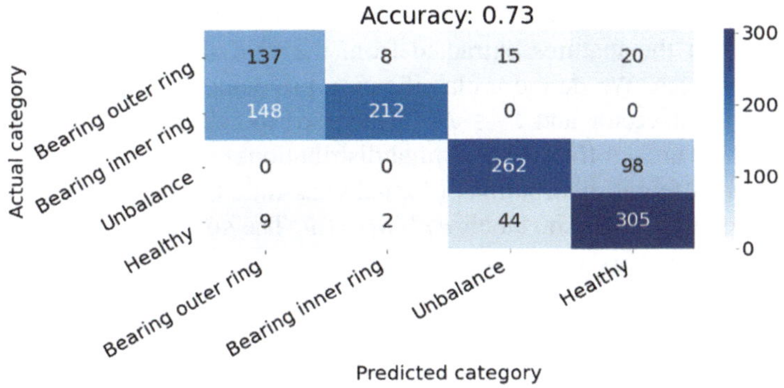

Fig. 4 Confusion matrix for damage classification using random forest model and statistical quantities as features

3 Data Set Energy and Domain Adaptation in Classification Problems

3.1 Introduction

Deploying AI in industry starts with the question of data set composition. Important questions are: "What is the quality of the data?" "What quantity of the data (number of data and dimension of feature vector) can be expected?" "What is the minimal complexity of the ML approach to handle the problem?" On the other hand, answering the questions above is closely related to the so-called domain adaptation problems. In many cases industrial data sets are coming from different domains. Differences in the data can be caused by air humidity and temperature in the production site, different machine operators, different manufacturers of the raw materials, and different wearing states of production plant components, e.g., different operation conditions and wearing status of rolling bearings installed. These differences can be modeled as data coming from different domains. Therefore, the questions regarding data quality and appropriate model choice are highly dependent on the domain in which the data have been collected.

In this section we will give a unifying view based on Shannon's information theory on both problems, i.e., data quality and domain adaptation.

3.2 Domain Divergence and Generalization Bound

In the first step, we will consider the domain adaptation approach proposed by [10]. The main requirement on domain adaptation and/or domain transfer approaches is the construction of the features (which are used for the classification; see previous sections) such that the generic correlation to the actual data domain is almost zero. This means that the features extracted from the data should not depend on the domain differences. We define a classification problem: let $X = \{X_1, ..., X_n\} \in R^{d\times n}$ be an input vector and $Y \in N^{l\times n}$ the corresponding set of labels. Assume further that there are u different but similar distributions over $X\times Y$, $\mathscr{D}_i$, $i = 1, ..., u$. From these u different distributions (domains) assume that only a subset u_L have labels and the rest u_U have no labels $u_L \cup u_U = u$. The goal of the domain-invariant classification machine is to build a mapping $f : X \rightarrow Y$ with a low risk on unlabeled validation data:

$$R_{\mathscr{D}_U}(f) = \sum_i P_{(x,y)\sim\mathscr{D}_i}\left(f(x) \neq y\right), i \in u_U \tag{6}$$

i.e., the sum of misclassification probabilities over all unlabeled data have to be minimized.

To quantify the similarity between a selected pair of domains, we use the definition of H-divergence given by [2]. Domain divergence uses so-called hypothesis classes. The hypothesis classes are defined by the VC dimension of the classification machines and are measuring the complexity of the learner [20]. VC dimension quantifies the maximum number of points that the classifier can separate in all possible ways, and it provides simultaneously the upper bound on the number of training samples needed for a training algorithm to achieve a specific level of performance. For example, the VC dimension of a linear classifier depends on the dimensionality of the feature space, while the VC dimension of the neural networks depends on the number of parameters, neurons and layers, and the type of neuron activation functions. The empirical way of computing H-divergence is the following formulation developed by [2]:

$$d_H^{ij} = 2\left(1 - min_{h\in\mathscr{H}}\left[\frac{1}{\#k_i}\sum_{k_i} I[h(x_{k_i}) = 1] + \frac{1}{\#k_j}\sum_{k_j} I[h(x_{k_j}) = 0]\right]\right) \tag{7}$$

where k_i and k_j are indexes of data from corresponding domains $i \in u_L$ and $j \in u_U$. $\mathscr{H}$ is the considered hypothesis class and I here denotes the indicator function with $I[h(x_{k_i}) = 1] = 1$ if $h(x_{k_i}) = 1$ and $I[h(x_{k_i}) = 1] = 0$ otherwise. The measure tells us that domains are close to each other (small d_H^{ij}) if the selection of classification machines with a particular complexity cannot differentiate between data coming from different domains. With this definition we obtain an H-divergence matrix of all labeled and unlabeled domain pairs. Let us now consider two domains: one source domain $\mathscr{D}_S$ with labeled data and one target domain $\mathscr{D}_T$ with unlabeled data. From [2] we know that the classification machine risk on the unlabeled target domain $\mathscr{D}_T$ is bounded from above by the following expression: As assumed let the training data be $S \in u_L$, then

$$R_{\mathscr{D}_T}(h) \leq R_{\mathscr{D}_S}(h) + g(d, n, \delta) + d_H^{TS} + C \tag{8}$$

here $R_{\mathscr{D}_S}(h)$ is the risk on the training data set, g is a continuous function depending on VC dimension d of the hypothesis class H, n is the number of samples in the training data set and $1 - \delta$ is the probability of choosing S or T. Interpreting the formula (8) tells us immediately that the learning machine generalization bound is a trade-off between training error, which estimates the empirical risk $R_{\mathscr{D}_S}(h)$, and the empirical H-divergence d_H^{TS}. C is the parameter which measures the ability of the hypothesis class to achieve low risk on both training and test data sets. As mentioned in [2], one can control the H-divergence by finding a representation of the input data set (e.g., parametric stochastic encoding R of an input sequence X) where a binary classifier of any VC dimension is not able to distinguish between representations generated by the source domain and the representations generated by the target domain data.

In [10] the authors added the approximated empirical H-divergence from formula (7) to the regularization term of a deep classification machine, for two domains by using adversarial learning approach known from [12].

In the next section, we will show the dependence of proper input (feature space) representation construction by deep unfolding of manifolds, with data quality requirements and domain adaptation problems.

3.3 Quantification of Data Information Content

In many industrial processes, data quality, which is closely related to the domain considerations from the previous section, plays a crucial role by answering questions regarding whether a ML approach will be successful or not. In [17] the authors defined the so-called Data Value Metric (DVM) that quantifies information content of large and complex data sets. The authors are using DVM as decision support to answer questions such as whether the augmentation of an existing data set may be beneficial in specific application domains or not. DVM is defined as follows:

$$DVM(R) = F(R) - P(h) \tag{9}$$

here $F(R)$ is the so-called fidelity which captures whether the original data set or its representation R is useful in order to approximate the parameters of a classification machine $h \in \mathscr{H}$ with chosen VC dimension. The second term in (9) penalizes the complexity h of the selected classification machine. In the following we concentrate on the fidelity term of $DVM(R)$. Based on the information bottleneck principle [1], the DVM fidelity for a representation R is defined as

$$F(R) := \frac{I(R; Y) - \beta I(X; R|Y)}{I(X; Y)}. \quad (10)$$

Here β is a weight hyperparameter, and $I(R; Y)$ is the mutual information between the latest representation R estimated by a learning machine (e.g., deep neural network) and class label vectors Y. X is the original feature vector. Mutual information is one of the most powerful dependency quantification methods between random variables. While commonly used (Pearson) correlation only identifies linear relationships, mutual information detects any type of dependency. Mutual information in terms of Kullback-Leibler divergence D_{KL} is defined as follows:

$$I(R; Y) = D_{KL}(p_{R,Y}|p_R p_Y) = \sum_{r,y} p_{R,Y}(r, y) \log \frac{p_{R,Y}(r, y)}{p_R(r) p_Y(y)} \quad (11)$$

Thus, Eq. (10) rewards high mutual information between the representation R and the class label Y which makes sense since we want the representation R to be useful for classification. Further, it penalizes mutual information between the input data vector X and the representation R conditionally on Y seen. This, again, is high if entropy of X given Y is high. This makes sense since we do not want any information in R which is not useful for the classification, for example, information about the domain.

We illustrate the fidelity effect on learning machine performance by using an example from a plastic extrusion process. In this particular example, we used data from a plastic extrusion process to train a 3-layer SAE (Stacked Auto-Encoder) with 16-dimensional input and binary (domain separation) categorical output. Here we added Gaussian noise to 25% of the data with the corresponding SNR (Signal to Noise Ratio) value in order to simulate a two-domain scenario where the test data is coming from a more noisy environment.

Figure 5 shows the fact that in low-noise environments (high SNR value), the classification machine cannot differentiate between domains (over 44% of the data have been classified as the wrong class by using any representation). Furthermore the deeper (entangled) representation (yellow line in Fig. 5) is more robust (leads to poorer classification performance) for data with lower SNR value (high-noise environments). This is beneficial from the domain adaptation point of view (see Eq. 8).

This can be explained by the so-called manifold hypothesis (MHT). MHT assumes that the probability distributions over, e.g., images, are highly concentrated, i.e., complex data manifolds are intrinsically low dimensional. For instance the MNIST data are represented by the 784-dimensional vector of pixels; however, the images can be characterized by few factors of variations like orientation and shape. The second finding is that from [3], we know that the regularized auto-encoders tend to disentangle the complex manifold structures in the data by unfolding and volume expansion [4]. In order to estimate the disentanglement process by the a(a=0.1)-

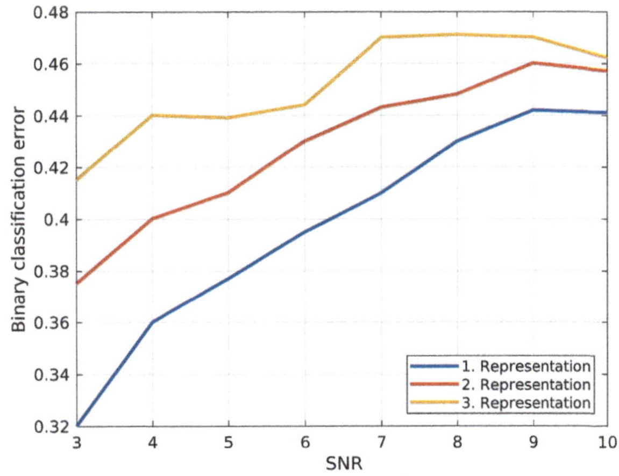

Fig. 5 Development of classification error for different SNR and representations estimated by SAE. "1. Representation" corresponds to the first auto-encoder layer and so on. Relation between deepness of representation and classification error is less pronounced for higher SNR values with decreasing magnitude

trimmed mean, we follow the idea from [4] and estimate the difference between geodesic distance $G_M(i, j)$ (sum of all edge lengths on the shortest path traveled on the manifold surface between two points) and Euclidean distance $G_E(i, j)$ for each point pair of a particular class by

$$G(i, j) = \frac{|G_M(i, j) - G_E(i, j)|}{|G_E(i, j)|}, \forall i, j \in \{1, ..., N_c\}$$

and the disentanglement is estimated by

$$D_c = \frac{1}{N_c(1 - 2a)} \sum_{i,j \in J_a} G(i, j) \tag{12}$$

where J_a denotes the set of arguments (i, j) of G which remain if we remove the smallest and the largest $100a\%$ of the values $G(i, j)$. A small value of D_c indicates the more disentangled manifolds in the class c data space.

From MHT point of view, the linear in-parameter classifier should be more precise for deeper layers of the SAE representations. This is not the case for the domain adaptation problem (binary output) (see Fig. 5) but is true for the original classification problem.

This observation supports the hypothesis that deeper representations are more focused on the separation of data itself and are ignoring multi-domain environments; see Fig. 6.

Figure 6 shows the dependency of the estimated D_c value (measure of disentanglement strength; see Eq. (12)) on the SNR value and the representation depth. We see for all classes that after an initial growth, the D_c steadily decreases with growing representation depth. The decrease seems also to be invariant with respect to the SNR value, which is in accordance with our findings. Below we look at the problem from a more formal point of view.

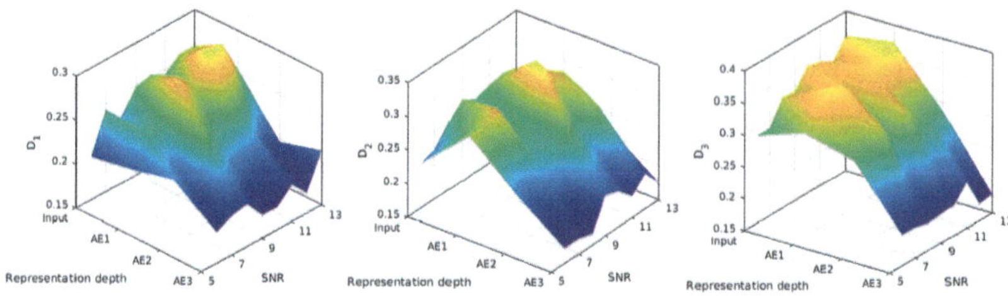

Fig. 6 Dependency of manifold entanglement measure from SNR and representation depth for randomly chosen three classes

Now we want to discuss the empirical finding from above from the Fano's inequality point of view [7] which combines the error probability with the concepts of entropy/mutual information. Using Fano's inequality and result from [19], the fact that $H(P_e) < 1$ for all $P_e \in [0, 1]$ (where H denotes the entropy) as well as the equality $H(Y|X) = H(Y) - I(X; Y)$, one arrives at the following: For two random variables X and Y where $h(X) = \hat{Y}$ is an estimator of Y, it holds that

$$0.5(H(Y) - I(Y; X)) \geq R_{\mathcal{D}_S}(h) \geq \frac{H(Y) - I(X; Y) - 1}{\log(|\mathcal{Y}|)}. \tag{13}$$

Here, $R_{\mathcal{D}_S}(h)$ is the error probability defined in (6) and $|\mathcal{Y}|$ is the support of Y.

We consider a general setting where we have a stacked auto-encoder learning machine with input X whose last (s-th) layer representation is R_s. Connected to that layer is the classification softmax layer with target Y. Also connected to R_s is the domain classification layer whose target is Y^d, the binary parameter describing the domain affiliation. In [10] a loss function for training a neural network for data from two domains was proposed. It simultaneously minimizes the risk on the source domain $R_{\mathcal{D}_S}(h)$ and the H-divergence between source and target domain d_H^{ST} with respect to the last layer of the network R_s.

In the following we will show a simple qualitative proof that the abovementioned modification of objective function is under special circumstances not necessary in accordance to the hypothesis formulated.

By Fano's inequality (13), we know that a small $R_{\mathcal{D}_S}(h)$ implies high $I(R_s; Y)$. On the other hand, a low d_H^{ST} leads to high error probability of the domain classifier (refer to (7)) resulting in a smaller $I(R_s; Y^d)$; see (13). The converse is not true here due to the minimization over all functions in the hypothesis class in the computation of d_H^{ST}. However, maximizing $I(R_s; Y)$ and minimizing $I(R_s; Y^d)$ is a reasonable way of trying to reduce the target risk (8). Intuitively this means that we want as much information as possible about the class Y to be in the extracted feature vector R_s. Simultaneously we want as little information about the domain as possible to be contained in R_s. In summary, the goals of the loss function in [10] are similar, but not equivalent, to maximizing $I(R_s; Y)$ and minimizing $I(R_S, Y^d)$.

Let us now combine these findings with the concept of DVM in order to find a target function for training a feature extraction neural network based on mutual information. We consider only the last layer of our network, R_s whose input is R_{s-1}. We define the fidelity of the s-th layer as follows:

$$F(R_s) = \frac{I(R_s; Y) - \beta I(R_{s-1}; R_s|Y)}{I(R_{s-1}; Y)} = \frac{(1+\beta)I(R_s; Y) - \beta I(R_s; R_{s-1})}{I(R_{s-1}; Y)}. \tag{14}$$

On the other hand, we can also define the fidelity for the domain classification problem:

$$F^d(R_s) = \frac{(1+\gamma)I(R_s; Y^d) - \gamma I(R_s; R_{s-1})}{I(R_{s-1}; Y^d)}. \tag{15}$$

As explained above, we want to maximize $I(R_s; Y)$ and minimize $I(R_S, Y^d)$. When choosing β and γ such that

$$\frac{\beta}{I(R_{s-1}; Y)} = \frac{\gamma}{I(R_{s-1}, Y^d)},$$

this is equivalent to the following problem:

$$T(R_s) = \max_{R_s} \left(F(R_s) - F^d(R_s) \right). \tag{16}$$

The problem (16) defines our proposed target function for the optimization. Finally we have the following qualitative dependency graph:

$$\begin{array}{c} I(R_s; Y) \uparrow \xrightarrow{(14)} F(R_S) \uparrow \xrightarrow{(13)} R_{\mathcal{D}_S}(h) \downarrow \xrightarrow{(MHT)} g(d, n, \delta) \downarrow \\ \downarrow \\ T(R_S) \uparrow \xleftarrow{(16)} F^d(R_S) \downarrow \xleftarrow{(15)} I(R_S; Y^d) \downarrow \xleftarrow{(13)} R_{\mathcal{D}_T}(h) \downarrow \xleftarrow{(8)} \end{array} \tag{17}$$

Equation (17) is a summary of qualitative dependency structure and can be read in the following manner: In first step the mutual information $I(R_S; Y)$ is improved by the training of the deep classification machine, which improves the fidelity of the representation $F(R_S)$. By Fano's inequality (13), the lower bound of the misclassification risk on training data set $R_{\mathcal{D}_S}(h)$ is lowered. By MHT the complexity penalty term $g(d, n, \delta)$ which is a continuous function depending on VC dimension d of the hypothesis class H is lowered and consequently by (8) the upper bound of misclassification risk on target data set is lowered. This implies by Fano's inequality that the mutual information $I(R_S; Y^d)$ and the fidelity $F^d(R_S)$ is lowered. Finally the overall objective function $T(R_S)$ is improved.

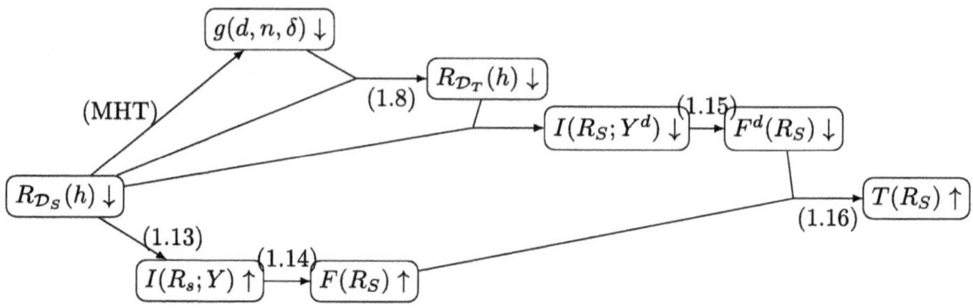

Fig. 7 Qualitative figure on the impact of deep neural network training

Figure 7 presents a qualitative dependency structure giving arguments for the following hypothesis.

Hypothesis 1 *Trained deep neural network generalization performance is invariant regarding multiple domain data scenarios and needs no special architecture or regularization terms to tackle domain adaptation problem.*

Reading graph 7 leads to the qualitative proof of Hypothesis 1. The target of deep neural network training is lowering the empirical risk $R_{\mathscr{D}_S}(h)$. This is done by estimating proper representation R_S. By the manifold hypothesis, the class manifolds get disentangled sequentially by deeper representations, leading to a lower VC dimension d and thus reduced $g(d, n, \delta)$ in the inequality (8). Moreover, since the difference between classes is, generally speaking, larger than the difference between target and source domain, data from both domains share the same class manifolds. This makes differentiating between the manifolds more difficult, resulting in a lower d_H^{ST} in inequality (8).

When combining this with the fact that $R_{\mathscr{D}_S}(h)$ decreases, (8) states that the upper bound on the risk on the target domain $R_{\mathscr{D}_T}(h)$ decreases too.

Since (upper bound of) the risk on both the source and target domain decreases, the representation that the classifier uses, i.e., R_S, has to lose information concerning the domain classification. Therefore, $I(R_S; Y^d)$ decreases. This, in turn, lowers $F^d(R_S)$ (15). Note here that $I(R_S, R_{S-1})$ should be the same for all R_S since R_S is computed with R_{S-1} as sole input.

On the other hand, if $R_{\mathscr{D}_S}(h)$ decreases, by Fano's inequality (13), the upper bound on $I(R_S; Y)$ has to increase and $F(R_S)$ increases as well (14). Combining this and the decrease of $F^d(R_S)$, we arrive at the conclusion that $T(R_S)$ increases.

In summary, according to our hypothesis, training a deep neural network should accomplish domain adaptation on its own. Domain adaptation is meant in the sense of the problem (16).

Let's redefine the fidelity from the Eq. (10) as follows:

$$F(R_s) = \frac{I(R_s; Y) - \beta I(R_{s-1}; R_s|Y)}{I(R_{s-1}; Y)} = \frac{(1+\beta)I(R_s; Y) - \beta I(R_s; R_{s-1})}{I(R_{s-1}; Y)} \tag{18}$$

Here R_s is the two-domain (very similar also for multi domain data) data representation generated by the s-th auto-encoder in SAE. Now let us define a general fidelity of the representation by incorporating domain information into the formula (18). From the domain adaptation theory, we know that a representation is superior if a classifier with a predefined complexity is unable to differentiate between data coming from different domains. Let's define representation fidelity for domain recognition problem:

$$F^d(R_s) = \frac{(1+\gamma)I(R_s; Y^d) - \gamma I(R_s; R_{s-1})}{I(R_{s-1}; Y^d)} \tag{19}$$

Here Y^d is the binary parameter describing the domain affiliation. Desired representation fidelity with high energy to discriminate between classes and low energy to differentiate data coming from different domains is then the result of the maximization problem of the difference between two:

$$F^O(R_s) = \max_{R_s} \left(F(R_s) - F^d(R_s) \right) \tag{20}$$

The problem (20) and the boundary (8) are equivalent in its structural goals. In (8) $R_{\mathscr{D}_T}(h)$ is minimized by minimizing empirical risk of the classifier $R_{\mathscr{D}_S}(h)$, by using Fano's inequality [7] we know that this is equivalent to maximizing the mutual information $I(R_s; Y)$ and consequently the fidelity $F(R_s)$ from Eq. (18). On the other hand, side low H-divergence d_H^{ij} between two domains in (8) also decreases the upper bound of empirical risk. Furthermore we know that low H-divergence implies low performance of the binary classifier and consequently by Fano's inequality low $F^d(R_s)$ from Eq. (19).

Furthermore, we can see that second terms in the denominator of formula (18) and formula (19) are the same up to a constant β and γ. The maximization problem (20) which summarizes two previous findings and equivalently minimization of the boundary for empirical risk from inequality (8) implies the minimization of $I(R_s; R_{s-1})$ from (18) and maximization of the same expression resulted from (19).

3.4 Conclusion

Summarizing previous finding we can see that better representation with high fidelity value $F(R_s)$ leads to low classification error and simultaneously reduces the fidelity of the representation with respect to the domain adaptation problem $F^d(R_S)$.

We have shown empirically growth of classification machine invariance with respect to the deeper representations for different SNR values of a SAE (see Fig. 5). Furthermore we measured class-specific manifold entanglement and have shown dependency of this measure on SNR and representation depth for each class (see Fig. 6). Quantitatively we have proved the hypothesis that the training of deep classification machines automatically lowers the fidelity of the representations regarding domain-specific data components and consequently leads to classification machines with invariance to the domain-specific aspects incorporated into the data.

References

1. Alemi, A. A., Fisher, I., Dillon, J. V., & Murphy, K. (2017). Deep variational information bottleneck. In *Proc. International Conference on Learning Representations* (pp. 1—19).
2. Ben-David, S., Blitzer, J., Crammer, K., & Pareira, F. (2006). Analysis of representations for domain adaptation. In *NIPS, Proc. International Conference on Neural Information Processing Systems* (pp. 137—144).
3. Bengio, Y., & Courville, A. (2013). Representation learning, a review and new perspectives. *IEEE Transactions on Software Engineering, 35*(8), 1798–1828.
4. Brahma, P., Wu, D., & She, Y. (2016). Why deep learning works: A manifold disentanglement perspective. *IEEE Transactions on Neural Networks and Learning Systems, 27*(10), 1997–2008.
5. Cheng, H., Zhang, Y., & Yang, Z. (2015). An extensive review of vibration modelling of rolling element bearings with localised and extended defects. *Journal of Sound and Vibration, 357*, 300–330.
6. Cheng, H., Zhang, Y., & Yang, Z. (2019). *Research on ball bearing model based on local defects* (vol. 1). Springer Nature Applied Sciences.
7. Cover, T. M., & Thomas, J. A. (2006). *Elements of information theory*. John Wiley & Sons, Inc.
8. Doguer, T. (2013). Simulation von Wälzlagerschäden unter Berücksichtigung variabler Betriebsbedingungen. Ph.D. thesis.
9. FAG. (2000). *Wälzlagerschäden, Schadenserkennung und Begutachtung gelaufener Wälzlager*.
10. Ganin, Y., Ustinova, E., Ajakian, H., Germain, P., Larochelle, H., Laviolette, F., Marchand, M., & Lempitsky, V. (2016). Domain-adversarial training of neural networks. *Journal of Machine Learning Research, 17*, 1—35.
11. Gasch, R., Nordmann, R., & Pfützner, H. (2002). *Rotordynamik*. Berlin: Springer.
12. Goodfellow, I., Pouget-Abadie, J., Mirza, M., Xu, B., Warde-Farley, D., Ozar, S., Courville, A., & Bengio, Y. (2014). Generative adversarial nets. In *Adv. Neural Information Processing Systems (NIPS)*.
13. Ho, D., & Randall, R. (2000). Optimisation of bearing diagnostic techniques using simulated and actual bearing fault signals. *Mechanical Systems and Signal Processing, 14*, 763–788.
14. Huang, N., & Attoh-Okine, N. O. (2005). *The Hilbert-Huang transform in engineering*. London: Taylor & Francis.
15. McFadden, P., & Smith, J. (1984). Model for the vibration produced by a single point defect in a rolling element bearing. *Journal of Sound and Vibration, 96*, 69–82.
16. Norton, M. P., & Karczub, D. (2003). *Fundamentals of noise and vibration analysis for engineers*. Cambridge University Press.
17. Noshad, M., Choi, J., Sun, Y., Hero, A., & Dinov, I. D. (2021). A data value metric for quantifying information content and utility. *Journal of Big Data, 8*(82), 1—23.
18. Ruan, D., et al. (2022). Dynamics modeling of bearing with defect in modelica and application in direct transfer learning from simulation to test bench for bearing fault diagnosis. *Electronics, 11*.
19. Tebbe, D., & Dwyer, S. (1968). Uncertainty and the probability of error (corresp.). *IEEE Transactions on Information Theory, 14*(3), 516–518. https://doi.org/10.1109/TIT.1968.1054135
20. Vapnik, V. N. (1998). *Statistical learning theory*. London: John Wiley & Sons, Inc.

Detecting Healthcare Fraud Using Hybrid Machine Learning for Document Digitization

Yannick Becker, Elisabeth Leoff, Henrike Stephani, and Thomas Weibel

Abstract Fraud and abuse are significant and growing problems for healthcare systems worldwide. In Germany, nursing services perform a variety of billable activities, which are invoiced using various paper documents. Some of these are standardized but not firmly formatted; most of them combine printed and handwritten texts. In the event of anomalies or fraud, all this information must be combined. In many cases this is still done manually, which is immensely time-consuming. Our aim is to support German law enforcement agencies detecting and persecuting these cases. We developed a semi-automatic method for document digitization that combines various state-of-the-art machine learning techniques to not only give reliable results but enable user interference and traceability for nonexperts. We evaluate our solution with the help of legal authorities as well as investigators and apply it to data from real criminal investigations.

1 Introduction

Since 1999, the number of people in need of care in Germany has more than doubled, reaching about five million in 2022. Insurance providers for long-term care pay out 25.2 billion euros per year for outpatient care alone—that is, for treating patients in their own homes. Due to several factors (e.g., demographic change and longer life expectancies), this number is expected to rise. At the same time, the number of police investigations concerning billing fraud in nursing services is rising steadily as well. The enormous amount of money involved in healthcare systems, their sheer size, and often complicated, decentralized billing procedures make them an attractive target for fraud. It is estimated that about 10% of healthcare expenditure is wasted due to fraud and abuse [2]. Not only is the monetary loss a problem for the

Y. Becker (✉) · E. Leoff · H. Stephani · T. Weibel
Fraunhofer ITWM, Financial Mathematics Department, Kaiserslautern, Germany
e-mail: yannick.becker@itwm.fraunhofer.de; elisabeth.leoff@itwm.fraunhofer.de; henrike.stephani@itwm.fraunhofer.de; thomas.weibel@itwm.fraunhofer.de

J. Franke, A. Schöbel (eds.), *Statistical Machine Learning for Engineering with Applications*, Lecture Notes in Statistics 227,
https://doi.org/10.1007/978-3-031-66253-9_5

financial stability of the whole social security system, fraud and abuse also reduce the quality of provided services. Patient's health can even be directly endangered due to unqualified care workers, unnecessary procedures, or negligence. In a healthcare system such as in Germany, where the service providers are reimbursed with money paid into health insurance by their insured patients, illegal outflow of money from the insurance companies directly impacts the resources available for services for all patients. Thus, detecting and preventing fraud is an important concern within public and private insurance companies.

Detecting fraud in healthcare faces unique challenges. The billing procedures are complex, often decentralized processes involving service providers, hospitals, accounting service companies, insurance companies, and patients themselves. They are country specific and sometimes even insurance specific. Thus, methods developed for certain data sets and countries are hard to generalize to other settings.

Frequently, not all necessary records and documents are even available digitally. This is especially true for outpatient care in Germany. A variety of billable services are performed and then invoiced monthly to the patient's insurance company using a paper document called "Leistungsnachweis" (service confirmation document). Based on this document, the insurance company calculates the invoice amount and pays the nursing service. Thus, the service confirmation document is usually the basis for fraud detection and criminal investigation in this area. But at the same time, its information is typically not digitally available during police investigations. Other paper documents such as itineraries are also highly relevant. Often times, the fraudulent behavior can only be identified by combining the information from different document classes, which are not digitally available either.

Our Contribution

Fraud detection and investigation in the area of nursing services must still be done manually in many cases, which is immensely time-consuming. Law enforcement agencies investigating a nursing service for fraud have to audit all paper documents by hand to prove the fraudulent behavior and quantify the damage. Comparing the costs of the investigation to the possible damage often leads to dropping the case or reducing its scope. To support the German law enforcement agencies detecting and prosecuting these cases, we developed a semi-automatic investigation approach.

In a first step, we propose a modular method for digitizing typical nursing service documentation. After digitizing the data, it can be investigated to provide legal proof of fraud. We developed the necessary data analysis tools, queries, and visualization in close collaboration with police detectives and public prosecutors. For various instances of possible fraud, we formalized the relevant data fields from several document classes, (cross)relationships between them, and the anomalies of interest. All inference can be done semiautomatically, making the investigation much more time-efficient. This is especially true for a real-life case as presented in Sect. 2.2, where 300,360 documents had to be digitized, combined, and analyzed by hand.

With our solution we aim at building a method that combines various state-of-the-art techniques to not only give reliable results but also enable user interference and traceability for nonexperts. These features are necessary to make results admissible in court, which can be challenging in real-world applications of machine learning techniques. The algorithms are developed and trained on real data coming from actual police cases. The results are very promising. Even if errors occur, e.g., in the structure recognition, due to extensive data augmentation, a small number of reannotations suffice to retrain the algorithms. All in all, the use of our application reduces the investigation expense which we show for a real-world example. Thus, more healthcare fraud can be brought to light.

The rest of this work is organized as follows: First, we give a short overview over relevant literature and methods for fraud detection and machine learning for image processing. In Sect. 2 we then describe the billing system and all relevant documents in the area of nursing services. We pay special attention to the complex relationship between the various actors, documents, and data sources, and give examples how these can be utilized to prove fraudulent behavior. In Sect. 3 we present our semi-automatic method for document digitization. We then outline the fraud detection workflow for the investigator in Sect. 4, including the necessary user interactions, data analysis, and reduction of investigation expense.

Literature on Fraud Detection, Machine Learning, and Image Processing
Fraud detection and deterrence is an important risk management activity. It is extensively applied in various contexts in business and government and covered by a large body of literature (see, e.g., [3] for an overview of common applications). Electronic records and sophisticated data mining or machine learning algorithms allow for a wide range of applications in fraud detection (see, e.g., [14] or [9] for extensive surveys or [8] for data mining approaches in healthcare).

With modern methods, the focus is often on (partially) automated systems that can be combined with statistical and expert knowledge [18]. For some types of fraud, especially in settings such as the healthcare sector, the fraudulent behavior can only be detected by combining different data sources [12]. Established computational approaches such as statistical models, data analysis, and data and process mining typically need data that is digitally available, which is not the case for the paper documents in our application.

Although there are many methods for handwritten or printed character recognition [10], they usually require a first step of identifying each character's position. This is not possible in our application. But existing methods of deep learning (see, e.g., [5, 15]) can be combined to recognize structure elements of each document with only very little training data necessary. Specific content can be identified based on classical image processing, pattern recognition, deep learning, or optical character recognition (OCR) [1, 7, 10, 17].

2 Reducing the Investigation Expense of Complex Healthcare Data Using Data Management

In this section, we describe the complex system of performance documentation and billing regarding home-care, as well as the investigation expense for typical fraud cases. Usually, nursing services[1] already organize and staff their activities (at least partially) using electronic tools. Yet they still have to invoice their services to insurance providers based on individual paper documents. Consequently, an investigation into fraudulent invoices is highly time-consuming. But the documents have logical connections, which can be used in electronic data management to link them and allow for semi-automatic data analysis. To illustrate our application, we first introduce two typical examples of home-care fraud that we will refer to later on.

Example 1 (Invoiced Home Care Performed by Unqualified Nurses) In Germany, each home-care service is grouped into one of four classes according to the Code of Social Law: SGB V, SGB XI, SGB XI 39, and SGB XI 45a. Nurses have different qualification levels. Only nursing specialists are allowed to perform all SGB V services,[2] but nurses with a higher qualification level are paid higher salaries.

We consider a real-life scenario of fraudulent activities by a nursing service operating in Saxony (Germany). Here, invoiced home care was often performed by nurses with insufficient qualifications. A large amount of SGB V services was invoiced, but only a few nurses with the necessary qualifications were employed. This was reported to the local police department by the insurance company. According to insurance companies and law enforcement, this incident is no isolated case.

Example 2 (Invoiced Home Care That Was Not Performed) Consider another typical fraudulent activity: Home care that was invoiced but not performed. One indication may be that, according to the invoiced confirmation documents, a single nurse visited several patients at the same time. Based only on the confirmation documents, we cannot say which one is true. Therefore, we take nurse itineraries into consideration. Services to patients that are not mentioned as a stop in the nurses' itinerary are questionable and can be used as evidence in court.

---

[1] In the following, we always refer to the context of home-care when mentioning nurses or nursing services.

[2] SGB V services belong to 1 of the 5 service groups, SGB XI services to one of around 20 service classes. The differentiation into groups or classes comes from different pricings.

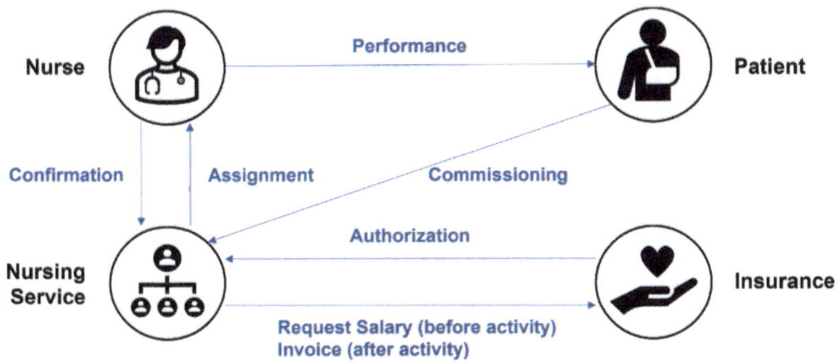

Fig. 1 The complex healthcare system in Germany and links between all roles regarding nursing activities. For reasons of simplicity, we leave out the role of a doctor

2.1 Complexity of the Healthcare System

Healthcare systems involve interactions between professionals, patients, and the organizational systems in care delivery [4]. For home-care services and their invoices to insurance companies, the system of interest involves the following components:[3] patients, nurses, nursing services, and insurance companies. To describe and detect healthcare fraud in our application, the relationships between these roles are important; see Fig. 1.

Complex systems do not derive directly from elementary components and their properties, but from the patterns of their interactions with no central control [11]. Thus, the healthcare system is a complex system in this sense. Since fraud detection in our case relies on paper billing documents, in Sect. 2.2 we will focus on their complexity. We outline all relevant components and their relation to the billing of home-care services. In Sect. 4.2.2, we then use these classifications to quantify and illustrate the reduction of investigation expense that we achieve using our methodology.

2.2 Complexity of Relevant Data and Investigation Expense

Nursing services organize home care using several classes of documents: service confirmation documents or nurses' itineraries. Service confirmation documents are used to invoice services to the insurance companies, whereas the others are used for internal organization. A single service confirmation document provides all information about invoiced services to one patient in 1 month. Even though nursing

[3] SGB V services require a doctor's prescription, which a patient has to request. Since we do not suspect a doctor of fraud, his role and relations to the other involved roles are not considered here.

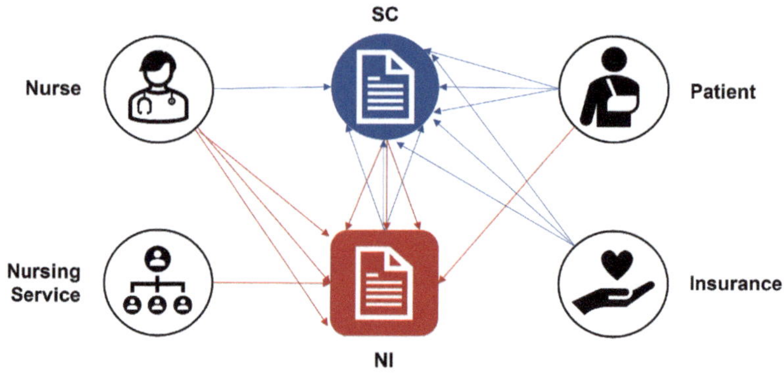

Fig. 2 Links from all roles to all involved document classes (one to one or one to many) relevant for healthcare fraud detection

services use their itineraries only internally, these are potentially relevant for fraud detection as seen in Example 2. If fraud is suspected, law enforcement agencies are allowed to confiscate all documents, even internal ones, within the context of home care services. Thus, there are two classes of data sources relevant for the analysis of home-care services:

(SC) Monthly Service Confirmations
(NI) Nurses' Daily Itineraries

The links between all involved roles and the two document classes are document specific; see Fig. 2. One SC is linked to many nurses but only to a single patient, while the opposite is true for NIs. Furthermore one SC is linked to many NIs and one NI is linked to many SCs. These links are highly relevant for our analysis since we are now able to compare the corresponding information in these two document classes to each other (e.g., finding all NIs for nurses of interest based on signatures from SCs). The following remark shows the volume of SCs and NIs based on their links to the roles involved.

Remark 1 The number of documents to be evaluated directly depends on the time span that is investigated. The amount of NIs further depends on the number of employees and routes included, whereas the volume of SCs further depends on the number of patients. One NI for one employee and day generally has five to seven pages. Regarding SCs, there are potentially two documents for one patient and month due to different codes of Social Law.[4] One SC usually has one page. These features lead to potentially enormous volumes of documents to be investigated by hand. Therefore, public prosecutors regularly have to limit the investigation by limiting the time span, number of patients, or service groups to be investigated.

[4] One SC is always limited to services of either SGB V or SGB XI, SGB XI 39, and SGB XI 45a. Patients often receive services from more than one category.

Example 3 (The Power of Databases for Large Document Volumes) We consider the real-life scenario from Example 2 for an investigation period of 2 years. The criminal investigation included 20 patients out of 120 and 60 nurses. There are potentially 960 SCs and 50,000 pages of NIs. Without loss of generality, each SC has one page and each NI has six. This leads to a document volume of 960 pages for SCs and 300,000 for NIs that had to be looked into by police officers.

One SC or NI is a single document and can therefore be understood as one data source. Here, traditional investigation included 300,960 data sources. Digitizing the documents allows for database storage of all information contained. For this, we created one table set for all SC data and another one for all NI data. Tables within each set are connected, so that one set can be understood as one data source. Thus, in the considered investigation, the transition from paper to electronic storage reduces the number of data sources from 300,960 to 2.

Remark 2 Consider again the scenario from Example 1, where invoiced home care was performed by unqualified nurses. The required data consists of all SCs for the respective time, the list of nurses including qualification levels, and the list of services which contains information about the required qualifications.

The traditional investigation method usually already starts the investigation by reducing its scope, looking, e.g., only at 1 week and nurse instead of the whole data set. The investigation of 1 week potentially relies on 240 SCs (2 SCs for each of the 120 patients). Every single paper SC has to be looked at; the relevant information must be identified and entered into an electronic document by hand. The nurse's qualification and the required qualification have to be looked up in separate documents every time or kept in mind. And lastly, the investigator has to locate the specific information that is actually relevant for the fraud scenario between all the superfluous data on the paper (e.g., the entire month of April 2017 and services that are not restricted by qualification levels). Consequently, the traditional investigation method is very time-consuming and does not scale well with the number of documents.

Remark 3 Not all investigations require both classes of documents. Their relevance depends on the specific case of fraud. For example, in Example 1, NIs were irrelevant, and the fraudulent behavior could be detected by combining data from the SCs with information about the qualifications. For other fraud phenomena, such as services that were not performed, both classes of documents are relevant.

In an investigation, both document classes are provided by the nursing services suspected of fraud, typically on paper. These include sensitive information about the following: patients, nursing services, nurses, and insurance companies. As core data, we use lists of patients (provided by the insurance company or created by investigators), nurses (provided by the nursing service), and German insurance companies. Furthermore, we closely collaborated with home-care nurses to create a standardized list of all billable services, where all synonyms are mapped onto a benchmark service.[5] In total, we distinguish between six different data classes within healthcare fraud detection:

(1) SC
(2) NI
(3) List of patients
(4) List of nurses (including their qualifications and data about the nursing service as a whole)
(5) List of insurance companies
(6) List of services: Standardized terms, synonyms, and required qualifications

SCs are individual for one patient and 1 month; NIs for one nurse and 1 day. However, data (3) and (4) are individual for a nursing service, but global for all patients and nurses within this nursing service. Data (5) and (6) are globally provided as core information since they are identical for all nursing services.

Recall Example 3, where we reduced 300,960 data sources down to 2 using a database. Adding data (3) to (6) is not only about adding four data sources. Since these are semi-core or core information, they are almost always important to consider whenever one of the other 300,960 data sources is viewed. Thus, the number of document views including data (1) or (2) gets multiplied and not added by data (3) to (6). We will address the number of views later when we describe the investigation expense within fraud detection in Sect. 4.2.2.

3 Digitizing Nursing Service Documentation

Optical character recognition (OCR) is a very mature technique for printed texts and in this application domain habitually used by investigators for "string" searching, i.e., searching for certain keywords [10]. However, the OCR recognized content is not sorted (in a stable manner) and has limited context. Because of that, we cannot automatically analyze the documents. We therefore propose a two-step process as described in Fig. 3.

In a first step, we identify the *structure* of the document, i.e., we segment the document according to its content. In a second step, depending on the content type, various methods for *content recognition* are employed. By proceeding this way, the process is modular and adaptable to different kinds of layouts and possible contents. This is especially important, since different nursing services use different layouts for their documents. Structural differences in the content representations are one of the great difficulties for the digitizing process. Both steps, structure segmentation and content identification, will be explained in the following sections.

[5] This list is globally valid for all nursing services and investigations since in Germany there is a predefined set of services which can be billed to the insurance companies.

Fig. 3 The process of digitizing the two different types of documents SC and NI consists of a structure detection and a content detection step. Input/output in blue, processing steps in green

3.1 Structure Detection

Both the NIs and SCs have no universal forms but are layouted by each nursing service individually. Nevertheless, the SCs are invoice documentation that have to fulfill a number of different content requirements (by law). For example, each single service (as described in Sect. 2.2) has to be noted with time and identifiable signature of the respective nurse. Therefore, a table-like structure is the central element of all SCs (see Fig. 4). Similarly, NIs have a common structure because they are itineraries of patients (usually with information on the respective services that need to be performed).

For about 10 years, (deep) convolutional neural networks (CNNs) have become a standard in image processing. While traditional image processing techniques are still used, CNNs provide generic and robust solutions for many different machine learning problems [13]. We therefore propose to learn the typical structures. We will explain this approach for SCs; the approach for NIs is analogous.

Good data is the basis for any machine learning task. However, as we want to keep the amount of manual labor to a minimum, our approach is to train on a very small data basis. For this purpose, a small number of example documents (e.g., 20) need to be annotated manually in the way displayed in Fig. 4. Training with such a small number of samples can easily lead to overfitting. Therefore, we use extensive image augmentation based on classical image processing such as rotation, geometric transforms, and noise addition. At the same time, we take care to only use label-preserving transforms (e.g., no vertical flipping for numbers and signatures, [16]).

We use this data to train a multichannel U-net [15]. This is a so-called encoder-decoder CNN architecture, which was first developed for biomedical image segmentation. In this sector, as in many practical applications, the amount of training data is limited. It has proven to be successfully adaptable to many different image processing tasks. We use the output classes to identify each document's regions, according to their content. As an example, the "day-columns" can be intersected with the "signature-lines" to assign a specific signature cell the correct time context. After this structuring of the document, the next step is the analysis of each individual cell's content.

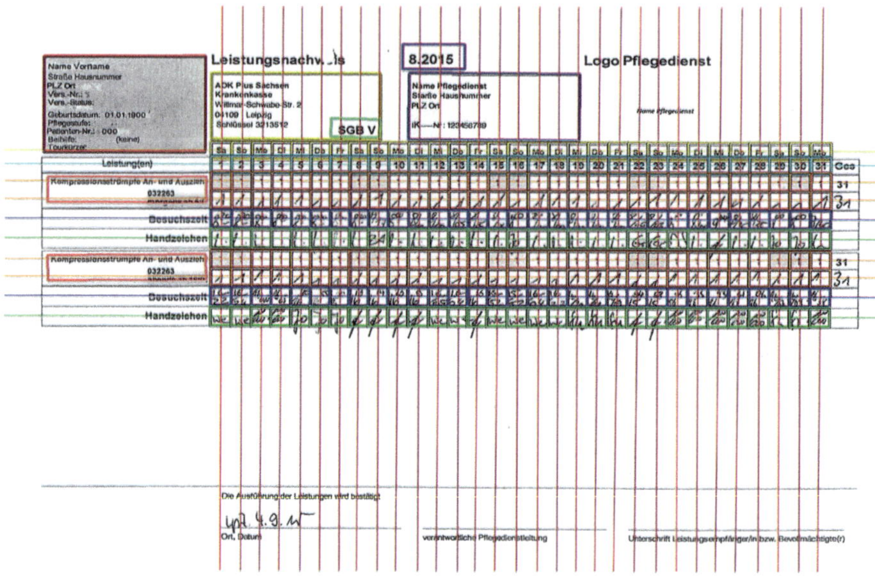

Fig. 4 The image shows how the different structural elements are annotated. Each content class is displayed in a different color

3.2 *Content Recognition*

We present the different kinds of content according to the method used to digitize them. As displayed in Fig. 3, we employ OCR, "global" deep learning (DL) recognition and investigation-specific DL recognition.

OCR

OCR is used on printed data. Fields containing printed data are the field with patient information, insurance information, and service description. To ensure a stable recognition, we furthermore apply a string matching to a given dictionary of these kinds of information (as described in Sect. 2.2). The output of this string matching is a confidence measure that helps the investigator decide if the digitized information is reliable enough to be imported into the database. In Sect. 4.1 we present the necessary user interaction in more detail.

"Global" DL Recognition

The next category is the so-called global recognition category. Global recognition means that we can train machine learning models that are applicable to different investigations and do not have to be retrained for each new case. Very easy examples are the printed day of the week and month. Both could be recognized by using OCR. However, training a simple CNN is more stable. A more difficult example of this category are the handwritten service times. Although OCR is very successfully applied to handwritten digit recognition [10], the single characters need to be

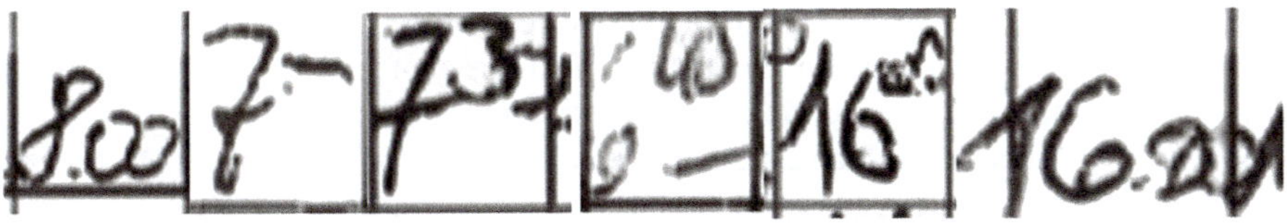

Fig. 5 Different forms of handwritten times that can be found on SCs

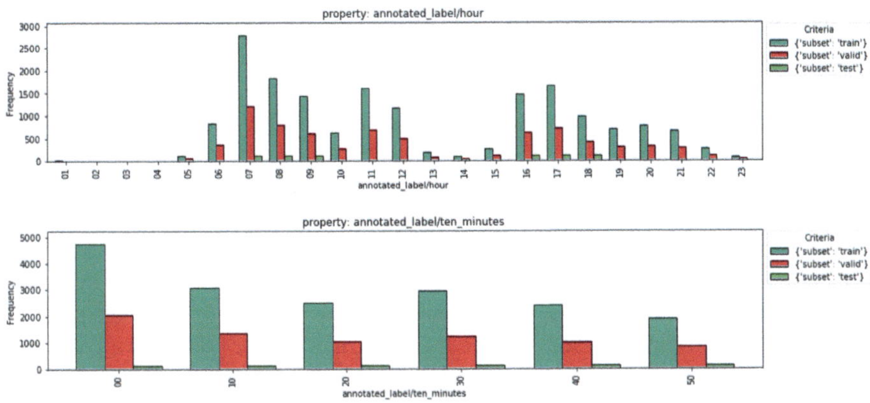

Fig. 6 Training, testing, and validation data set for service time classification

separable for this to work. This is not feasible due to the limited resolution as well as the high variance of possible notations—see Fig. 5 for examples.

Therefore, we use CNNs for the recognition of service times. The variance of the handwritten times is very high. Thus, in contrast to the structure recognition, the necessary amount of training data for this classification task is high. Additionally, the training data needs to represent this variance. To achieve such a high number of training data, we use all available SCs and annotate the service times. The annotation effort is quite large; however, as the aim is a global service time model, any annotation will benefit future solutions and investigations. The accuracy of the classifier will thereby improve over time.

Currently we have about 25,000 annotated times in our solution. To limit the number of classification categories, we train two models: one model that predicts the hour and one model that predicts 10-minute intervals. In Fig. 6 the distribution over the various classes can be seen. As can be expected, the classes are not balanced. We again limit this shortcoming by image augmentation as described before—however, augmentation is difficult as the variance of the data cannot easily be created by classical image transform and the likes. An additional measure to improve the amount of training data and to control its distribution is to generate the data. This can be done by providing a form where handwritten times need to be noted at specific positions and distribute it widely (no annotation and therefore no additional manual labor is needed in this case).

We train on a basis of a keras implementation of the Efficient Net architecture [5, 17]. The model provides a confidence measure that can be used to transfer data over a certain confidence measure directly into the database, while everything below this confidence has to be checked manually. We globally set the confidence to 95% which leads to a relatively low recall—i.e., between 60 and 70%—but a high precision of 95%. For this application not classifying wrongly is very important to build trust in the solution. The recall will improve with more data balance.

Investigation-Specific DL Recognition
Next to the service times, the signature of a nurse is one of the most important things to be digitized. However, in contrast to times, the recognition cannot be done globally, as the signatures (or rather signature initials) are highly personal. The data basis for training a signature classifier has to be generated each time a new investigation takes place. For this reason, to generate a first data set, we used unsupervised classification or clustering [1]. A rough list of employed nurses can be provided by the investigators as well as an example for each signature—both lists are usually not complete. Nevertheless, the list of nurses provides an estimation of the clusters of signatures that should be in the data set. We cluster the signatures of each new nursing service by using the features that were calculated in the previous investigation. The signature clusters can then be sorted and named easily, minimizing the annotation effort to 1–2-day work.

Using the data set created in this way, we again train an Efficient Net [17] architecture. Analogously to the time classification explained above, we use the confidence measure as a threshold for direct usage in the database. As the data is much more self-similar and has less variance than the times, the recall at 95% confidence is much higher. Depending on the specific data set, it lies above 90%. The rest of the data is checked manually as detailed in the next section.

4 Fraud Detection Workflow and User Interaction

In this section, we describe the fraud detection workflow from source to analysis and the user's interaction with our application. The data is handled in two consecutive streams: First the extraction (from source to database), followed by the evaluation (from database to final data interpretation); see Fig. 7.

4.1 User Interaction During Digitization and Data Import

The image processing step includes structure recognition as well as content recognition and can require user interactions for both; see Fig. 7. The structure recognition results in suggested annotations for each document and a corresponding confidence level as presented in Sect. 3. If this is below 95%, users have to

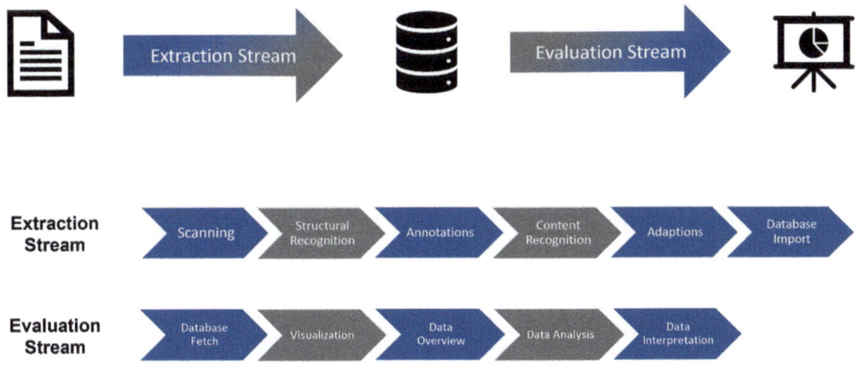

Fig. 7 Pipeline of data handling: From source through database storage to visualization. The database divides the pipeline into two streams: extraction and evaluation of data. Both streams are divided into sub-streams which contain either user interactions (blue) or server-side processes (gray)

correct the annotations for the respective content if applicable. The confirmed annotations can then be used in the algorithmic pipeline to improve the pre-trained weights for further runs. Once the algorithm is familiar with the current document structure, the structure recognition will perform well for all documents of the current investigation.

The content recognition follows in the next step, using different techniques depending on the respective content (cf. Sect. 3). The algorithms suggest allocations for the content and corresponding confidences where relevant, but a person can still change the allocation. For user input concerning low-confidence results, we have to distinguish two cases: *data of choice* and *free data*.

Data of choice means that there is a predefined, finite selection of possible terms (e.g., the list of services). The output of the image processing is matched to this finite list using a string matching procedure, which computes scores following the method *longest common substring* of the R package *stringdist* [19]. The procedure performs rather well since we could adapt it to the predefined lists during the development of the tool. If there is one single suggestion with the best score, this suggestion is directly allocated by the program. In case of more than one plausible suggestion, the user has to decide on the correct allocation by choosing from the predefined list. Contrary to that, *free data* indicates that no global fixed list of possible options is available. An example of this would be the nurses' signatures or the patient names. All possible classes for these have to be either provided by the investigator or created by clustering algorithms during the image processing steps. These classes are then offered as options for the user during data import. Depending on the data, string matching might be employed as well. But since free data depends on the currently investigated nursing service, it is not possible to adapt the string matching method beforehand, so the results are less reliable than for data of choice.

Once all data is allocated, the final results are stored in the database to be available for visualizations and evaluations; see Fig. 7. The user can then proceed with the actual data analysis to detect fraudulent behavior, which we will describe in the following section.

4.2 Fraud Detection Workflow and Reduction of Investigation Expense

4.2.1 Semi-Automatic Data Analysis

Following [6], data visualizations can increase efficiency and effectiveness of fraud detection. When detecting fraudulent home-care invoices, SCs and NIs are the two most important document classes. Therefore, our application has individual analysis sections for both. We closely collaborated with active home-care nurses as well as police officers investigating home-care fraud to develop several relevant data visualization formats. Different table views represent information from all six data classes outlined in Sect. 2.2. Tables enable semi-automatic views into the data. *Semi-automatic* means that the user still needs to filter the tables for anomalies on his own, but is able to view the data aggregated from all SCs or NIs.

The power of the application is about displaying, filtering, limiting, organizing, or, in short, *structuring* data while going through the tables step by step with different aggregation levels. For SCs the aggregation spans from yearly overviews to an entire SC or just single services. For NIs there is a yearly and a monthly view. Furthermore, we joined data from the list of nurses and services (e.g., the qualification levels of nurse and services). The structured data allows the investigator to draw first conclusions, spot single anomalies quickly, and eventually start more advanced fraud investigation.

Example 4 (Detecting Invoiced Home Care Performed by Unqualified Nurses) To illustrate this process, consider again Example 1 dealing with services performed by unqualified nurses. While considering all SCs with data from the 1 week of interest and all available nurses results in 2229 entries in the table view, filtering criteria enable the user to exclude irrelevant weeks and nurses resulting in only 100 actually relevant entries; see Fig. 8.

Example 5 (Detecting Invoiced Home Care That Was Not Performed) An investigation into the scenario from Example 2 requires both the SC and NI data. In the application, the user filters the SC tables to find all services performed by the nurse of interest and week. Via including the NI tables, these services can now be compared to those in the NI data. The tool joins the SC and NI data to enable a deeper investigation. For this we added an investigation or tracing page to the application that we describe in Sect. 4.2.2.

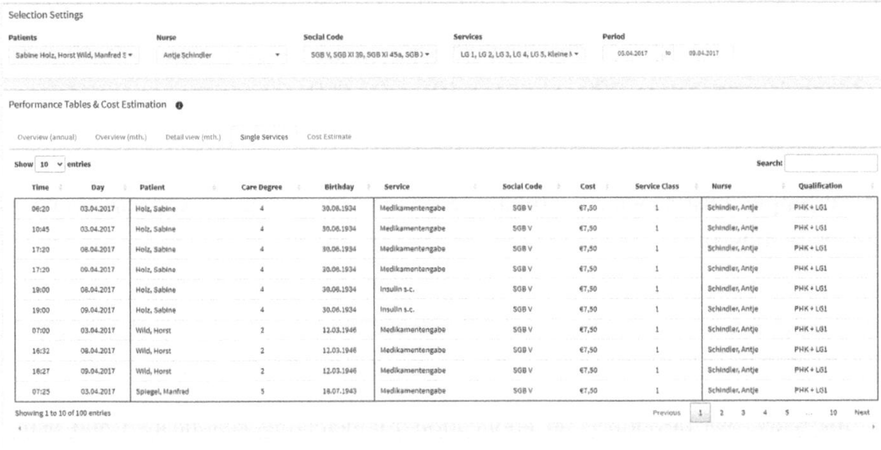

Time	Day	Patient	Care Degree	Birthday	Service	Social Code	Cost	Service Class	Nurse	Qualification
06:20	03.04.2017	Holz, Sabine	4	30.06.1934	Medikamentengabe	SGB V	€7,50	1	Schindler, Antje	PHK + LG1
10:45	03.04.2017	Holz, Sabine	4	30.06.1934	Medikamentengabe	SGB V	€7,50	1	Schindler, Antje	PHK + LG1
17:20	08.04.2017	Holz, Sabine	4	30.06.1934	Medikamentengabe	SGB V	€7,50	1	Schindler, Antje	PHK + LG1
17:20	09.04.2017	Holz, Sabine	4	30.06.1934	Medikamentengabe	SGB V	€7,50	1	Schindler, Antje	PHK + LG1
19:00	08.04.2017	Holz, Sabine	4	30.06.1934	Insulin s.c.	SGB V	€7,50	1	Schindler, Antje	PHK + LG1
19:00	09.04.2017	Holz, Sabine	4	30.06.1934	Insulin s.c.	SGB V	€7,50	1	Schindler, Antje	PHK + LG1
07:00	03.04.2017	Wild, Horst	2	12.03.1946	Medikamentengabe	SGB V	€7,50	1	Schindler, Antje	PHK + LG1
16:32	08.04.2017	Wild, Horst	2	12.03.1946	Medikamentengabe	SGB V	€7,50	1	Schindler, Antje	PHK + LG1
18:27	09.04.2017	Wild, Horst	2	12.03.1946	Medikamentengabe	SGB V	€7,50	1	Schindler, Antje	PHK + LG1
07:25	03.04.2017	Spiegel, Manfred	5	18.07.1943	Medikamentengabe	SGB V	€7,50	1	Schindler, Antje	PHK + LG1

Fig. 8 The table view for SCs using the selection of one nurse (Antje Schindler) and 1 week (3rd to 9th April, 2017). All names represented were anonymized

Once the investigation implies a specific, punishable offense, overview representations of the data are not sufficient anymore, and in-depth investigations start. Rather than searching the tables for all instances of the offense, this task is automated by our tool.

4.2.2 Automatic Data Analysis

Our application has a tracing section to detect predefined cases of fraud as in Example 1. Since the specific kind of fraud and the data of interest need to be selected by the user, the fraud detection could still be seen as semi-automatic. But the analysis routines then scan all relevant data automatically for instances of this fraud, which speeds up the investigation considerably. An investigator often needs to go through a single document multiple times. In Example 2, the investigator cannot memorize all services of a nurse from one SC by just one single view to the document. Without loss of generality, we consider one document view for each information processed (e.g., one view for one signature on a SC). Thereby, we only count relevant information depending on the respective fraud.

Example 6 Again, we consider Example 4 where the investigation comprises three different document classes (and volumes): SC (100; see Fig. 8), list of nurses (1), and list of services (19). In our setting of healthcare fraud, we determine an **investigation expense** as the number of relevant document views. In the case of traditional investigation, we multiply the numbers of views to different document classes, since each entry of one class gets compared to an entry of another class.

However, in the other cases, data of different document classes is joined and thus provided in one representation. This implies:

- Manual investigation: $100 \cdot (1 + 1 + 19) = 21{,}000$
- Semi-Automatic Investigation (Table Views): 100
- Automatic Investigation (Tracing Section): 1

This example illustrates the enormous reduction of investigation expense achieved with our application.

5 Conclusion

In this work, we developed an application for detecting healthcare fraud using hybrid machine learning, where we combined different machine learning techniques and classical image processing, followed by data analysis. We pointed out the complexity in both content and structure of the billing data. Since this data is paper-based, we used deep learning, among other methods, for document digitization. Then, we described the evaluation stream within the application. In particular, we implemented single and aggregated data representations as well as data analysis for certain investigation cases. The necessary analysis and visualizations were developed in close collaboration with law enforcement agencies.

The digitization of the paper-based documents is not completely transferable between those of different nursing services, due to individual layouts and signatures. Making the thus necessary adjustments to the method more efficient and automatic could be researched further in the future. Another possible extension is to automatize sub-processes to a greater extent in order to provide a more general data analysis. Since our developed tool successfully reduced the investigation expense of police officers investigating healthcare fraud, we expect to have a broader interest group including insurance companies in the future.

References

1. Bishop, C. M., & Nasrabadi, N. M. (2006). *Pattern recognition and machine learning*. Springer.
2. Button, M., & Gee, J. (2015). *The financial cost of healthcare fraud*. University of Portsmouth.
3. Cendrowski, H., Petro, L. W., Martin, J. P., & Wadecki, A. A. (2006). *The handbook of fraud deterrence*. Wiley.
4. Chandler, J., Rycroft-Malone, J., Hawkes, C., & Noyes, J. (2015). Application of simplified complexity theory concepts for healthcare social systems to explain the implementation of evidence into practice. *Journal of Advanced Nursing, 72*, 09.
5. Chollet, F., et al. (2015). *Keras*. https://keras.io.
6. Dilla, W. N., & Raschke, R. L. (2015). Data visualization for fraud detection: Practice implications and a call for future research. *International Journal of Accounting Information Systems, 16*, 1–22.

7. Gonzalez, R. C., & Woods, R. E. (2018). *Digital image processing* (Fourth edition, Global edition). New York: Pearson.
8. Joudaki, H., Rashidian, A., Minaei, B., Mahmoud, M., Geraili, B., Nasiri, M., & Arab, M. (2015). Using data mining to detect health care fraud and abuse: A review of literature. *Global Journal of Health Science, 7*, 194–202.
9. Li, J., Huang, K.-Y., Jin, J., & Shi, J. (2008). A survey on statistical methods for health care fraud detection. *Health Care Management Science, 11*(3), 275–287.
10. Memon, J., Sami, M., Khan, R. A., & Uddin, M. (2020). Handwritten optical character recognition (ocr): A comprehensive systematic literature review (slr). *IEEE Access, 8*, 142642–142668.
11. Mitchell, M., & Toroczkai, Z. (2010). Complexity: A guided tour. *Physics Today, 63*, 47.
12. Ortega, P. A., Figueroa, C. J., & Ruz, G. A. (2006). A medical claim fraud/abuse detection system based on data mining: A case study in chile. In *International Conference on Data Mining*.
13. O'Mahony, N., Campbell, S., Carvalho, A., Harapanahalli, S., Hernandez, G. V., Krpalkova, L., Riordan, D., & Walsh, J. (2019). Deep learning vs. traditional computer vision. In *Science and Information Conference* (pp. 128–144). Springer.
14. Phua, C., Lee, V., Smith, K., & Gayler, R. (2010). A comprehensive survey of data mining-based fraud detection research. In *Intelligent Computation Technology and Automation (ICICTA)* (pp. 50–53).
15. Ronneberger, O., Fischer, P., & Brox, T. (2015). U-net: Convolutional networks for biomedical image segmentation. In *International Conference on Medical Image Computing and Computer-Assisted Intervention* (pp. 234–241). Springer.
16. Shorten, C.. & Khoshgoftaar, T. M. (2019). A survey on image data augmentation for deep learning. *Journal of Big Data, 6*(1), 1–48.
17. Tan, M., & Le, Q. (2019). Efficientnet: Rethinking model scaling for convolutional neural networks. In *International Conference on Machine Learning* (pp. 6105–6114). PMLR.
18. Thornton, D., Mueller, R. M., Schoutsen, P., & van Hillegersberg, J. (2013). Predicting healthcare fraud in medicaid: A multidimensional data model and analysis techniques for fraud detection. *Procedia Technology, 9*, 1252–1264.
19. van der Loo, M. P. J. (2014). The stringdist package for approximate string matching. *The R Journal, 6*, 111–122.

Cracks in Concrete

Tin Barisin, Christian Jung, Anna Nowacka, Claudia Redenbach, and Katja Schladitz

Abstract Finding cracks in images of concrete and properly classifying each pixel to belong either to a crack or to the complement of the crack system, i.e., segmenting the image into crack system and background, is a challenging task. Cracks are thin and rough, and being air filled they yield a very weak contrast in 3D images obtained by computed tomography. Enhancing and segmenting dark lower-dimensional structures is already demanding. The heterogeneous concrete matrix and typical computed tomography images having more than 8 million voxels further increase the complexity. Machine learning methods have proven to solve difficult segmentation problems when trained on enough and well-annotated data. However, so far, there is not much 3D image data of cracks available at all, let alone annotated. Interactive annotation is error-prone as humans can easily tell cats from dogs or roads without from roads with cars but have a hard time deciding whether a thin and dark structure seen in a 2D slice continues in the next one. Training networks by synthetic, simulated images is an elegant way out, yet bears its own challenges. In this contribution, we describe how to generate semisynthetic image data to train CNN like the well-known 3D U-Net or random forests for segmenting cracks in 3D images of concrete. The thickness of real cracks varies widely, even within the same sample and within one crack. The segmentation method should therefore be invariant with respect to scale changes. We introduce the so-called

T. Barisin · C. Jung · C. Redenbach
RPTU Kaiserslautern-Landau, Kaiserslautern, Germany
e-mail: christian.jung@rptu.de; claudia.redenbach@rptu.de

A. Nowacka
Fraunhofer ITWM, Kaiserslautern, Germany

RPTU Kaiserslautern-Landau, Kaiserslautern, Germany
e-mail: anna.nowacka@itwm.fraunhofer.de

K. Schladitz (✉)
Fraunhofer ITWM, Kaiserslautern, Germany
e-mail: katja.schladitz@itwm.fraunhofer.de

J. Franke, A. Schöbel (eds.), *Statistical Machine Learning for Engineering with Applications*, Lecture Notes in Statistics 227,
https://doi.org/10.1007/978-3-031-66253-9_6

RieszNet designed for exactly this purpose. Finally, we discuss how to generalize the ML crack segmentation methods to other concrete types.

1 Why Do We Look for Cracks in Concrete?

Concrete is a brittle material; cracks thus occur naturally. Where and at which load they appear as well as their shape, size, and orientation bear valuable information on the particular concrete under investigation. Observations on the surface of concrete samples are state of the art.

Computed tomography (CT) has proven to yield much richer information as it adds the third dimension and images crack regions that are sure not to have been affected by the sample preparation [1, 27]. However, wide application in civil engineering research has been hampered by the conflict between the high lateral resolution needed to properly observe the cracks and the concrete's microstructure on the one hand and the large size of representative samples on the other hand. A dedicated, custom-built CT device promises to resolve this conflict [30].

2 Why Is It Difficult to Segment Cracks in Images of Concrete?

Cracks are thin, essentially rough surfaces in 3D images and paths in 2D images, and thus lower-dimensional. Cracks are filled by air and thus appear dark in images. They are therefore hard to perceive even visually on the image background of usually very heterogeneous concrete structure. See Fig. 1 for a cutout from an industrial optical image and Fig. 2 for 2D sections through reconstructed CT images of several concrete types. In CT images, cracks yield a low X-ray absorption contrast. Moreover, concrete contains pores as well. Being air filled, too, the pores feature approximately the same gray value distribution as the cracks.

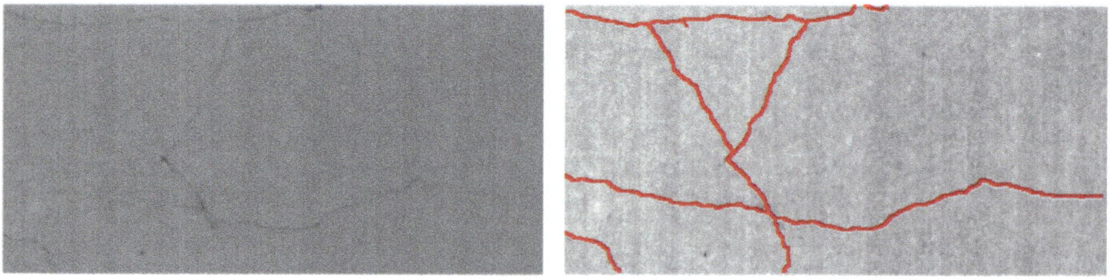

Fig. 1 Example of cracks in concrete as they appear in optical images. Small cutout from a $14{,}000 \times 32{,}000$ pixel 2D image of a concrete panel of dimensions 1.2×2.0 m. The 500 by 250 pixel cutout shown here covers approximately 4 cm by 2 cm. Left: original. Right: crack as segmented by CrackNet [24] see Sect. 3

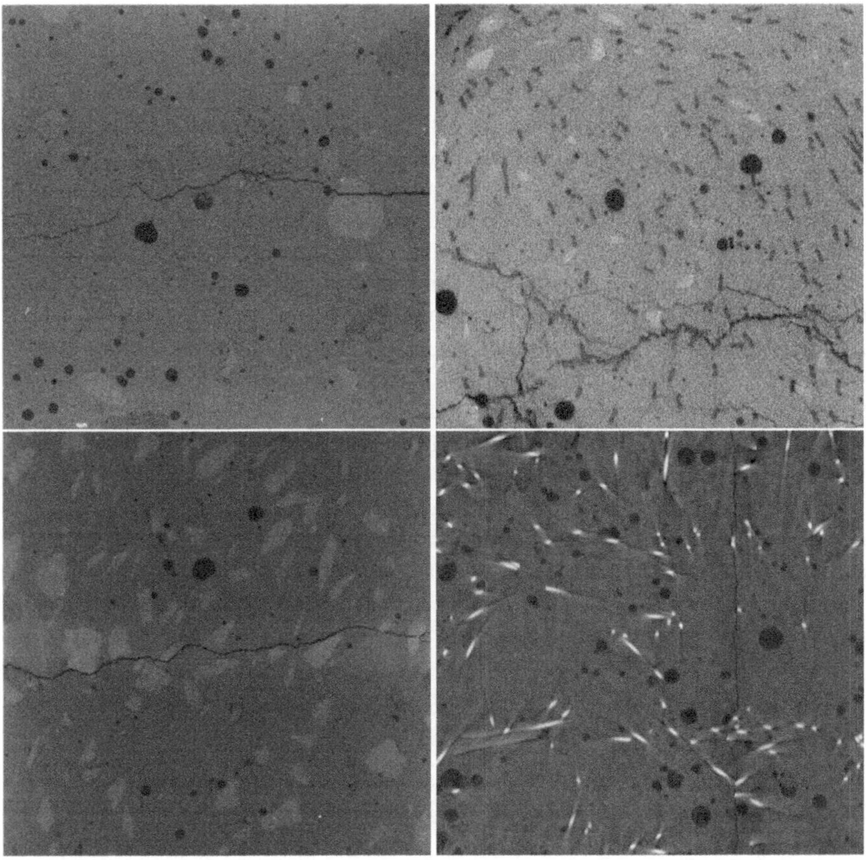

Fig. 2 Examples of cracks in concrete as they appear in CT images. 2D slices from the 3D images. All samples created by University of Kaiserslautern-Landau (RPTU), Civil Engineering. CT scans by Fraunhofer ITWM, except for top right scanned by Fraunhofer EZRT using the setup [30]. Left: Normal concrete. Right: Fiber (bottom steel, top polymer) reinforced concrete. Section images contain 1200 × 1200 (top, bottom left) and 900 × 900 pixels (bottom right) covering squares of edge lengths between 2.25 and 4.3 cm

Within these noisy gray value images, likely to feature imaging artifacts, we want to assign each image pixel or voxel the class crack or the class background. That is, a semantic segmentation problem has to be solved.

Deep learning (DL) methods for solving it are additionally challenged by the problem immanent class imbalance. Obviously, the vast majority of pixels do not belong to the crack system. Hence, besides a lack of training data featuring cracks, the common quality measures do not differentiate well.

3 ML for Crack Segmentation in 2D Images

In civil engineering, manual marking of cracks on surfaces is still rather the rule than an exception. Attempts to automate the task by classical image processing have been made [9], but either do not detect the cracks sufficiently reliably or find far too many false positives.

Machine learning (ML) has been used to segment cracks for at least a decade by now [9]. CNN approaches are abundant, for example [22, 32, 35] (see also the review [17]), but random forests and other tree-based decision rules are applied successfully, too [10, 31]. All of these methods solve the task given sufficiently large and consistent training data. Current research focuses on training with as few annotated crack images as possible. Semi-supervised [23, 32, 35] or weakly supervised [22] learning is explored to increase robustness.

In [24], the well-known SegNet [2] CNN architecture is adapted to segmenting the very fine surface cracks on industrial concrete panels shown in Fig. 1. In only 353 small subimages of size 224×224 pixels, the cracks are marked interactively. Augmentation by rotating and flipping the images yields an eightfold larger data set consisting of 2824 images. Histogram equalization ensures robustness wrt illumination variations. The class imbalance is taken care of by weighting the cross-entropy by the reciprocals of the class sizes as suggested for SegNet [2]. The validation and comparison based on 424 images shows that augmentation is crucial for the prediction quality. The weighting has a significant yet smaller impact, too.

4 Challenge 1: Segment 2D Cracks in 3D Images

In [13], methods for segmenting cracks in 3D images of concrete obtained by computed tomography are reviewed. The so-called Hessian-based percolation is generalized and adapted to 3D. A shape criterion based on the Hessian matrix of second-order gray value derivatives is used to identify dark, locally sheetlike structures. These candidate regions are allowed to grow as long as the local shape does not change too strongly. For more details see [13] and [3]. The latter, more recent comparison includes 3D U-Net and a random forest. Much care is taken to compare the methods in a fair and objective way. Both training ML methods and quantitatively comparing them require data with underlying ground truth. In our case, these are 3D gray value images with known crack structure. Real CT images with known crack pixels are hard to come by. The only way to generate them is interactive annotation of the crack. This is a hard and tedious task, and it is nearly impossible to guarantee spatial consistency when annotating slice wise. Semisynthetic images combining simulated crack structures with the concrete matrix from real CT images therefore play a crucial role. Two ingredients are needed—geometric models for the crack structure on the one hand and a way to generate the heterogeneous background on the other. We discuss two approaches to crack structure modeling as well as how to embed these synthetic cracks in realistic CT images in Sect. 4.2.

4.1 Convolutional Neural Networks for Crack Segmentation

U-Net [29] is a dedicated CNN architecture for semantic segmentation with a well-described 3D extension [8]. U-Net has been used for crack segmentation in 2D images already [18]. In [18] and following the same idea also in [17], patches including crack pixels are chosen with much higher probability during training to counteract the class imbalance. We follow another strategy, closer to the weighting originally suggested in [29], giving crack pixels a higher weight instead of the whole patches. That is, let p_0 and p_1 be the proportion of background (class 0) and crack pixels (class 1), respectively. The losses of crack pixels are weighted by $w = p_0/p_1 > 1$, whereas the loss of background pixels keeps weight $w = 1$.

In semantic segmentation, it is common to allow deviations by a couple of pixels. Applying such a tolerance is particularly reasonable in a case like the cracks where the structure to be segmented is very thin.

4.2 Synthetic Image Data for Training Crack Segmentation CNN

CT imaging is simulated for dimensional measuring where so-called traceability is demanded. Several software packages like ARTIST [5], the ASTRA toolbox [34], and the CT module of CIVA [14] are available; see [6] for a wider overview. However, so far, the results do not reflect the heterogeneity of a typical concrete matrix, where material density and chemical composition locally vary enormously.

Therefore, for the time being, a hybrid technique as described in [3] is favored: Simulated cracks are imposed on CT images of crack-free concrete samples. The gray value distribution in the cracks is derived from the gray value distribution in pores. Edges are smoothed slightly to mimic the partial volume effect in CT images: Gray values in voxels at the boundary of two material components are a weighted mean of the gray values of these components. As a consequence, voxels at the edge of a crack are brighter than those completely within the crack but still darker than concrete matrix voxels.

Fractional Brownian Surface

The fractional Brownian surface is a model for crack structures readily available as a MATLAB function [7]. Its only parameter—the Hurst index H—controls the roughness of the surfaces. The index takes values in [0, 1] with the surfaces getting smoother with increasing H. Finally, the output of the generator [7] is discretized to yield a binary (black and white) image of the crack; see [3] for details.

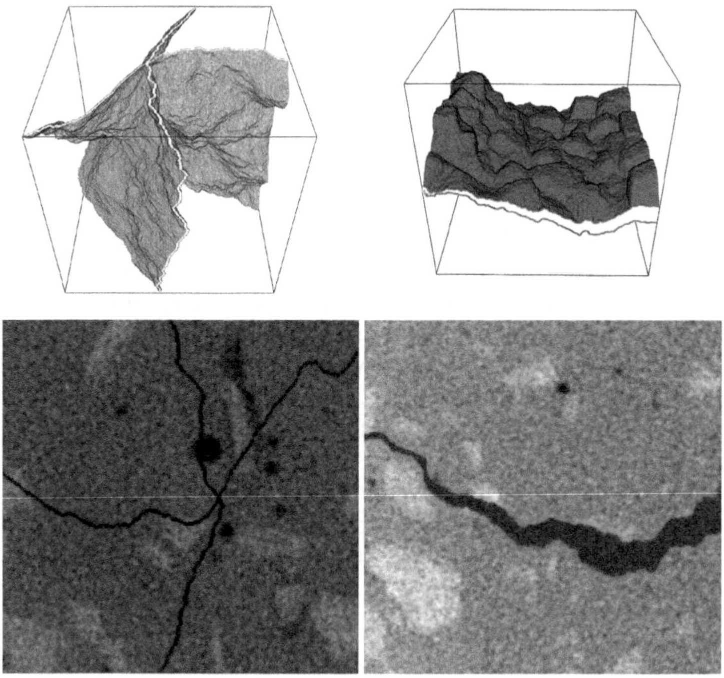

Fig. 3 Examples of simulated cracks in $256 \times 256 \times 256$ voxel images. Left: two fractional Brownian surfaces, both widened to constant thickness 3 pixels and with Hurst index 0.97. Right: a minimal surface from a spatial Voronoi tessellation, width varying according to a Bernoulli random walk with parameter $p = 0.1$. Top: volume renderings. Bottom: 2D slices of the 3D images of the cracks superimposed on CT images of high-performance concrete, scanned with a voxel size of 23.5 μm. The visualized cubes and slices have therefore edge length 6 mm

Combining realizations allows for crossing cracks. An example is shown on the left side of Fig. 3. The local thickness can be altered by superposition with a suitable function. Superimposing simulated cracks and CT images of concrete yields image data for both purposes (training and validation) as shown in [3].

We generate 60 images of size 256^3 voxels with synthetic cracks of constant thickness 1, 3, and 5 voxel edge lengths, 20 images of each.

Minimum Weight Surfaces in 3D Voronoi Diagrams

A new, more versatile model has been suggested recently [21] to better capture features of real cracks like branching, variation of roughness within one crack, and sudden changes of direction. A minimal surface is extracted from a realization of a random spatial Voronoi tessellation. As in the case of the fractional Brownian surfaces, the stochastic nature of the model enables fast generation of as many crack surfaces as needed and naturally captures their heterogeneity.

In its simplest version, generated by a homogeneous Poisson point process, just one parameter drives the tessellation, namely, the intensity of the point system. The denser the point system, the smaller the Voronoi cells and consequently also the facets finally forming the crack surface.

Given the realization of the tessellation, assign weights to all arcs and facets by respective functions. Now consider the cuboidal window given by the image. Intersecting the facets of the tessellation and the four vertical facets of the cuboid yields a set of line segments on these cuboid facets and a set of vertices on each of the vertical cuboid edges. From these segments and vertices, construct a minimal cycle on the surface of the cuboid by first choosing one vertex on each vertical edge and second connecting the four vertices by line segments using Dijkstra's algorithm. Finally, find a minimum-weight surface bounded by the cycle formed by these paths by solving an integer program [21].

Modeling the crack by a system of facets opens a variety of additional degrees of freedom by marking and nesting. Moreover, the crack surface being a subsystem of a spatial tessellation, 2D or 1D branches, can be generated rather naturally.

Locally Varying Thickness by Adaptive Dilation
The locally varying thickness featured by real cracks can be captured by adaptive dilation of the crack. That is, a morphological dilation with size of the structuring element varying according to a size map is applied. Currently, in each 2D x-slice of the 3D image of the crack, the foreground is dilated N_i times by a 2 by 2 pixel square structuring element, where i is the slice index. This results in the crack in slice i being dilated to a $(N_i + 1) \times (N_i + 1)$ pixel square. The $N_i,\ i = 0, \dots, n$ increase with slice index. That is, $N_{i+1} = N_i + k_i$, with increments k_i being Bernoulli distributed with parameter $0 < p < 1$. See Fig. 3, right side for an example.

4.3 Results: 3D U-Net on Synthetic Cracks

We use a 3D U-Net with exactly the original architecture [8] and the same activation function. As in [29], a dropout layer at the bottleneck between decoder and encoder helps to prevent overfitting. The dropout probability is 0.5. The learning rate decays by 0.5 after every fifth epoch. As in the 2D case, we weight the binary cross entropy loss by the reciprocals of the class sizes in order to cope with the class imbalance.

We split the 60 images described above into 9 for training, 3 for validation, and 48 for evaluation. The 256^3 voxel images are divided into patches of 64^3 voxels overlapping by 14 voxels to avoid edge effects. Augmentation by rigid motions, cropping and zooming, sharpening and blurring, as well as gray value distortions and gray value variations triples the number of patches yielding 768 for training. Gray values are normalized.

The 3D U-Net trained this way segments the synthetic cracks very well, keeping mean recall and precision above 0.9 even for crack width 1 and no tolerance. In particular for these thin cracks, it thereby performs better than all classical competitors [3].

5 Challenge 2: Segment Real Cracks

The most striking difference of real cracks in comparison with the fixed width synthetic cracks considered so far is their widely varying thickness. A range of 1–20 voxels is not at all unusual. The example featured in Table 1 shows that 3D U-Nets trained solely on fixed width cracks fail to generalize to these multiscale cracks.

5.1 *Multiscale 3D U-Net*

Image pyramids are a fast and straightforward approach to processing multiscale structures. This image representation is based on downsampling the image size by fixed factors, here 2. The new voxel gray values are interpolated, here by splines of order 3. Iteration yields a sequence of images starting with the original one and getting coarser with each step. We use just three levels, the original image, one with doubled, and one with quadruplicated voxel edge length.

Downsampling shrinks thicker cracks to thinner ones in the number of voxel edge lengths. A crack width 20 on the original level is way out of the range the U-Net has been trained on. However, this original crack width corresponds to 5 after downsampling twice by factor 2. Hence, thin cracks can be segmented at the lower levels, thicker cracks at the higher levels of the image pyramid.

Our multiscale 3D U-Net consists of a 3D U-Net on each level of the pyramid. The result is upsampled to the original image size. The results of all levels are fused by a simple voxelwise maximum. More precisely, the same threshold is applied on each level separately, and the voxelwise maximum is applied to the resulting binary images. Consequently, a voxel is segmented as crack voxel as soon as it is found to be a crack voxel on one pyramid level. This strategy has proven to work [20], it is however costly.

5.2 *3D U-Net Fine-Tuned Using Multiscale Cracks*

The synthetic Voronoi tessellation-based multiscale cracks from Sect. 4.2 enable an alternative strategy, closer to a natural learning process: We take our 3D U-Net trained on the images featuring constant width cracks from Sect. 4.3 and further train it on 5 of these multiscale crack images.

Table 1 Two real crack systems, segmented by 3D U-Net trained on cracks of thickness 1, the multiscale 3D U-Net, the fine-tuned 3D U-Net, and the Riesz network. 2D slices through the 3D images

CNN	Normal concrete	hpc
Original	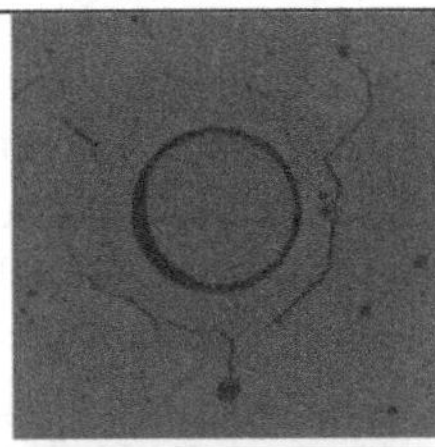	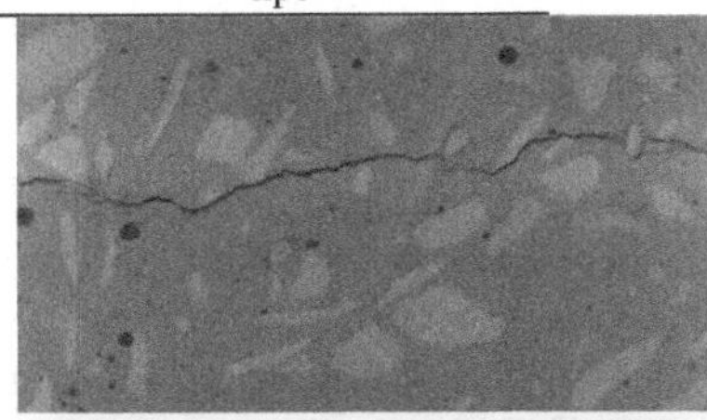
Multiscale 3D U-Net trained on 1 voxel wide cracks	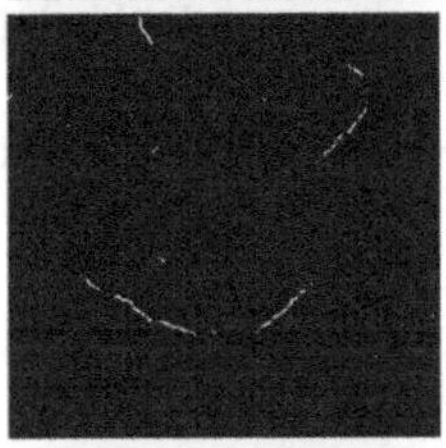	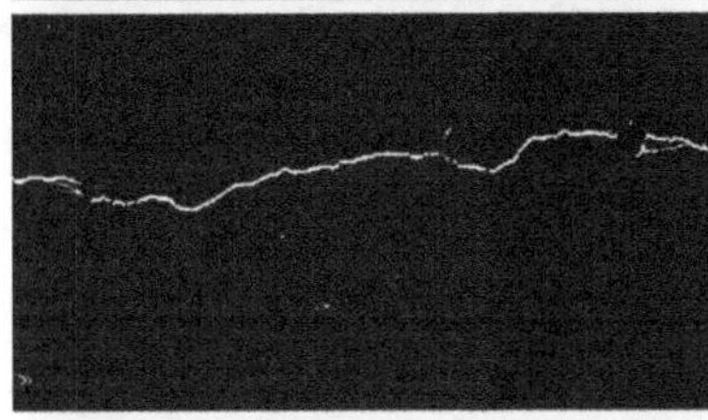
Multiscale 3D U-Net from 5.1	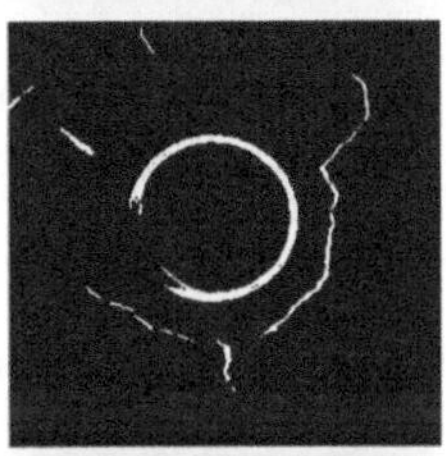	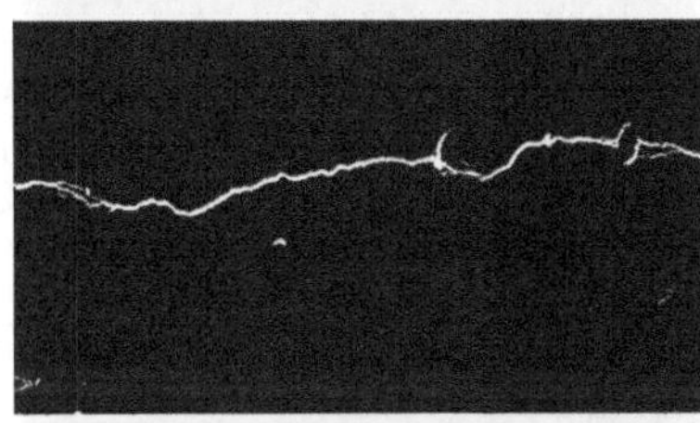
Fine-tuned multiscale 3D U-Net from 5.2	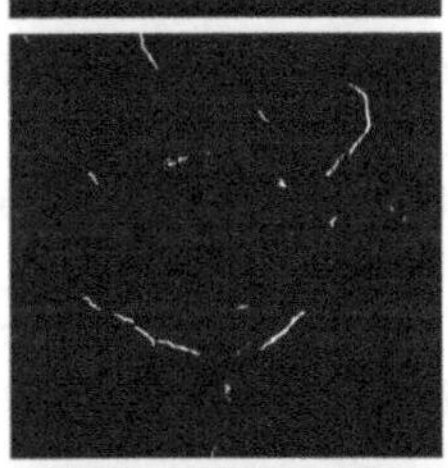	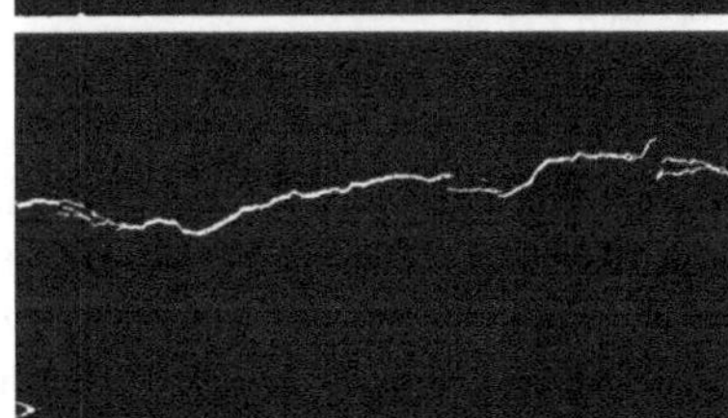
Riesz network as described in 5.3	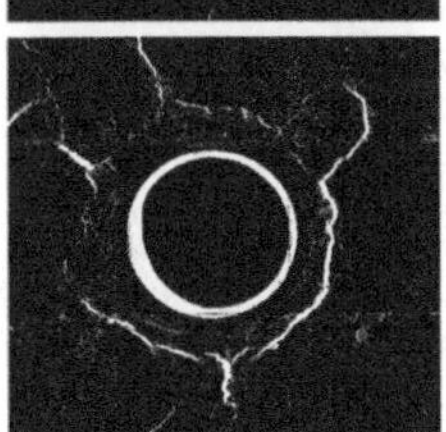	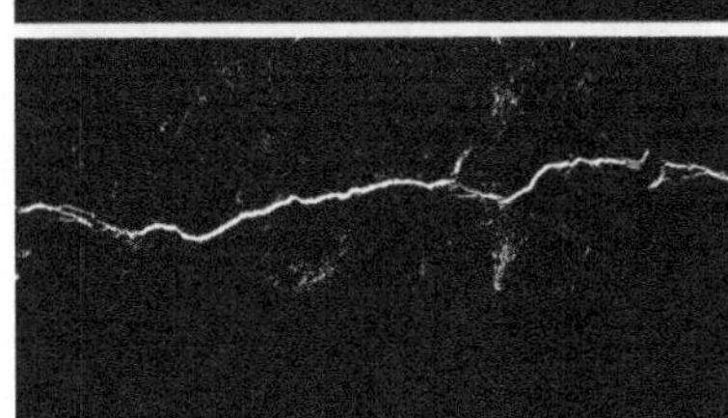

5.3 A Dedicated Network: The Riesz Network

A far more elegant solution is to impose scale invariance on the neural network as devised by Barisin [4]. The main idea is to replace the convolutions in the encoder by Riesz transforms whose scale equivariance is known and has been exploited in image processing before [12, 33].

First- and second-order Riesz transforms have proven to be useful as low-level features in analyzing and classifying textures [11], and estimating orientations [28] in 2D images. Unser [33] built steerable wavelet frames, so-called Riesz-Laplace wavelets based on the Riesz transform. These are the first ones utilizing the scale invariance property of the Riesz transform, and have inspired the quasi monogenic shearlets [16]. Recently, Riesz transforms have been applied in DL as supplementary features to improve robustness [19].

The first-order Riesz transform $\mathscr{R} = (\mathscr{R}_1, \cdots, \mathscr{R}_d)$ of a d-dimensional image $f : \mathbb{R}^d \to \mathbb{R}$ is defined as

$$\mathscr{R}_j(f)(x) = C_d \lim_{\epsilon \to 0} \int_{\mathbb{R}^d \setminus B_\epsilon} \frac{y_j f(x-y)}{|y|^{d+1}} dy,$$

where $C_d = \Gamma((d+1)/2)/\pi^{(d+1)/2}$ is a normalizing constant and B_ϵ the ball of radius ϵ centered at the origin. Higher-order Riesz transforms are derived as sequences of first-order Riesz transforms:

$$\mathscr{R}^{(j_1, j_2, \ldots, j_d)}(f)(x) := \mathscr{R}_1^{j_1}(\mathscr{R}_2^{j_2}(\cdots(\mathscr{R}_d^{j_d}(f(x))))$$

for $j_1, j_2, \ldots, j_d \in \{1, \cdots, d\}$. Here, $\mathscr{R}_i^j$ refers to applying the Riesz transform $\mathscr{R}_i$ sequentially j times and $\mathscr{R}_i^0(f) = f$ for $i \in \{1, \cdots, d\}$ and $j \in \mathbb{N}$.

The Riesz transform has several desirable properties like translation invariance and an all-pass filter property [33]. Denote by $L_a(f)(x) = f(\frac{x}{a})$ a dilation or rescaling operator with size parameter $a > 0$. The scale equivariance of the Riesz transform expresses itself as commutation with scaling, formally

$$\mathscr{R}_j(L_a(f))(x) = L_a(\mathscr{R}_j(f))(x).$$

The base layer of a Riesz network consists of a linear combination of first- and second-order Riesz transforms. So far, systematic investigation and comparison of the Riesz network's abilities [4] are restricted to the planar case $d = 2$. Its extension to the spatial case $d = 3$ is nevertheless straightforward: In the Kth layer, the ith input channel is linked to the jth output channel by the linear combination:

$$J_K^{(j,i)}(f) = C_0^{(j,K)} + \sum_{k=1}^{3} C_k^{(j,i,K)} \mathscr{R}_k(f) + \sum_{\substack{k,\ell,m \\ k+\ell+m=2}} C_{k,\ell,m}^{(j,i,K)} \mathscr{R}^{(k,\ell,m)}(f), \quad (1)$$

where the parameters $\{C_0^{(j,K)}, C_k^{(j,i,K)}, C_{k,\ell,m}^{(j,i,K)} | k, \ell, m \in \{0, \cdots, 3\}, k + \ell + m = 2\}$ are learned during training. The offset parameters $C_0^{(j,K)}$ are the same for all input channels. Additionally, each pair of input and output channels generates $3+3\cdot 2 = 9$ trainable parameters resulting in $c(K+1) + c(K)c(K+1)9$ trainable parameters for the whole layer, where $c(K)$ is the number of input channels and $c(K+1)$ the number of output channels of the Kth layer. Assume now that the Kth network layer takes input:

$$F^{(K)} = \left(F_1^{(K)}, \cdots, F_{c(K)}^{(K)}\right) \in (\mathscr{L}_2(\mathbb{R}^d))^{c(K)}$$

with $c(K)$ feature channels and generates output

$$F^{(K+1)} = \left(F_1^{(K+1)}, \cdots, F_{c(K+1)}^{(K+1)}\right) \in (\mathscr{L}_2(\mathbb{R}^d))^{c(K+1)}$$

with $c(K+1)$ channels. Then, the output in channel $j \in \{1, \cdots, c(K+1)\}$ is

$$F_j^{(K+1)} = \sum_{i=1}^{c(K)} J_K^{(j,i)}\left(F_i^{(K)}\right). \tag{2}$$

The Riesz network is built from layers consisting of the transformation batch normalization, Riesz layer, and ReLU (rectified linear unit) activation, in this order. Batch normalization improves the training capabilities and avoids overfitting, ReLUs introduce nonlinearity, and the Riesz layers extract scale equivariant spatial features. A network with $K \in \mathbb{N}$ layers can be simply defined by a K-tuple specifying the channel sizes, e.g., $(c(1), \cdots c(K))$. An additional final layer linearly combines the features from the previous layer followed by a sigmoid function yielding the desired scale-invariant probability map as output.

We apply a three-layer Riesz network with channel sizes (16, 16, 32). It has $(16 + 1\cdot 16\cdot 9) + (16 + 16\cdot 16\cdot 9) + (32 + 16\cdot 32\cdot 9) + (1 + 32\cdot 1) = 7153$ trainable parameters. This is most remarkable when comparing to U-Net with more than 2 million parameters.

5.4 Results

We tested the three approaches described above on the remaining 15 images (numbered 1–15) featuring Voronoi tessellation-based multiscale cracks. We report precision, recall, and F1 score for tolerances 0 and 1 in Table 2. As had to be expected, results for tolerance 2 are even better than those for 1. However no qualitative differences occur. All three methods perform worst on image 7 whose gray value distribution is indeed strangely distorted; see the original in the first row

Table 2 Results of the multiscale and the fine-tuned 3D U-Net and the three-layer Riesz network on 15 images with synthetic multiscale cracks for tolerances 0 and 1. The images yielding the minimal values are named in parentheses

Tolerance 0									
Method	3D U-Net, multiscale			3D U-Net, fine-tuned			Riesz network		
Measure	Min	Mean	Median	Min	Mean	Median	Min	Mean	Median
Precision	0.822 (10)	0.920	0.931	0.880 (10)	0.976	0.984	0.671 (10)	0.848	0.854
Recall	0.182 (7)	0.859	0.929	0.046 (7)	0.878	0.963	0.805 (7)	0.918	0.920
F1 score	0.307 (7)	0.869	0.918	0.088 (7)	0.897	0.970	0.773 (10)	0.872	0.898
Tolerance 1									
Method	3D U-Net, multiscale			3D U-Net, fine-tuned			Riesz network		
Measure	Min	Mean	Median	Min	Mean	Median	Min	Mean	Median
Precision	0.962 (12)	0.989	0.992	0.991 (4)	0.997	0.998	0.846 (10)	0.923	0.917
Recall	0.374 (7)	0.923	0.984	0.096 (7)	0.913	0.989	0.942 (7)	0.983	0.985
F1 score	0.544 (7)	0.946	0.986	0.175 (7)	0.930	0.994	0.912 (10)	0.957	0.961

of Table 3. It is however remarkable how robust the Riesz network reacts compared to the two 3D U-Nets (Table 2).

6 Challenge 3: Fiber-Reinforced Concrete

Traditionally, the brittle material concrete is reinforced by thick steel rebars. Mixing much smaller fibers directly into the concrete matrix to improve mechanical strength and allow for freer design has a long history, too. Attempts to replace the rebars completely are more recent and subject of current research. Examples of steel and polymer fiber-reinforced concrete samples are shown on the right side of Fig. 2.

These fibers increase the complexity of our segmentation problem even further. Polymer fibers are thin, dark structures and thus easily mistaken to be cracks. Steel fibers absorb X-rays much stronger than the other microstructure components and thus require adapted imaging parameters which in turn result in more similar gray values for pores, cracks, and concrete matrix. Even worse, often, steel fibers feature a thin dark halo either due to having been coated by a resin enhancing bonding or due to imaging artifacts arising from the high local difference in absorption. As a consequence, our 3D U-Nets trained on concrete without fibers fail; see the central column of Fig. 4.

In order to train a CNN robust with respect to reinforcing fibers, too, we generate synthetic crack images with the background acquired from CT images of fiber-reinforced concrete samples. We train a 3D U-Net as described in Sect. 4.1, now on 24 images featuring one or two cracks of widths 1, 3, and 5 on backgrounds of normal and high-performance concrete, as well as concretes reinforced by polypropylene and steel fibers. Detailed results are presented in [26]. Here, we demonstrate the effect using two examples in Fig. 4.

7 Discussion and Conclusion

Overall, ML is definitely able to solve our segmentation problem. This is however only possible, thanks to the simulated training data. A wide variety of new research topics arise from this. Most prominent is the question which properties of the synthetic training images enable generalization to their real-world counterparts. The Riesz network is a new, promising alternative to the abundant CNN. Already in the very experimental form presented here, it can compete with 3D U-Net. The strikingly low number of parameters that have to be fit during training renders it very attractive in all cases where scale invariance is a natural requirement and annotated training data are scarce.

For the practical problem to be solved, challenges remain, too. First of all, generalization to unseen concrete formulations. Does it just require additional training using the well-known simulated cracks and one CT image of the new

Table 3 Examples of segmented synthetic cracks. 2D slices through the 3D images. Image 7 is the one where the 3D U-Nets seem to fail whereas the Riesz network still reaches considerable accuracy. Image 10 is also a hard one according to the accuracy values in Table 2 yet by far not as difficult as 7

CNN	Image 7	Image 10
Original	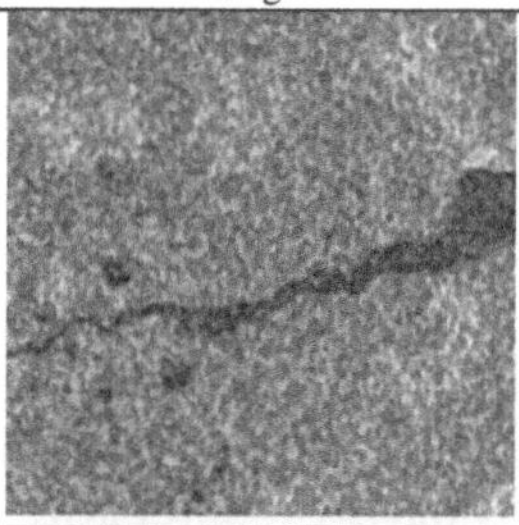	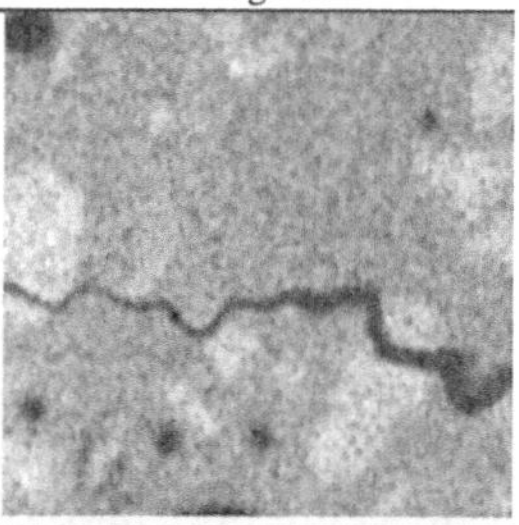
Multiscale 3D U-Net trained on fixed width cracks, 5 voxels (left) and 1 voxel (right)	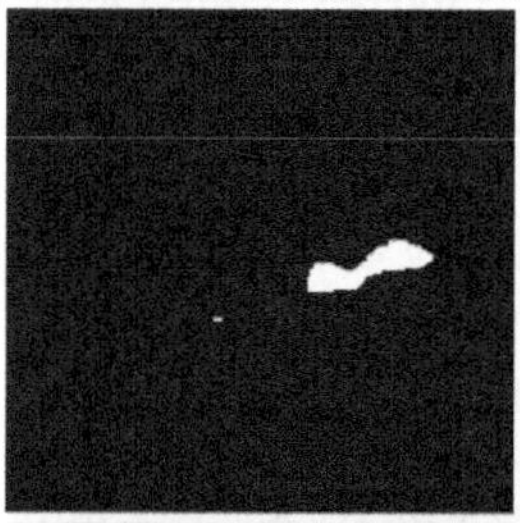	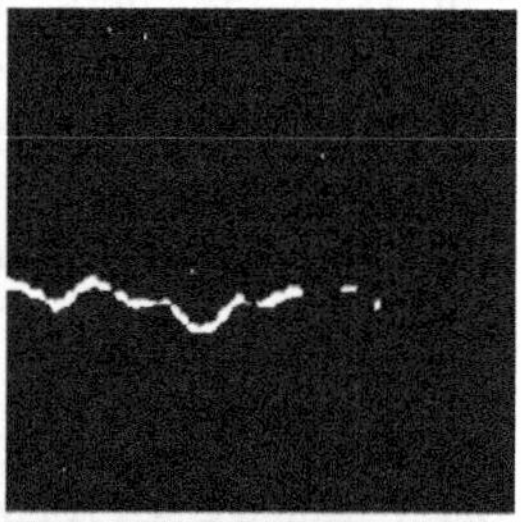
Multiscale 3D U-Net from 5.1	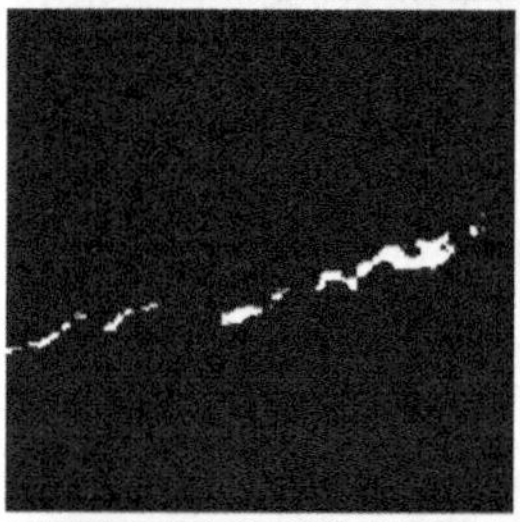	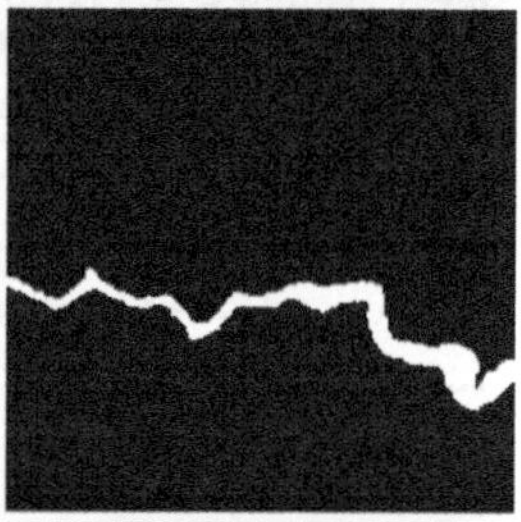
Fine-tuned multiscale 3D U-Net from 5.2	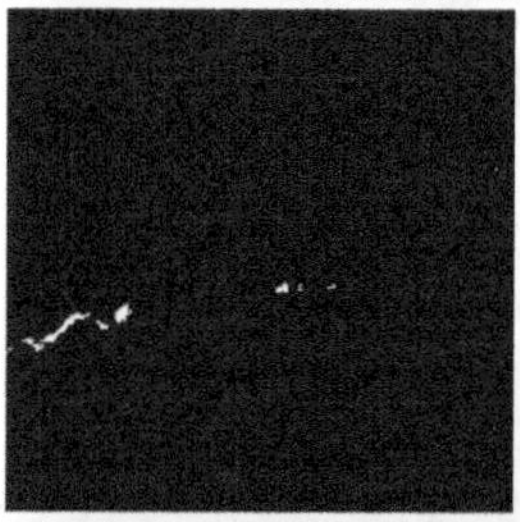	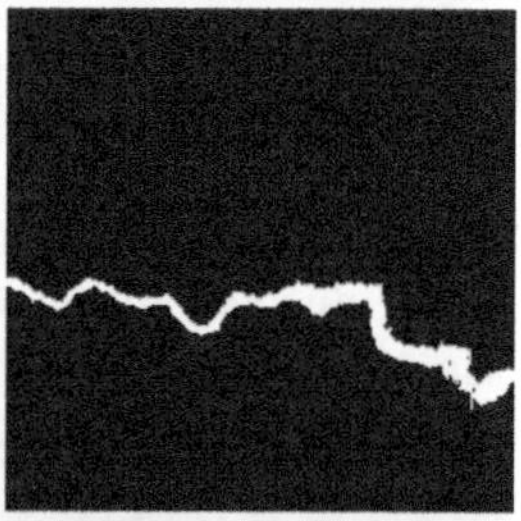
Riesz network as described in 5.3	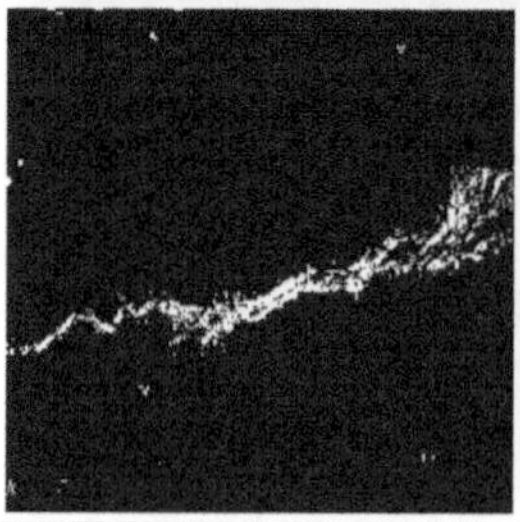	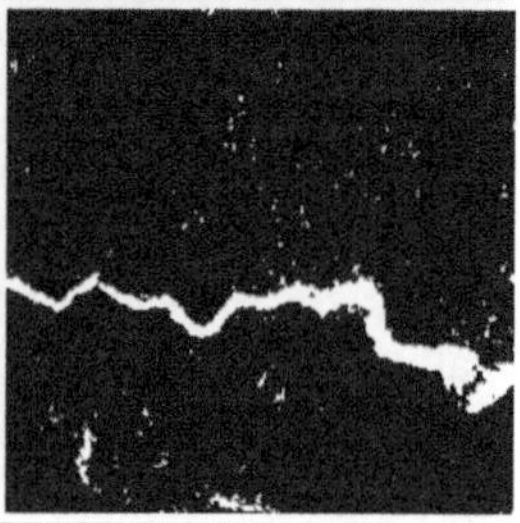

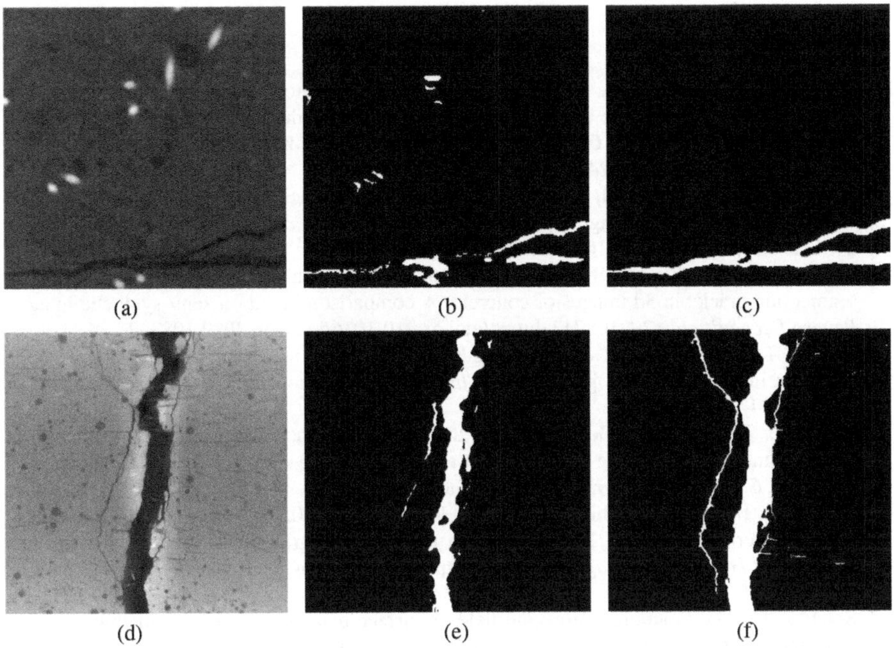

Fig. 4 Examples of segmentation results for fiber-reinforced concrete. The 3D U-Net not trained on FRC mistakes dark regions close to the bright steel fibers as cracks (**b**) and does not detect the thin branches of the crack in the PPFRC (**e**). The fine-tuned 3D U-Net performs much better although classifying some pp fibers as crack (**f**). (**a**) SFRC, original. (**b**) 3D U-Net. (**c**) 3D U-Net, trained on FRC. (**d**) PPFRC, original. (**e**) 3D U-Net. (**f**) 3D U-Net, trained on FRC

concrete in the spirit of a calibration? Or is it possible to train a model that finds cracks of all shapes in images of all types of concrete? In the latter case, how is correctness of this automatic generalization ensured?

Clearly, new opportunities open up once the cracks are correctly segmented. Geometric characterization of the crack structures helps to better understand the material concrete as well as material systems involving concrete [15]. In situ four-point bending tests within the XXL CT device Gulliver [30] will allow to observe crack initiation and evolution in concrete beams of a length of more than 1 m. Analysis of the induced local motion fields [25] may even unveil cracks too thin to be segmented based on just a static image. Finally, the geometric modeling can be refined, maybe in combination with multiscale modeling of the concrete matrix.

Acknowledgments This work was supported by the German Federal Ministry of Education and Research (BMBF) [grant number 05M2020 (DAnoBi)]. We thank Matthias Pahn and Szymon Grzesiak from CE of RPTU for samples and experimental design, Franz Schreiber from Fraunhofer ITWM and Michael Salamon from Fraunhofer EZRT for CT imaging, and Ali Moghiseh from Fraunhofer ITWM for input on CrackNet.

References

1. Ansell, A., Nordström, E., & Guarin, A. (2018). Laboratory investigation of steel fibre reinforced sprayed concrete using a computed tomography method. In *Eighth International Symposium on SPRAYED CONCRETE - Modern Use of Wet Mix Sprayed Concrete for Underground Support* (pp. 24–38).
2. Badrinarayanan, V., Kendall, A., & Cipolla, R. (2017). SegNet: A deep convolutional encoder-decoder architecture for image segmentation. *IEEE Transactions on Pattern Analysis, 39*(12), 2481–2495. https://doi.org/10.1109/TPAMI.2016.2644615
3. Barisin, T., Jung, C., Müsebeck, F., Redenbach, C., & Schladitz, K. (2022). Methods for segmenting cracks in 3d images of concrete: A comparison based on semi-synthetic images. *Pattern Recognition, 129*, 108747. https://doi.org/10.1016/j.patcog.2022.108747.
4. Barisin, T., Redenbach, C., & Schladitz, K. (2024). Riesz networks: scale invariant neural networks in single forward pass. *J Math Imaging Vis 66*, 246–270. https://doi.org/10.1007/s10851-024-01171-4
5. Bellon, C., Burger, K., & Gollwitzer, C. (2019). Virtual ct acquisition and reconstruction of complex and noisy scanning trajectories in aRTist. In *Proceedings of the 9th Conference on Industrial Computed Tomography*, Padova, Italy.
6. Binder, F., Bellon, C., Wohlgemuth, F., & Hausotte, T. (2021). Ein praxisnaher Leitfaden für computertomografische Untersuchungen mit der radiografischen Simulationsumgebung aRTist. In *DGZfP Jahrestagung 2021*. https://jahrestagung.dgzfp.de/portals/jt2021/bb176/inhalt/p4.pdf
7. Botev, Z. (2022). Fractional Brownian field or surface generator. https://de.mathworks.com/matlabcentral/fileexchange/38945-fractional-brownian-field-or-surface-generator. (accessed 02 February 2022)
8. Çiçek, Ö., Abdulkadir, A., Lienkamp, S. S., Brox, T., & Ronneberger, O. (2016). 3D U-Net: Learning dense volumetric segmentation from sparse annotation. In *Lecture Notes in Computer Science* (pp. 424–432). https://doi.org/10.1007/978-3-319-46723-8_49
9. Chambon, S., & Moliard, J. M. (2011). Automatic road pavement assessment with image processing: review and comparison. *International Journal of Geophysics, 2011*, 1–20. https://doi.org/10.1155/2011/989354
10. Chun, P., Izumi, S., & Yamane, T. (2020). Automatic detection method of cracks from concrete surface imagery using two-step light gradient boosting machine. *Computer-Aided Civil and Infrastructure Engineering, 36*, 1–12. https://doi.org/10.1111/mice.12564
11. Depeursinge, A., Foncubierta-Rodriguez, A., Van de Ville, D., & Müller, H. (2012). Multiscale lung texture signature learning using the Riesz transform. In N. Ayache, H. Delingette, P. Golland, & K. Mori (Eds.) *Medical Image Computing and Computer-Assisted Intervention – MICCAI 2012* (pp. 517–524). Berlin, Heidelberg: Springer. https://doi.org/10.1007/978-3-642-33454-2_64
12. Dobrovolskij, D., Persch, J., Schladitz, K., & Steidl, G. (2019). Structure detection with second order Riesz transform. *Image Analysis & Stereology, 38*(1), 107–119. https://doi.org/10.5566/ias.1964
13. Ehrig, K., Goebbels, J., Meinel, D., Paetsch, O., Prohaska, S., & Zobel, V. (2011). Comparison of crack detection methods for analyzing damage processes in concrete with computed tomography. In *International Symposium on Digital Industrial Radiology and Computed Tomography*.
14. EXTENDE. (2023). CIVA, NDT simulation software. https://www.extende.com/computed-tomography-with-civa
15. Grzesiak, S., Barisin, T., Schladitz, K., & Pahn, M. (2023). Analysis of the bond behaviour of a GFRP rebar in concrete by in-situ 3D imaging test. *Materials and Structures, 56*, article number 163.

16. Häuser, S., Heise, B., & Steidl, G. (2014). Linearized Riesz transform and quasi-monogenic shearlets. *International Journal of Wavelets, Multiresolution and Information Processing, 12*(03), 1450,027. https://doi.org/10.1142/S0219691314500271
17. Hsieh, Y. A., & Tsai, Y. J. (2020). Machine learning for crack detection: Review and model performance comparison. *Journal of Computing in Civil Engineering, 34*(5), 04020,038. https://doi.org/10.1061/(ASCE)CP.1943-5487.0000918
18. Jenkins, M. D., Carr, T. A., Iglesias, M. I., Buggy, T. W., & Morison, G. (2018). A deep convolutional neural network for semantic pixel-wise segmentation of road and pavement surface cracks. In *2018 26th European Signal Processing Conference (EUSIPCO)* (pp. 2120–2124).
19. Joyseeree, R., Otálora Montenegro, J., Müller, H., & Depeursinge, A. (2019). Fusing learned representations from Riesz filters and deep CNN for lung tissue classification. *Medical Image Analysis, 56*. https://doi.org/10.1016/j.media.2019.06.006
20. Jung, C., Müsebeck, F., Barisin, T., Schladitz, K., Redenbach, C., Kiesche, M., & Pahn, M. (2022). Towards automatic crack segmentation in 3d concrete images. In *11th Conference on Industrial Computed Tomography (iCT 2022)*, Wels, Austria.
21. Jung, C., & Redenbach, C. (2023). Crack modeling via minimum-weight surfaces in 3D Voronoi diagrams. *Journal of Mathematics in Industry, 13*(10). https://doi.org/10.1186/s13362-023-00138-1
22. König, J., Jenkins, M., Mannion, M., Barrie, P., & Morison, G. (2021). Weakly-supervised surface crack segmentation by generating pseudo-labels using localization with a classifier and thresholding. https://arxiv.org/pdf/2109.00456.pdf (accessed 02 February 2022).
23. Li, G., Wan, J., He, S., Liu, Q., & Ma, B. (2020). Semi-supervised semantic segmentation using adversarial learning for pavement crack detection. *IEEE Access, 8*, 51,446–51,459. https://doi.org/10.1109/ACCESS.2020.2980086
24. Müller, O., Moghiseh, A., Stephani, H., Rottmayer, N., & Huang, F. (2018). Application of deep learning for crack segmentation on concrete surface. Forum Bildverarbeitung 2018. https://doi.org/10.5445/KSP/1000085290
25. Nogatz, T., Redenbach, C., & Schladitz, K. (2022). 3D optical flow for large CT data of materials microstructures. *Strain, 58*(3), e12,412. https://doi.org/10.1111/str.12412.
26. Nowacka, A., Schladitz, K., Grzesiak, S., & Pahn, M. (2024). Segmentation of cracks in 3d images of fiber reinforced concrete using deep learning enabling image based characterization. *Journal of Soft Computing in Civil Engineering*, To appear, https://doi.org/10.22115/scce.2024.414873.1712
27. Paetsch, O., Baum, D., Ehrig, K., Meinel, D., & Prohaska, S. (2012). Automated 3D crack detection for analyzing damage processes in concrete with computed tomography. In *Proceedings Conference on Industrial Computed Tomography* (pp. 321–330).
28. Reinhardt, M., Bernstein, S., & Heise, B. (2022). Multi-scale orientation estimation using higher order Riesz transforms. *International Journal of Wavelets, Multiresolution and Information Processing, 20*(03), 2040,007. https://doi.org/10.1142/S021969132040007X
29. Ronneberger, O., Fischer, P., & Brox, T. (2015). U-net: Convolutional networks for biomedical image segmentation. In *Lecture Notes in Computer Science* (pp. 234–241). https://doi.org/10.1007/978-3-319-24574-4_28
30. Salamon, M., Reims, N., & Hanke, R. (2018). High energy X-ray tomography for large concrete structures. In M. Pahn, C. Thiele, C. Glock, & J. Schnell (Eds.) *Vielfalt im Massivbau: Festschrift zum 65. Geburtstag von Prof. Dr.-Ing. Jürgen Schnell* (pp. 595–609). Ernst & Sohn.
31. Shi, Y., Cui, L., Qi, Z., Meng, F., & Chen, Z. (2016). Automatic road crack detection using random structured forests. *IEEE Transactions on Intelligent Transportation Systems, 17*(12), 3434–3445. https://doi.org/10.1109/TITS.2016.2552248
32. Shim, S., Kim, J., Cho, G. C., & Lee, S. W. (2020). Multiscale and adversarial learning-based semi-supervised semantic segmentation approach for crack detection in concrete structures. *IEEE Access, 8*, 170,939–170,950. https://doi.org/10.1109/ACCESS.2020.3022786
33. Unser, M., & Van De Ville, D. (2010). Wavelet steerability and the higher-order Riesz transform. *IEEE Transactions on Image Processing, 19*(3), 636–652. https://doi.org/10.1109/TIP.2009.2038832

34. van Aarle, W., Palenstijn, W. J., De Beenhouwer, J., Altantzis, T., Bals, S., Batenburg, K. J., & Sijbers, J. (2015). The ASTRA toolbox: A platform for advanced algorithm development in electron tomography. *Ultramicroscopy, 157*, 35–47. https://doi.org/10.1016/j.ultramic.2015.05.002
35. Wang, W., & Su, C. (2021). Semi-supervised semantic segmentation network for surface crack detection. *Automation in Construction, 128*, 103,786. https://doi.org/10.1016/j.autcon.2021.103786

Machine Learning Methods for Prediction of Breakthrough Curves in Reactive Porous Media

Daria Fokina, Pavel Toktaliev, Oleg Iliev, and Ivan Oseledets

Abstract Reactive flows in porous media play an important role in our life and are crucial for many industrial, environmental, and biomedical applications. Very often the concentration of the species at the inlet is known, and the so-called breakthrough curves, measured at the outlet, are the quantities which could be measured or computed numerically. The measurements and the simulations could be time-consuming and expensive, and machine learning approaches can help to predict breakthrough curves at lower costs. Methods like Gaussian processes and fully connected neural networks, as well as the cross-approximation tensor method, are well suited for predicting breakthrough curves (Fokina et al., On the performance of machine learning methods for breakthrough curve prediction (2022)). In this paper, we demonstrate their performance in the case of pore-scale reactive flow in catalytic filters.

1 Introduction

Reactive flows in porous media play an important role in our life and are crucial for many industrial, environmental, and biomedical applications. Therefore, it is

D. Fokina · P. Toktaliev
Fraunhofer ITWM, Kaiserslautern, Germany

Technical University Kaiserslautern, Kaiserslautern, Germany
e-mail: daria.fokina@itwm.fraunhofer.de; pavel.toktaliev@itwm.fraunhofer.de

O. Iliev (✉)
Fraunhofer ITWM, Kaiserslautern, Germany

Bulgarian Academy of Sciences, Sofia, Bulgaria
e-mail: oleg.iliev@itwm.fraunhofer.de

I. Oseledets
Skolkovo Institute of Science and Technology, Moscow, Russia

Artificial Intelligence Research Institute, Moscow, Russia
e-mail: i.oseledets@skoltech.ru

J. Franke, A. Schöbel (eds.), *Statistical Machine Learning for Engineering with Applications*, Lecture Notes in Statistics 227,
https://doi.org/10.1007/978-3-031-66253-9_7

important to understand how the reactive transport works and how one can control or improve it. Note that chemical reactions occur at pore scale; however, the measurements are usually done at macroscopic scale. Very often the concentration of the species at the inlet is known, and the so-called breakthrough curves are the quantities which could be measured or computed numerically. Breakthrough curves represent the concentration of the species at the flow outlet. In problems like parameter identification, optimizations, etc., breakthrough curves for many of the input parameters are needed. The measurements and simulations could be time-consuming and expensive, and machine learning approaches can help to predict breakthrough curves at lower costs. Machine Learning (ML) methods, such as Gaussian processes and fully connected neural networks, as well as the cross-approximation tensor method, are well suited for predicting breakthrough curves [2]. In this paper, we demonstrate their performance in the case of pore-scale reactive flow in catalytic filters. The computational geometry for a piece of the filtering membrane could come from 3D CT image, or can be generated by proper software. The governing equation is the convection-diffusion-reaction equation, and linear and nonlinear reactions are considered. The computational experiments are followed by discussions about the required size of the training set, applicability, and the performance of each of the methods for different flow regimes, namely, diffusion dominated and reaction dominated. We expect that the considered methods will show similar performance for predicting breakthrough curves for other industrial or environmental problems, supposing similar reactions and flow regimes are considered.

2 Problem Formulation

We consider at microscale a reactive flow through a piece of a catalytic filter membrane. A sample of the geometry can be seen in Fig. 1a. The red color is reserved for the solid inert particles, the green color denotes the active washcoat particles, and the microscale pores are transparent. There is no transport through the inert particles. Reactions occur within the washcoat; they occupy the domain indicated by Ω_w. The resolved microscale pores occupy a domain denoted as Ω_f. Thus, the computational domain is $\Omega = \Omega_w \cup \Omega_f$. Mathematically, the process is modeled by the following transport equation:

$$\frac{\partial c}{\partial t} - D\,\Delta c + \langle \mathbf{v}, \nabla c \rangle + r(c) = 0, \quad x \in \Omega, t > 0, \tag{1}$$

where $c(t, x)$ denotes the concentration at location x and time t, D is the diffusion coefficient which is different in the pores and in the washcoat, $\mathbf{v}$ is the stationary velocity field which is computed in advance (see below), $r(c(t, x)) = 0$, $x \in \Omega_f$, $t > 0$. For the reaction in the washcoat, we consider linear and nonlinear reactions. Further in the text we refer only to the processes in the washcoat. If the

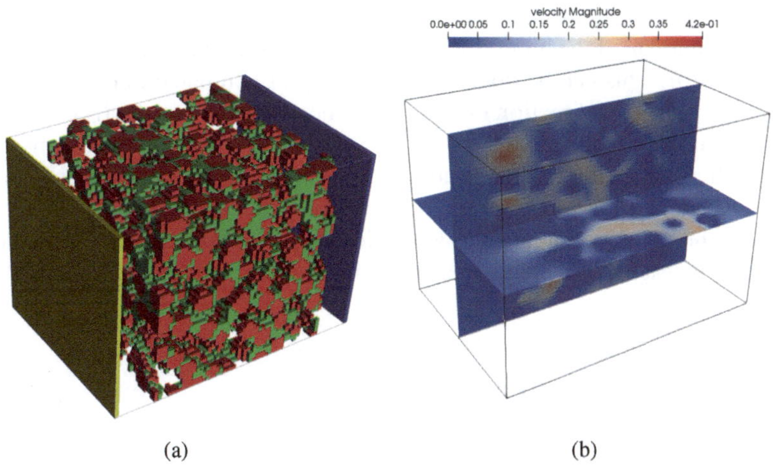

Fig. 1 Geometry and flow field for Eqs. 1–3: (**a**) is the considered geometry of the porous medium: yellow denotes the inlet, blue is the outlet, red are the solid particles, green is the washcoat material, and (**b**) is the velocity field inside the medium

reaction is linear: $r(c) = kc$, then in a dimensionless form, the equation can be written as follows:

$$\frac{\partial c}{\partial t} - \mu_{\mathrm{Da}}\, \Delta c + \langle \mathbf{v}, \nabla c \rangle + \mu_{\mathrm{Pe}}\, c = 0, \tag{2}$$

where $\mu_{\mathrm{Da}} = \frac{kL}{D}$ is the Damköhler number, $\mu_{\mathrm{Pe}} = \frac{u_{in} L}{D}$ is the Peclet number, L is the length of the computational domain, and u_{in} is the velocity at the inlet.

The initial and boundary conditions are specified as follows:

$$\begin{aligned} c(x, 0) &= 0, \quad x \in \Omega, \\ c(x, t) &= 1, \quad x \in \Gamma_{\mathrm{inlet}}, t > 0, \\ \frac{\partial c}{\partial \mathbf{n}}(x, t) = \langle \nabla c, \mathbf{n} \rangle &= 0, \quad x \in \Gamma_{\mathrm{outlet}} \cup \Gamma_{\mathrm{walls}}, \end{aligned} \tag{3}$$

where Γ denotes the domain boundary and $\mathbf{n}$ is a normal to the boundary. Further we use the notation: $\mu = (\mu_{\mathrm{Da}}, \mu_{\mathrm{Pe}})$.

The quantity of interest is the time-dependent concentration averaged over the outlet area Ω_{outlet}, i.e., the breakthrough curve:

$$s(t; \mu) = s_{\mu}(t) = \int\limits_{\Omega_{\mathrm{outlet}}} c(x, t; \mu) dx. \tag{4}$$

We use *supervised* machine learning methods to approximate the dependency of $s_\mu(t)$ as a function of the input parameters μ. For that we need the values s_μ for a set of values of μ. This can be done experimentally or computed numerically. In both cases we do not get a function of time $s_\mu(t)$, but vectors $\boldsymbol{s}_\mu = (s_\mu(t_1), \ldots, s_\mu(t_N))$ for fixed moments in time $t_1, \ldots, t_N$. To obtain $\boldsymbol{s}_\mu$ we use PoreChem software [9], which provides the results of numerical simulations. Based on these numerical values, we train a machine learning model and then choose the best configuration based on the results on new data, validation data, yet not seen by the model. The third set, the test set, is used to present the final results of the best models.

3 Considered Methods

In the scope of this paper, we compare two machine learning methods—namely, Gaussian processes and fully connected neural networks—and one tensor method, namely, cross approximation. In the following, we discuss these methods in detail. Please note that for our problem, other machine learning methods could be considered as well.

3.1 *Machine Learning Methods*

The considered machine learning methods considered are the so-called *regression* methods. This means that we have a function $f : \Omega \subset \mathbb{R}^n \to \mathbb{R}^m$, which we would like to approximate from several data points: $\mathscr{D} = (X, Y)$ which consists of pairs $\{(\mathbf{x}_i, \mathbf{y}_i = f(\mathbf{x}_i))\}, i = 1, \ldots, N$, $\mathbf{x}_i$ is the input, and $\mathbf{y}_i$ is the target value. The approximation is represented as a parametrized mapping $\hat{f}_\theta$; θ are the parameters. Normally optimal parameters θ^* that give the best fit to the data points are chosen by minimizing mean squared error (MSE):

$$l(\theta; \mathscr{D}) = \sum_{(\mathbf{x}_i, \mathbf{y}_i) \in \mathscr{D}} \|\hat{f}_\theta(\mathbf{x}_i) - \mathbf{y}_i\|^2 \to \min_\theta . \tag{5}$$

This functional is often used in neural networks.

3.1.1 Gaussian Processes

The first method to consider is a Gaussian process. For simplicity of the description, consider a function $f : \Omega \to \mathbb{R}$. Then set of given target values $Y = \mathbf{y}$ is a vector. This is a Bayesian method, which gives the conditional distribution $p(y_*|x_*, X, Y)$ of $y_* = f(x_*)$ at a new point x_*, given a training data set (X, Y). Consider a Gaussian process $f(x) \sim GP(0, K)$, where K is a kernel function, or simply a kernel. This means that for any vector $\mathbf{y}$, $y_i = f(x_i)$, $i = 1, \ldots, N$, has a normal distribution $\mathcal{N}(\mathbf{0}, K(X, X))$, $K(X, X)_{ij} = K(x_i, x_j), i, j = 1, \ldots, N$. When several values of f are known for the set of points $X \subset \Omega$, the posterior distribution of $y_* = f(x_*)$ for the new point $x_* \in \Omega$ is the following:

$$y_*|\mathbf{y}, X, x_* \sim \mathcal{N}(K(x_*, X)K(X, X)^{-1}\mathbf{y}, \\ K(x_*, x_*) - K(x_*, X)K(X, X)^{-1}K(X, x_*)). \tag{6}$$

The approximation is the mean of this distribution, i.e., $\hat{f}(x_*) = K(x_*, X) K(X, X)\mathbf{y}$.

We consider three kernel functions:

- Radial Basis Function (RBF) kernel: $K(x, z) = \exp\left(-\frac{d(x,z)^2}{2l^2}\right)$,
- Matern kernel: $K(x, z) = \frac{1}{\Gamma(\nu)2^{\nu-1}}\left(\frac{\sqrt{2\nu}d(x,z)}{l}\right)^l K_\nu\left(\frac{\sqrt{2\nu}d(x,z)}{l}\right)$,
- Rational quadratic (RQ) kernel: $K(x, z) = \left(\frac{1+d^2(x,z)}{2\alpha l^2}\right)^{-\alpha}$,

where K_ν is the modified Bessel function, $\Gamma(\nu)$ is the Gamma function, and $d(x, z)$ is the Euclidean distance function.

3.1.2 Neural Networks

Another class of ML models which has demonstrated a good efficiency are artificial neural networks (NNs). In our considerations we use a classical fully connected neural network. For this kind of neural network, the approximation is the following:

$$f_\theta = L_d \circ g \circ \ldots \circ L_2 \circ g \circ L_1, \tag{7}$$

where $L_k(\mathbf{x}) = \mathbf{x}W_k + \mathbf{b}_k$, $W_k \in \mathbb{R}^{h_{k-1} \times h_k}$, $\mathbf{b}_k \in \mathbb{R}^{h_k}$, $\theta = \{(W_k, \mathbf{b}_k\}_{k=1}^d$, $h_0 = n$, $h_d = m$, and g is a nonlinear scalar function; applied element-wise to each component of the input vector, this function is also called an activation function. The parameters are found by solving the minimization problem (5). This is usually done by stochastic gradient descent or its modifications. In this work we use Adam optimizer [5]. We experiment with three activation functions:

- Rectified linear unit (ReLU): $g(x) = \max(0, x)$,
- Leaky ReLU: $g(x) = \begin{cases} x, & \text{if } x > 0 \\ \alpha x, & \text{if } x \le 0 \end{cases}$, $\alpha = 0.01$,
- Exponential linear unit (ELU): $g(x) = \begin{cases} x, & \text{if } x > 0 \\ \alpha \exp(x - 1), & \text{if } x \le 0 \end{cases}$, $\alpha = 0.01$.

3.2 Cross Approximation

3.2.1 Discrete Case

Unlike the previous methods, cross approximation [2] is a tensor method. As it follows from the name, it operates on tensors, i.e., on n-dimensional matrices. We also note that it is not a well-known cross-approximation method for tensor compression [7]. We describe this method applied to a discretized parameter domain, and then comment on how we extend it to the continuous case.

The input of the model is a vector $\boldsymbol{\mu} = (\mu_{\mathrm{Da}}, \mu_{\mathrm{Pe}})$, each component of which is located within a closed interval of interest: $\mu_{\mathrm{Da}} \in \gamma_1 = [p_1^1, p_2^1]$, $\mu_{\mathrm{Pe}} \in \gamma_2 = [p_1^2, p_2^2]$. At the first step, the space of input parameters $\gamma_1 \times \gamma_2$ is restricted to a $k \times k$ grid $A \times B$. This is done by uniformly discretizing each of intervals γ_1 and γ_2. The notation is the following:

$$\mu_{\mathrm{Da}} \in \{A\} := \{A_i\}_{i=1}^k \text{ with } A_1 = p_1^1 \text{ and } A_k = p_2^1$$
$$\mu_{\mathrm{Pe}} \in \{B\} := \{B_i\}_{i=1}^k \text{ with } B_1 = p_1^2 \text{ and } B_k = p_2^2$$

If we compute the curves s_μ for each element on the grid, we get a three-dimensional tensor $\mathsf{S} \in \mathbb{R}^{k \times k \times N_T}$, where N_T is the number of simulated time steps. We suppose that S can be represented in a low-rank form using the Tucker decomposition [6, 10]:

$$\mathsf{S} = \mathsf{G} \times_1 U \times_2 V, \tag{8}$$

$G \in \mathbb{R}^{r_1 \times r_2 \times N_T}$, $U \in \mathbb{R}^{r_1 \times k}$, $V \in \mathbb{R}^{r_2 \times k}$. The multiplication $\times_m$ is performed along the axis m. For $\mathsf{C} = \mathsf{A} \times_m B$, where $\mathsf{A} \in \mathbb{R}^{d_1 \times \ldots \times d_n}$, $B \in \mathbb{R}^{d_m \times l}$, its elements are expressed as

$$C_{i_1,\ldots,i_{m-1},j,i_{m+1},\ldots,i_n} = \sum_{i_m} A_{i_1,\ldots,i_n} B_{i_m,j}. \tag{9}$$

Cross approximation provides an algorithm to choose U, V and G. Denote slices of S for some k_1 and k_2 as $P_1 := S[k_1, :, :]$, $P_2 := S[:, k_2, :]$. For these matrices we compute a singular value decomposition:

$$P_1 = \hat{U} S_1 \Phi_1; \; P_2 = \hat{V} S_2 \Phi_2, \tag{10}$$

with $U^T = \hat{U}[:, : r_1] \in \mathbb{R}^{k \times r_1}$ and $V^T = \hat{V}[:, : r_2] \in \mathbb{R}^{k \times r_2}$ are the leftmost singular vectors r_1 and r_2, respectively. To find G a MaxVol [3] algorithm is used. MaxVol is a greedy algorithm looking in a matrix for a submatrix of maximum volume (for a square matrix $\mathrm{vol}(A) = |\det(A)|$). This algorithm gives us a set of indices, for which we need to compute values of S, to get the tensor G, and, respectively, to reconstruct the whole tensor S. The elements necessary for computing S are indicated in red and blue in Fig. 2.

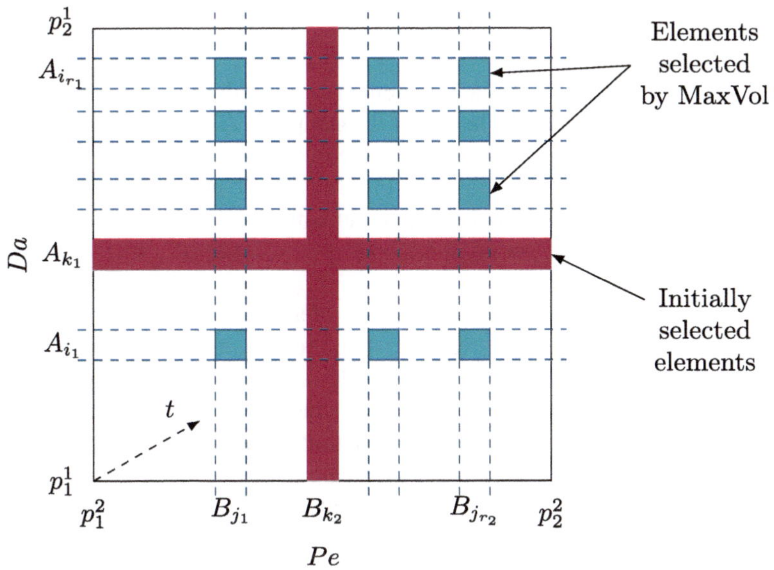

Fig. 2 Elements on the parameter grid, required for cross approximation: in red are the elements required for computing U and V in Eq. (8), in blue are the elements selected by MaxVol, required to find G in Eq. (8)

3.2.2 Continuous Case

We consider the tensor S in the Tucker decomposition; this means that its elements can be represented as

$$S_{mnt} = \sum_{i,j=1}^{r_1,r_2} G_{ijt} U_{im} V_{jn}, \quad m = 1, \ldots, k, \ n = 1, \ldots, k, \ t = 1, \ldots, N_T. \tag{11}$$

The value of U_{im} depends only on the value A_m of μ_{Da}; the value of V_{jn} depends only on the value B_n of μ_{Pe}. In the continuous case, it would be written as

$$S_t(\mu_{\mathrm{Da}}, \mu_{\mathrm{Pe}}) = \sum_{ij} G_{ijt} U_i(\mu_{\mathrm{Da}}) V_j(\mu_{\mathrm{Pe}}). \tag{12}$$

We call $U_i(\mu_{\mathrm{Da}})$ and $V_j(\mu_{\mathrm{Pe}})$ basis functions and find them by interpolating the vectors U_i and V_j correspondingly.

4 Computational Experiments

We conduct the experiments on a 3D geometry—a random packing of spheres. Its size is $50 \times 50 \times 50$ voxels. The length of the sample is $L = 0.0001$ m, the diameter of the spheres is $l = 6$ µm, the velocity on the inlet is $u_{\mathrm{in}} = 0.047$, and the porosity is $\phi = 0.741$. The geometry is shown on Fig. 1a with the precomputed velocity field on Fig. 1b. To calculate the velocities, Geodict software tool [1] was used. For each value of input parameters, we compute concentration for 200 time steps (N_T) for $t \in [0.00001, 0.002]$.

We start with different regimes for the case of a linear reaction and then consider the case of a strong nonlinear reaction. For the training of Gaussian processes and neural networks, we uniformly randomly sample the input parameters from the domain of interest. In each of the cases, the size of the training data set for Gaussian processes is equal to 100, and for neural networks, it is 600. For the cross-approximation method, we need parameter values for the middle cross section of the parameter domain and also at positions found by the MaxVol algorithm. For the evaluation of these three methods, we randomly sample 100 points for validation and separately 100 points for the test data set.

The validation data set is used for the choice of hyperparameters. In our case, however, it is the choice of the kernel function for the Gaussian process and the choice of the activation function for the neural network. The kernel parameter selection is implemented in the scikit-learn Python package [8]. It is done automatically and is based on likelihood maximization. For the neural network width and depth, we have found out that shallow (5 layers in our case) and wider (100 neurons per layer) neural networks perform better. The test set is used to represent the final errors.

For each of the sampled parameters, we numerically compute the breakthrough curves, which are the target values for the models. To compare the results, we compute two errors for each data set (training, validation, test):

$$e_{l_2} = \sqrt{\frac{\sum\limits_{\mu \in X} \sum\limits_{t=1}^{N_T} (\hat{s}_t(\mu) - s_t(\mu))^2}{\sum\limits_{\mu \in X} \sum\limits_{i=1}^{m} s_i(\mu)^2}}, \tag{13}$$

$$e_{\max} = \max_{\mu \in X} \sqrt{\frac{\sum\limits_{t=1}^{N_T} (\hat{s}_t(\mu) - s_t(\mu))^2}{\sum\limits_{t=1}^{m} s_t(\mu)^2}}, \tag{14}$$

where $s_t(\mu)$ is the target breakthrough curve value from definition (4) for the parameter $\mu = (\mu_{\mathrm{Pe}}, \mu_{\mathrm{Da}})$, $\hat{s}(\mu)$ is the predicted value from , e_{l_2} is a relative mean square root error taken over the data set X, and $e_{\max}$ is maximum of relative l_2 errors between the curves taken over the set X.

4.1 Linear Reaction

As mentioned above, we begin with a linear reaction $r(c) = k\,c$. There are two regimes considered:

- Diffusion dominated with $\mu_{\mathrm{Da}} \in [0.01, 1.1]$, $\mu_{\mathrm{Pe}} \in [0.01, 1.1]$,
- Strong reaction with $\mu_{\mathrm{Da}} \in [800, 1200]$, $\mu_{\mathrm{Pe}} \in [0.05, 0.1]$.

4.1.1 Diffusion-Dominated Case

The results for Gaussian processes for different kernels are provided in Table 1, for neural networks for different activation functions are in Table 2, and for cross approximation—in Table 3. In all tables, we highlight in bold font the validation and test errors for the method with the smallest error $e_{\max}$. Table 4 provides the test errors for each method. Figure 3 compares breakthrough curves predicted by the models (solid line) with numerically computed target values (dotted line) for several values of input parameters (μ_{Da}, μ_{Pe}). The predicted values are provided for the best configuration of each method. The difference between predicted and target values is not seen by a human eye. In a zoomed-in figure, a difference can be noticed, though it is also very small.

As one can see from Table 4, the best result is achieved by a neural network with ELU activation function. The cross approximation gives a similar result, but with a smaller amount of training data.

To give an idea of the approximate training and prediction times of the methods, we give them for the diffusion-dominated regime in Table 5.

Table 1 Linear reaction, diffusion-dominated case: error values from Eqs. (13), (14) on training and validation data sets for Gaussian processes for three kernel functions. Gaussian process with Matern kernel provides the best result

	Train error		Validation error	
Kernel	e_{l_2}	$e_{\max}$	e_{l_2}	$e_{\max}$
RBF	0.00161	0.00489	0.04281	0.22586
Matern	1.0×10^{-7}	5.96×10^{-7}	**0.04777**	**0.20501**
RQ	1.67×10^{-6}	8.71×10^{-6}	0.06546	0.33635

Table 2 Linear reaction, diffusion-dominated case: error values from Eqs. (13), (14) on training and validation data sets for neural networks for three activation functions. Neural network with ELU activation function has the smallest error

	Train error		Validation error	
Activation function	e_{l_2}	$e_{\max}$	e_{l_2}	$e_{\max}$
ReLU	0.00011	0.00040	0.00166	0.00905
Leaky ReLU	0.00038	0.00136	0.00248	0.01771
ELU	0.00084	0.00475	**0.00113**	**0.00428**

Table 3 Linear reaction, diffusion-dominated case: error values from Eqs. (13), (14) on training and validation data sets for cross-approximation and two interpolation methods. Piecewise linear interpolation gives a better result

	Train error		Validation error	
Interpolation	e_{l_2}	$e_{\max}$	e_{l_2}	$e_{\max}$
Polynomial	0.01026	0.03127	0.01041	0.03135
Piecewise linear	0.00210	0.01820	**0.00380**	**0.01974**

Table 4 Linear reaction, diffusion-dominated case: error values (13, 14) on the test set. The neural network outperforms other methods

	Test error	
Method	e_{l_2}	$e_{\max}$
Gaussian process	0.01716	0.13296
Neural network	**0.00055**	**0.00126**
Cross approximation	0.00046	0.00219

4.1.2 Strong Reaction

In the case of a strong linear reaction (large Damköhler number), the results are presented in Table 6 for Gaussian processes, Table 7 for neural networks and Table 8 for cross approximation. In Table 9 test errors are presented. We also plot the breakthrough curve samples, numerically computed, and predicted by the considered methods.

In this case, the best performance shows cross approximation with polynomial interpolation (Table 9). For Gaussian processes with rational quadratic kernel, the error is bigger; however, the difference to cross approximation is minor. Neural networks have the worst result, although they use the most of the data. On Fig. 4 we can notice the difference between the values predicted by the model and numerically simulated data for neural networks, but this difference is noticed only at a closer look at the figure.

4.2 Nonlinear Reaction

Here we consider a nonlinear reaction

$$r(c) = -\frac{k \cdot c}{(1 + k_{\text{inhib}} c)^2}, \tag{15}$$

$k_{\text{inhib}} = 4\ \text{m}^3/\text{mol}$.

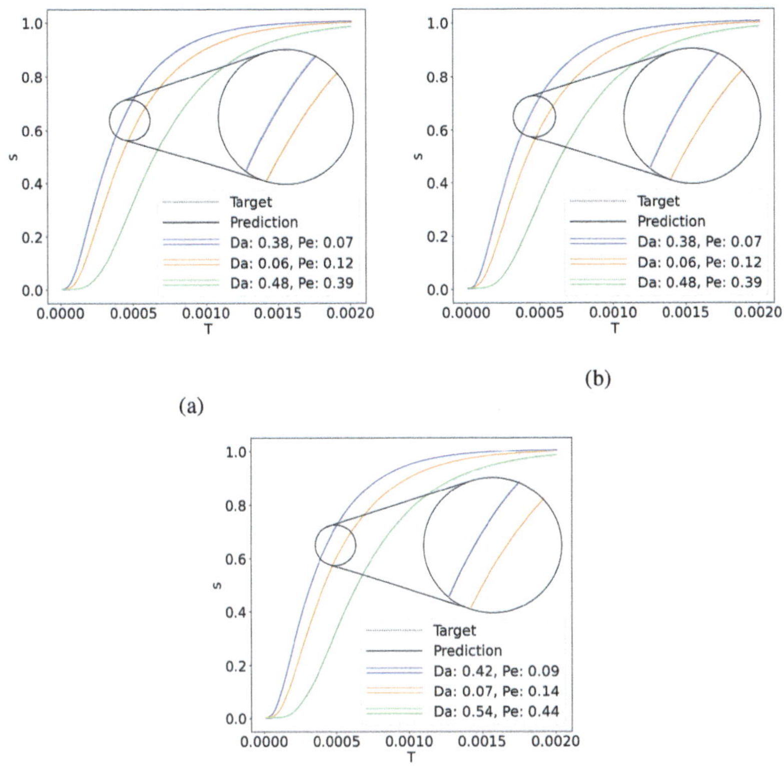

(c)

Fig. 3 Linear reaction, diffusion-dominated case. Solid lines correspond to several examples of breakthrough curves predicted by: the Gaussian process (**a**), the neural network (**b**), the cross approximation (**c**). The target values for comparison are shown by dotted lines

Table 5 Linear reaction, diffusion-dominated case: computing times for training and inference (prediction) of different methods. For cross approximation we provide times for each step of the method: building the basis U, V and G (Eq. 8), interpolation of U and V

	Computing time, s			
Method	Train			Inference
Gaussian process	0.11698			0.00386
Neural network	1213.3			0.00156
	Basis	Interpolation	Inference	
Cross approximation	0.005936	0.000626	0.001030	

Like in the case of the linear reaction, we introduce here in the same way Da and Pe numbers. The parameter intervals for this case are the same as for the strong linear reaction: $\mu_{\text{Da}} \in [800, 1200]$, $\mu_{\text{Pe}} \in [0.05, 0.1]$.

Table 6 Linear strong reaction: error values from Eqs. (13), (14) on training and validation data sets for Gaussian processes for three kernel functions. Rational quadratic kernel provides the smallest error on the validation data set

Kernel	Train error		Validation error	
	e_{l_2}	$e_{\max}$	e_{l_2}	$e_{\max}$
RBF	6.79×10^{-6}	3.57×10^{-5}	9.32×10^{-6}	4.04×10^{-5}
Matern	1.83×10^{-5}	0.00017	3.97×10^{-5}	0.00028
RQ	2.05×10^{-6}	9.41×10^{-6}	$\mathbf{8.37 \times 10^{-6}}$	$\mathbf{7.35 \times 10^{-5}}$

Table 7 Linear strong reaction: error values from Eqs. (13), (14) on training and validation data sets for neural networks for three activation functions. A neural network with leaky ReLU activation function gives the best result on the validation data set

Activation function	Train error		Validation error	
	e_{l_2}	$e_{\max}$	e_{l_2}	$e_{\max}$
ReLU	0.00297	0.00661	0.00322	0.00573
Leaky ReLU	6.14×10^{-5}	0.00016	**0.00055**	**0.00011**
ELU	0.00109	0.00234	0.001163	0.00186

Table 8 Linear strong reaction: error values from Eqs. (13), (14) on training and validation data sets for cross-approximation and two interpolation methods. Polynomial interpolation gives a better result

Interpolation	Train error		Validation error	
	e_{l_2}	$e_{\max}$	e_{l_2}	$e_{\max}$
Polynomial	2.24×10^{-6}	4.24×10^{-6}	$\mathbf{2.20 \times 10^{-6}}$	$\mathbf{3.48 \times 10^{-6}}$
Piecewise linear	4.48×10^{-6}	9.92×10^{-6}	4.32×10^{-6}	8.86×10^{-6}

Table 9 Linear strong reaction: error values from Eqs. (13), (14) on the test set. Cross approximation outperforms other methods

Kernel	Test error	
	e_{l_2}	$e_{\max}$
Gaussian process	8.3×10^{-6}	4.13×10^{-5}
Neural network	8.83×10^{-5}	0.00027
Cross approximation	$\mathbf{2.11 \times 10^{-6}}$	$\mathbf{4.03 \times 10^{-6}}$

We present errors of the models in Tables 10, 11, and 12 for Gaussian processes, neural networks, and cross approximation correspondingly. Test errors can be found in Table 13. In this case we also plot samples of predicted and precomputed breakthrough curves as well (Fig. 5). The results of the methods are similar to the results in the linear case. Here it should be noted that breakthrough curves in both linear and nonlinear cases are quite similar and can sometimes coincide. This depends on the parameter values, i.e., ratio Da/Pe. As it follows from Table 13, cross approximation outperforms other methods.

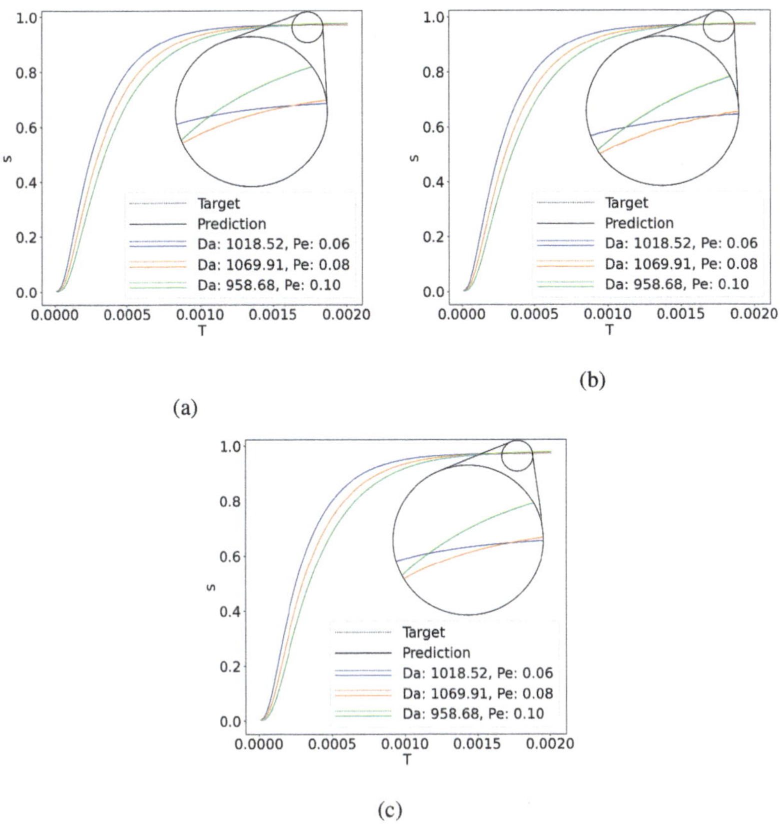

Fig. 4 Linear strong reaction. Solid lines correspond to several examples of breakthrough curves predicted by the Gaussian process (**a**), the neural network (**b**), the cross approximation (**c**). The target values for comparison are shown by dotted lines

Table 10 Nonlinear reaction: error values from Eqs. (13), (14) on training and validation data sets for Gaussian processes for three kernels. RBF kernel gives the smallest error on the validation data set

	Train error		Validation error	
Kernel	e_{l_2}	$e_{\max}$	e_{l_2}	$e_{\max}$
RBF	9.43×10^{-6}	7.24×10^{-5}	$\mathbf{7.20 \times 10^{-6}}$	$\mathbf{2.76 \times 10^{-5}}$
Matern	6×10^{-5}	0.00075	2.15×10^{-5}	0.00013
RQ	1.12×10^{-5}	0.00013	7.91×10^{-6}	6.42×10^{-5}

5 Conclusion

We have performed a comparison study of two machine learning methods and one tensor method extended by interpolation when applied to predict breakthrough curves for a catalytic filter. The training sets were obtained by simulating at pore-

Table 11 Nonlinear reaction: error values from Eqs. (13), (14) on training and validation data sets for neural networks for three activation functions. Leaky ReLU activation function gives us the smallest error on the validation data set

	Train error		Validation error	
Activation function	e_{l_2}	$e_{\max}$	e_{l_2}	$e_{\max}$
ReLU	6.91×10^{-5}	0.00031	8.48×10^{-5}	0.00031
Leaky ReLU	3.77×10^{-5}	8.87×10^{-5}	$\mathbf{5.35 \times 10^{-5}}$	**0.00021**
ELU	0.00045	0.00055	0.00045	0.00054

Table 12 Nonlinear reaction: error values from Eqs. (13), (14) on training and validation data sets for cross-approximation and two interpolation methods. Polynomial interpolation gives a better result

	Train error		Validation error	
Interpolation	e_{l_2}	$e_{\max}$	e_{l_2}	$e_{\max}$
Polynomial	1.89×10^{-6}	3.64×10^{-6}	$\mathbf{1.82 \times 10^{-6}}$	$\mathbf{3.59 \times 10^{-6}}$
Piecewise linear	4.06×10^{-6}	9.78×10^{-6}	4.05×10^{-6}	9.92×10^{-6}

Table 13 Nonlinear reaction: error values from Eqs. (13), (14) on the test data set. Cross approximation outperforms other methods

	Test error	
Kernel	e_{l_2}	$e_{\max}$
Gaussian process	6.55×10^{-6}	2.15×10^{-5}
Neural network	4.68×10^{-5}	0.00016
Cross approximation	$\mathbf{1.83 \times 10^{-6}}$	$\mathbf{3.44 \times 10^{-6}}$

scale 3D reactive flow using the software tool PoreChem. All the methods show good performance, though for different regimes the order of error differs. In the case of a strong reaction, the errors obtained are much smaller than in the diffusion-dominated case. This may be due to the fact that the curves changed less for the chosen intervals of parameters. The methods requiring a small amount of data (Gaussian processes and cross approximation) are more beneficial in this case, as numerical simulations for larger geometries require hours of running to compute just one point. In our case, it took several minutes to compute one sample.

In the case of cross approximation, we have empirically shown that the selected basis vectors for the low-rank approximation together with the data selected by the maximum volume algorithm provide a good approximation to the considered tensor. Further interpolation of the basis vectors gives us a good extension to the continuous case.

Future work includes applying the considered methods for cases of more complex reactions and more complex geometries. For a problem with more input parameters, the cross approximation with the chosen basis may have poor performance. A different basis choice can then be considered. The investigation, how the choice of the parameter space influences the model performances, is a matter of further studies. Additionally, these approaches can be considered for similar tasks in other areas, e.g., decline curves in oil industry [4].

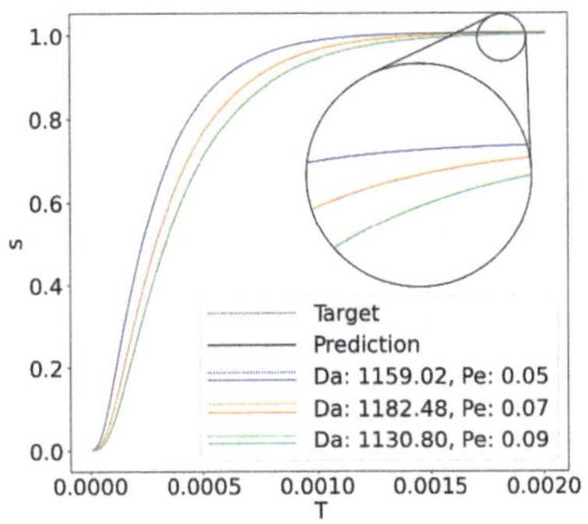

(a)

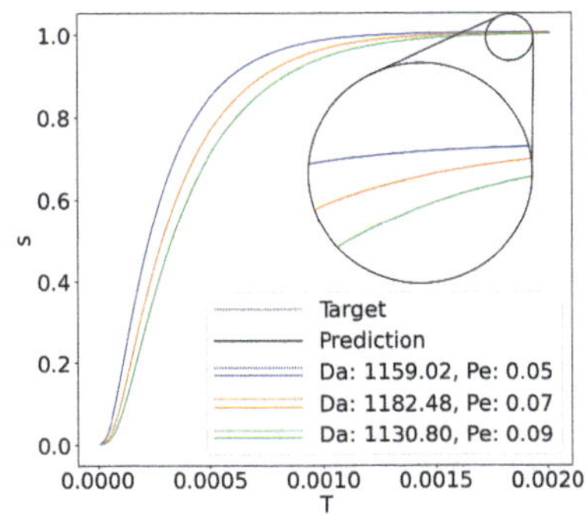

(b)

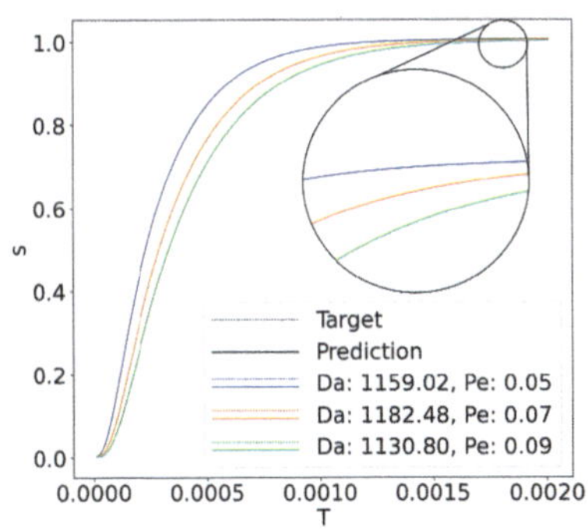

(c)

Fig. 5 Nonlinear reaction. Solid lines correspond to several examples of breakthrough curves predicted by the Gaussian process (**a**), the neural network (**b**), the cross approximation (**c**). The target values for comparison are shown by dotted lines

References

1. Becker, J., Biebl, F., Glatt, E., Cheng, L., Grießer, A., Groß, M., Linden, S., Mosbach, D., Wagner, C., Weber, A., & Westerteiger, R. (2021). Geodict (release 2022) [simulation software]. https://doi.org/10.30423/release.geodict2022
2. Fokina, D., Iliev, O., Toktaliev, P., Oseledets, I., & Schindler, F. (2022). On the performance of machine learning methods for breakthrough curve prediction. arXiv:2204.11719.
3. Goreinov, S. A., Oseledets, I. V., Savostyanov, D. V., Tyrtyshnikov, E. E., & Zamarashkin, N. L. (2010). How to find a good submatrix. In V. Olshevsky, & E. Tyrtyshnikov, (Eds.), *Matrix methods: theory, algorithms and applications* (pp. 247–256). World Scientific.
4. Kamari, A., Mohammadi, A. H., Lee, M., & Bahadori, A. (2017). Decline curve based models for predicting natural gas well performance. *Petroleum, 3*(2), 242–248.

5. Kingma, D. P., & Ba, J. (2015). Adam: A method for stochastic optimization. In *3rd International Conference on Learning Representations, ICLR 2015*, San Diego, CA, USA, May 7–9, 2015, Conference Track Proceedings.
6. Malik, O. A., & Becker, S. (2018). Low-rank tucker decomposition of large tensors using tensorsketch. In S. Bengio, H. Wallach, H. Larochelle, K. Grauman, N. Cesa-Bianchi, & R. Garnett, (Eds.), *Advances in neural information processing systems* (vol. 31). Curran Associates, Inc.
7. Oseledets, I., & Tyrtyshnikov, E. (2010). Tt-cross approximation for multidimensional arrays. *Linear Algebra and its Applications, 432*(1), 70–88.
8. Pedregosa, F., Varoquaux, G., Gramfort, A., Michel, V., Thirion, B., Grisel, O., Blondel, M., Prettenhofer, P., Weiss, R., Dubourg, V., Vanderplas, J., Passos, A., Cournapeau, D., Brucher, M., Perrot, M., & Duchesnay, E. (2011). Scikit-learn: Machine learning in Python. *Journal of Machine Learning Research, 12*, 2825–2830.
9. Toktaliev, P., & Iliev, O. (2023). Numerical modeling of reactive flow in porous media at the pore scale. *PAMM, 23*(1), e202200327.
10. Tucker, L. R. (1966). Some mathematical notes on three-mode factor analysis. *Psychometrika, 31*(3), 279–311.

Segmentation and Aggregation in Text Classification

Franziska Diez, Michael Trebing, and Stefanie Schwaar

Abstract In the context of AI systems for industry, analysis of text becomes an important part. One application is the classification of documents. While multiple classification algorithms are known, in application those suffer from the different length of the documents. One way to handle those is the segmentation setup. As a result, the classification algorithm gives a probability for each segmentation and every class. To classify the whole document, an aggregation method is needed. We investigate different approaches of aggregating and apply those methods to text documents for public administration.

1 Introduction

In this chapter, we explore different methods for automatically routing digital mail by analyzing the digital text and assigning it to the appropriate contact person or department. One challenge we face is the varying text length, which affects the structure of the text and poses difficulties for classification algorithms. To overcome this, a classical approach is to split the text into segments of equal size and classify each segment individually. However, we need to determine how to combine the results of these segments to classify the entire document. In this study, we investigate and analyze the advantages and disadvantages of different aggregation methods to identify the most suitable method for automatic routing of digital mail. Text classification is a topic of ongoing interest in natural language processing (NLP) with applications in various areas such as authorship classification and spam detection. Different approaches have been developed, ranging from statistics to machine learning and sentiment analysis (see, e.g., [5, 12, 14, 15]). An overview is given in [10].

F. Diez · M. Trebing · S. Schwaar (✉)
Fraunhofer Institute for Industrial Mathematics, Kaiserslautern, Germany
e-mail: franziska.diez@itwm.fraunhofer.de; michael.trebing@itwm.fraunhofer.de; stefanie.schwaar@itwm.fraunhofer.de

J. Franke, A. Schöbel (eds.), *Statistical Machine Learning for Engineering with Applications*, Lecture Notes in Statistics 227,
https://doi.org/10.1007/978-3-031-66253-9_8

In this study, we focus on word-based methods that use word frequencies as features for standard classifiers. The preprocessing of text, including the generation of features, is crucial for the model's performance as mentioned in [16]. We specifically focus on the term frequency-inverse document frequency (tf-idf) method (see, e.g., [9]; in detail see [11] for term frequency and for inverse document frequency [17]), which considers the frequency of a word within a text relative to its frequency in a collection of texts. While further developments have been made in this area (e.g., [6]), we concentrate on standard tf-idf in this study.

Our application example is based on the implementation of automatic routing of digital mail in German municipalities. These municipalities are divided into a maximum of 45 departments. Of those, 31 are present in our data set.

The varying text length in the input data presents a challenge here, as text length will affect the norm of the vectors[1] produced by tf-idf. Similar inputs differing only in norm, i.e., where one vector is a multiple of the other, may not be treated equally by standard classifiers, potentially leading to differences in classification. One approach to addressing this challenge is to segment or split the data into pieces of equal length, allowing the classification algorithms to be trained on comparable data. As a consequence of this segmentation scheme, predictions are initially obtained for individual text segments and need to be combined in order to derive predictions for the original whole documents. To achieve this, we define different aggregation methods and present the results of a comparative study that applies each of them to all the results from different combinations of preprocessing pipelines and classifiers.

Since the preprocessing is of high importance for the model performance (see [16]), we apply different combinations of preprocessing steps. Those steps are centered around the tf-idf method, and the results are then used to train and evaluate different classifiers.

This work is structured as follows: In Sect. 2 we describe the different aggregation methods in detail. After that, in Sect. 3, we lay out the experimental setup used to test the different approaches. Finally, we provide our key results in Sect. 4 and discuss them in Sect. 5.

2 Aggregation Methods

In dealing with the issue that our data contains documents of different lengths, we split the document into segments of similar length and classify every segment. For every segment, the classification algorithms produce vectors of probabilities. If we wanted to predict a class for the segment, we would choose the one with the largest probability value. However, in order to estimate the underlying true class of the

[1] The norm of a numeric vector is a value derived from its components which can be interpreted as its magnitude or the distance between the vector's start and end points.

whole document, we need to find some way to derive document-related values for all classes from the information we have on the segment level.

For aggregating values, different functions can be considered: we can choose the class with the highest:

- sum of probabilities,
- averaged probability, or
- weighted averaged probability.

Furthermore, different sets are possible on which these functions can be calculated. We can use only the predicted class and its corresponding probability, or we use the probabilities of all classes provided by the classification. This leads to seven different aggregation algorithms described in the following.

This chapter uses the following notation. The number of segments derived from one document is denoted by m. It differs between documents and depends on the document length and the fixed text segment length. The document or the text segments will be assigned to one of the n classes. When considering one particular combination of pipeline and classification algorithm, we denote by $p_j^{(i)}$, $i = 1, \ldots, m$ and $j = 1, \ldots, n$, the probability that the ith text segment is assigned to the jth class. The predicted class of the ith segment is the class with the maximum $p_j^{(i)}$.

2.1 Restricted Aggregation Algorithms

First, we restrict the set of possible predicted document classes to those which are predicted in at least one segment. We define the set K_j as containing the indices of those segments for which class j is the predicted class, i.e.,

$$K_j = \left\{ i \in \{1, \ldots, m\} : j = arg\,max_{l \in \{1,\ldots,n\}} \left(p_l^{(i)} \right) \right\}, \quad j = 1, \ldots, n.$$

Based on these sets, we obtain the following aggregation methods regarding the aggregation functions mentioned above.

Maximum Predicted Class (Max. Predicted Class)
This is the most simple algorithm considered here. We count the segments for which class j is predicted and choose the class with the maximum count. The predicted document class j^* is thus given by

$$j^* = arg\,max_{j \in \{1,\ldots,n\}} \left(|K_j| \right).$$

Ties are resolved by choosing the class which appears first when sorting the assignments by the original order of the segments. This amounts to a random choice. While it would be possible to base the decision on detailed prediction probabilities, we purposely avoid resorting to information other than the segment level prediction in order to better study the effect of aggregation. There will be more information on this aspect in Sect. 4.

Restricted Maximum Sum Probability (R-Max. Sum)
For every class only the probability values from the segments predicting this class are added up. Then, the class with the maximum sum is assigned to the document:

$$j^* = arg\,max_{j\in\{1,\dots,n\}}\left(\sum_{k\in K_j} p_j^{(k)}\right).$$

Restricted Maximum Average Probability (R-Max. Average)
Here we use the same values as in *Restricted Maximum Sum Probability*, but use the mean as aggregation function. Thus, the predicted document class is derived as

$$j^* = arg\,max_{j\in\{1,\dots,n\}}\left(\frac{1}{|K_j|}\sum_{k\in K_j} p_j^{(k)}\right).$$

Restricted Maximum Weighted Average Probability (R-Max. Weighted Average)
This is a modified version of *Restricted Maximum Average Probability* that uses a weighted mean where the weights are the segment lengths which can be defined by the number of characters or number of words. The predicted document class is then

$$j^* = arg\,max_{j\in\{1,\dots,n\}}\left(\frac{1}{\sum_{l\in L_j} w_l}\sum_{k\in K_j} w_k p_j^{(k)}\right).$$

2.2 Unrestricted Aggregation Algorithms

Here, we do not restrict the aggregation functions to the set K_j but make use of all probability values obtained from the classification. In combination with the three aggregation functions mentioned above, this yields three more methods. The Maximum Predicted Class is not affected by this change.

Maximum Sum Probability (Max. Sum)
For every class j, $j = 1, \dots, n$, the probabilities are summed up over all segments. The class with the maximum sum is chosen as the predicted document class j^*:

$$j^* = arg\,max_{j\in\{1,\dots,n\}}\left(\sum_{i=1}^{m} p_j^{(i)}\right).$$

Maximum Average Probability (Max. Average)
With the average as aggregation function in place of the sum, we obtain the document class via

$$j^* = arg\,max_{j \in \{1,\ldots,n\}} \left(\frac{1}{m} \sum_{i=1}^{m} p_j^{(i)} \right).$$

Maximum Sum Probability and *Maximum Average Probability* yield of course always the same value for all statistics because they always predict the same class. However, the probability assigned to the predicted class is different in both methods. In *Maximum Sum Probability*, the sum of the probabilities is used, while the mean of the probabilities is the quantity in *Maximum Average Probability*. Application settings are given where the probabilities are necessary. For example, in an automatically routing setting, we can determine a threshold for the probabilities, and all documents with a larger probability will be routed automatically whereas the other ones assigned to human treatment. The difference in the probabilities would then become relevant.

Maximum Weighted Average Probability (Max. Weighted Average)
Replacing simple by weighted average probabilities, the document class j^* becomes:

$$j^* = arg\,max_{j \in \{1,\ldots,n\}} \left(\frac{1}{\left(\sum_{l=1}^{m} w_l\right)} \sum_{i=1}^{m} w_i\, p_j^{(i)} \right).$$

3 Experiment

In German municipal administration, there exists a standard layout of 45 departments (see [1]), all of which may need to be able to receive secure e-mails in one common postbox (see [3]). As an example of text classification, we consider the task of distributing incoming mail to the correct agencies based on their contents.

In this section, we present the course of analysis in detail. After some remarks about the data used, we describe the experimental setup including the segmentation process, measures for balancing and vectorizing the data, further steps aimed at making inputs easier digestible for the classifiers, and the precise setup of the classifiers used including the steps taken for the optimization of their hyperparameters. Finally, we report on the cross-validation results obtained from training all our classifiers on our pool.

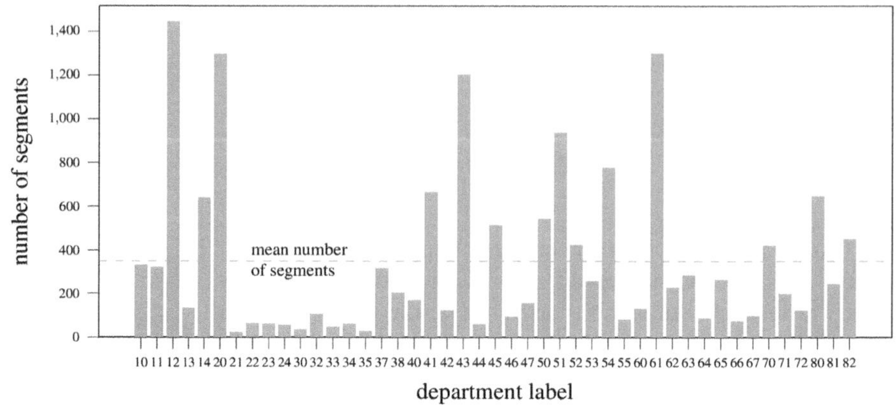

Fig. 1 Number of segments by department in original sample

3.1 Data

In our analysis we use a collection of 1462 labeled text examples extracted from pdf documents, provided by Governikus GmbH & Co. KG after retrieving them from municipal web pages. Some preprocessing steps have already been performed at the word level before we start our analysis: word stemming[2] via the CISTEM stemmer for the German language (see [18]) as well as the removal of names, numbers, and stop words.[3] These steps convert every document into a sequence of terms separated by blanks, where each term is composed of lowercase letters only. Due to the normalizing effect of the applied measures, there are much fewer different terms now than word shapes in the original material.

Additional preprocessing steps that operate on the entirety of each converted document rather than single words are part of our experimental setup, two of which aim at balancing the classes.

From the splitting procedure described in Sect. 3.2, we obtain a total of 15,802 text samples. There is, however, excessive imbalance with respect to the individual departments' representation in the collection, with the number of segments ranging from 24 to 1446, as shown in Fig. 1. Imbalanced data has the potential to severely degrade the performance of classification algorithms.

However, we expect that this problem can easily be avoided in actual applications by collecting more balanced (and larger) samples to train on. Even though incoming data may well be comparably imbalanced as ours, significant numbers of texts can

[2] Stemming algorithms aim at reducing words to a (less complex) form that is shared between all words belonging to the same root, e.g., the words "run", "runs", "running" and "runner" could all be stemmed to "run".

[3] Stop words are a priori expected to contribute little to the meaning of a text. For example, articles or conjunctions can be classified as stop words.

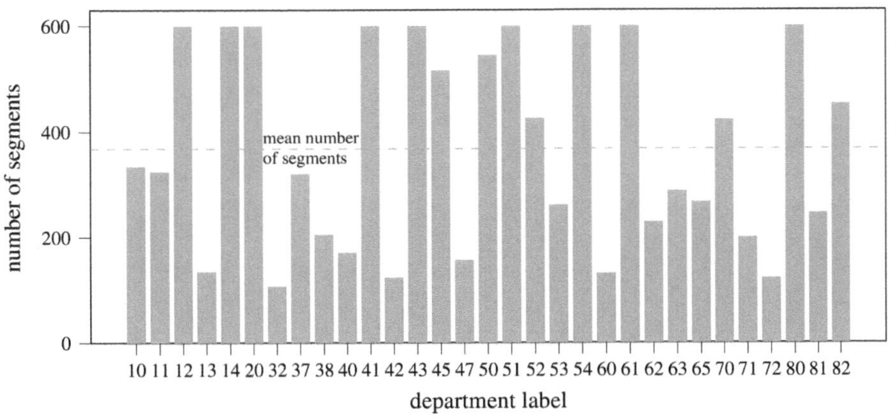

Fig. 2 Number of segments by department after filtering

be expected to accumulate over time even for the departments occurring less often here. Therefore, we deem it appropriate to perform some balancing measures on our data.

First, we exclude departments from the analysis that contribute less than 100 segments. On the other hand, we restrict the number of segments taken into consideration from any department to 600. To this end, for all departments owning a larger number of samples, we randomly eliminate segments from all documents possessing more than a given number of pieces and then gradually lower that value until the number in the department reaches the targeted threshold.[4] Finally, 11,386 segments stemming from 1169 documents remain. The remaining samples belong to just 31 departments, with the number of segments per department starting from 107 and averaging 367.3,[5] as illustrated by Fig. 2. There is a minimum of eight documents from every department (Fig. 3).

As this amount of imbalance may still lead to insufficient consideration of smaller classes in the learning process, all pipelines used in the analysis contain an oversampling step that increases the number of samples in all classes to that of the majority class. After boosting all classes to the maximum number of 600 observations, our pool contains 38.8% generated data. Had we relied on the oversampling exclusively for balancing our data, the share of generated data would have been 75.7%. Because the oversampler simply produces convex combinations of existing vectorized specimens, the generated data contributes no new information. Instead, classes may appear purer than they were before the oversampling effect. This is why we decided in favor of combining an elimination and an oversampling step.

[4] This way, no documents are eliminated completely because there are fewer than 600 documents from every class.

[5] with a corrected sample standard deviation of 187.5.

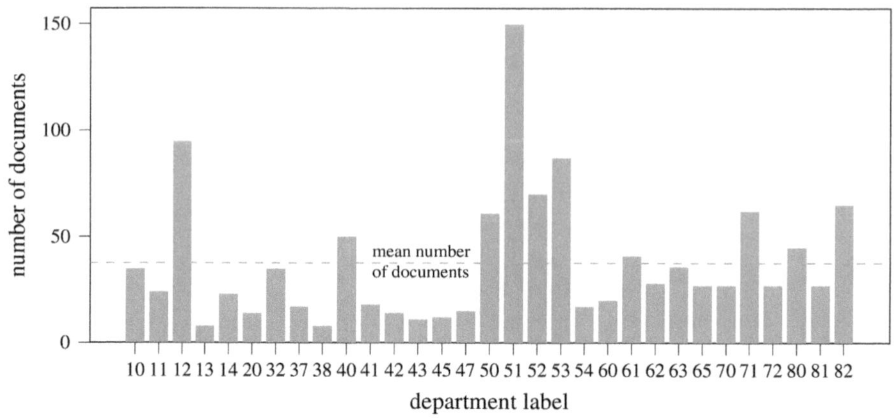

Fig. 3 Number of documents by department after filtering

3.2 Pipeline

An essential step that needs to be taken before classification algorithms can be applied is the preprocessing of the text. There is a wealth of possible measures that can be applied in this phase. We chose a set of preprocessing steps and combined them in four chains, or pipelines. The individual steps used are:

- Splitter: Split the text in segments with fixed length.
- Term frequency (tf) transformer: Transform the character sequence into a vector of numbers by counting term appearances.[6]
- Synthetic minority oversampler: Generate new examples of the minority classes.
- Term frequency-inverse document frequency[7] (tf-idf) converter: Take into account the number of documents a term appears in.
- Truncated singular value decomposition: Reduce the input dimension.
- L^2-Normalizer: Scale the vectors to equal L^2-norm.

We investigate four different experiment pipelines made up from these steps, as depicted in Fig. 4. The base version, our first preprocessing pipeline, contains all of the steps listed above.

First we split the document in segments with fixed character length. For this, we define a minimum number of 42 characters. All documents with a length in characters less than or equal to that minimum are removed from the data set. On

[6] Term frequency is the number of a term's occurrences in the text.

[7] Term frequencies are adjusted for the relevance of terms in the document collection by multiplying them with the inverse document frequency to reflect the assumption that terms are less significant if they occur in many documents. Inverse document frequency is calculated as the logarithm of the ratio of the total number of documents to the number of documents containing the respective term.

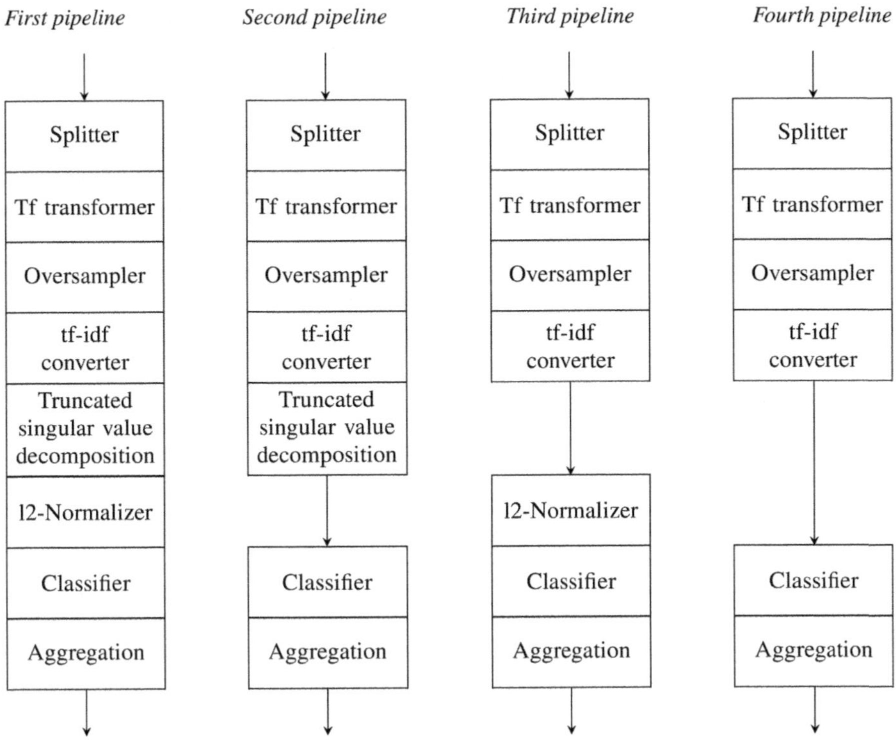

Fig. 4 The different experiment pipelines

the other hand, we determine the maximum length of a segment to 2048 characters. Documents longer than that are split into segments of 2048 characters each with the possibly shorter rest remaining in the data set as another segment.

After dividing the documents into segments with similar length, the segment texts are split on blanks (i.e., "tokenized"), yielding lists of terms, mostly the stemmed words contained in our input material. To transform these lists into vectors of term frequencies, every term occurring in any segment is assigned a unique vector position. For every segment, the number of the term's occurrences is entered at that position in the vector created for the segment. The term frequency transformer performs both steps, the tokenization and the transformation. The term frequency vectors obtained this way form the basis of the oversampling process. Synthetic examples of minority classes are generated until all classes contain the same number of data points. To synthesize a new sample in a minority class, first a data point from that class is chosen at random. Then one of the five closest data points in the same class is again picked randomly. Finally, the synthesized example is determined as a random point on the straight line between the two data points. The oversampling step is used exclusively in the training phase, not in inference. Based on the now increased data pool, the inverse document frequencies (idf) and from these the tf-idf scores are calculated.

The remaining two steps are only present in two versions each of the preprocessing part of the pipeline, resulting in a total of four variants. A reduction to 800 dimensions is achieved by replacing the tf-idf matrix with a reduced version of its singular value decomposition that calculates only the largest 800 eigenvalues. Finally, the vectors derived from all segments are scaled so that they all have the same L^2-norm. Pipeline two differs from the base version in that the L^2-normalizer is skipped, while pipeline three does not contain the truncated singular value decomposition. The fourth pipeline omits both, so that the classifier algorithms are trained on, and applied to, the entire tf-idf vectors directly.

We consider three different classifiers:

- logistic regression (see, e.g., [8]),
- random forest (see, e.g., [2]), and
- support-vector classifier (see, e.g., [4]).

In the inference step, after getting the predicted class and probabilities of the segments, we aggregate them to one document class by the aggregation algorithms described in Sect. 2. We define the weights used in *Restricted Maximum Weighted Average Probability* and *Maximum Weighted Average Probability* as the word number of the segment. In fact, in doing so we take into account an additional measure of segment length since we split the documents based on the number of characters.

3.3 Evaluation

We evaluate the result by means of a fivefold cross-validation. For this, we divide the split text segments into five partitions of nearly equal size under the restriction that all segments of one and the same document are in the same subset. Every partition is used as test set once with the remaining subsets forming the training set.

For comparing the models, we calculate several statistics using the data from all test sets. Except for accuracy, these are derived as averages of per-class evaluations weighted on class size:

- *Accuracy* is the ratio of correctly assigned items, calculated over all classes.
- *Precision* is calculated as the ratio of observations assigned to a class that actually belong there.
- *Recall* is obtained by dividing the number of correctly identified members of a class by the total number of true class members. The class-weighted average recall is thus the same as the accuracy statistic.
- The *F1-Score* for each class is as usual derived as the harmonic mean of precision and recall.

Table 1 Hyperparameter search space

Classifier	Hyperparameter	Search space
Logistic regression	Regularization parameter	$[1 \cdot 10^{-6}; 100]$
	Tolerance for stopping criteria	$[1 \cdot 10^{-6}; 1 \cdot 10^{-2}]$
Random Forest	Number of trees	[1; 1000]
	Maximum tree depth	[1; 1000]
Support-vector classifier	Regularization parameter	$[1 \cdot 10^{-6}; 100]$
	Kernel	{rbf; linear}
	Kernel coefficient	$[1 \cdot 10^{-6}; 1 \cdot 10^{-2}]$
	Tolerance for stopping criteria	$[1 \cdot 10^{-6}; 1 \cdot 10^{-2}]$

Besides model evaluation, cross-validation is also used when optimizing the hyperparameters of the classifiers. We use a sequential model-based hyperparameter optimization, namely, Bayesian hyperparameter search (see [13]). The mean of the accuracy over all test folds is the maximizing score. The hyperparameter search space for the different classifier is listed in Table 1 and the chosen hyperparameter of the different experiments depending on the classifier in Tables 5, 6, and 7 in the Appendix.

4 Results

We considered a total of 12 combinations out of 4 preprocessing pipelines and 3 classifiers. In this section, we highlight some patterns observable in the results, which are tabularized in the Appendix. For reference, an additional row, denoted by *Segment Level Values* in the aggregation algorithm column, is included for each pipeline. These rows contain results at the segment level, i.e., which have been calculated directly from the predictions returned by the various classifiers.

Aggregated values are in general higher by several percentage points than values for raw segments. The exception to this rule is given by the weighted and unweighted *Restricted Maximum Average Probability* aggregators which often yield only slightly better fits than those obtained at the segment level. On the other hand, the level of amelioration in many cases surpasses seven or even eight percentage points, with the highest gains apparently obtainable in the very cases where the quality on the segment level is already rather high. Hence, we conclude that the mode of aggregating segment level results heavily influences the overall performance of the classification pipeline.

Preprocessing Pipelines

Across all aggregation methods, the third preprocessing pipeline performs best independently of the classifier employed. For example, the highest accuracy for the logistic regression classifier is achieved using this pipeline and *Maximum Weighted*

Average Probability. It is 1.96% higher than the worst value obtained with the same aggregation method which stems from pipeline two. For the random forest with the same aggregation algorithm, the accuracy scores increases from 79.81% to 82.21% using the third pipeline instead of the first. The largest difference is given for the support-vector classifier and again the same aggregation method where the third pipeline beats the fourth by 8.64%. For this aggregation variant, it thus depends on the classifier which pipeline performs worst.

On closer inspection, we can observe that pipeline one actually produces higher accuracy values for the logistic regression and support-vector machines on segment level. The difference lies here in the dimensionality reduction step which is present in the first, but lacking in the third pipeline. In light of the discussion in Sect. 4, we can assume that the sample averages calculated in the aggregation methods cause a stronger development towards the normal distribution for the third pipeline. This suggests that the distribution of probability values for the most probable classes tends to have more outliers for these classifiers in the absence of the dimensionality reduction step. In fact, random forests are known to cope well with high-dimensional input spaces.

Precision values are higher than their recall counterparts for all pipelines considered here.

Classifiers

As far as classifiers are concerned, the best results are delivered by the logistic regression for all pipelines (see Tables 8, 9, and 10 in the Appendix). Tables 2, 3, and 4 show the minimum, mean, and maximum over the different pipelines and aggregation algorithms of the scores for every considered classifier. These tables show a 6.46% higher averaged accuracy for the logistic regression results as compared to the random forest and a 4.04% higher averaged accuracy when compared to the support-vector machine.

For the pipelines containing an L^2-normalization step, the logistic regression is closely followed by the support-vector classifier. This underscores the well-known (e.g., [7]) fact that support-vector machines profit from normalized inputs. The support-vector classifier also outperforms the random forest in most cases when using the second pipeline.

Aggregation Methods

There exists no single one aggregation method which delivers the best scores for all combinations of pipelines and classifiers. However, among the best 3 aggregation

Table 2 Minimum, mean, and maximum percentage over all pipelines and aggregation algorithms of the different scores for the logistic regression classifier

	Accuracy	Precision	Recall	F1-Score
Minimum	81.01	81.22	81.01	80.82
Mean	85.05	85.53	85.05	84.90
Maximum	87.25	87.89	87.25	86.99

Table 3 Minimum, mean, and maximum percentage over all pipelines and aggregation algorithms of the different scores for the random forest classifier

	Accuracy	Precision	Recall	F1-Score
Minimum	73.74	74.96	73.74	72.99
Mean	78.59	79.71	78.59	78.30
Maximum	82.21	83.07	82.21	82.00

Table 4 Minimum, mean, and maximum percentage over all pipelines and aggregation algorithms of the different scores for the support-vector classifier

	Accuracy	Precision	Recall	F1-Score
Minimum	75.96	78.60	75.96	75.86
Mean	81.01	82.56	81.01	80.80
Maximum	85.89	86.82	85.89	85.40

methods for all 12 test cases, there are only 3 different algorithms: *Maximum Weighted Average Probability*, *Restricted Maximum Sum Probability*, and *Maximum Average Probability / Maximum Sum Probability*.[8] *Maximum Weighted Average Probability* is the best aggregation algorithm in 7 of the 12 test cases. It produces the best results for the first and third pipeline and for the second pipeline in combination with the random forest classifier. For the second and fourth pipeline with logistic regression and for the fourth pipeline with the support-vector machine, *Restricted Maximum Sum Probability* performs best. In the latter setting, it is tied with *Maximum Average Probability*, the method which also produces the best aggregation results for the remaining two combinations.

For practical purposes, it can be of interest that *Restricted Maximum Sum Probability* can be applied under circumstances involving limited information at the segment level, in that it only requires knowledge of the winning classes and probabilities.

All aggregated values are higher than values based on segment levels, except for the values of *Restricted Maximum Average Probability* for the fourth pipeline and random forest. The improvement is caused by the variance of the sample mean. We will explain this effect using *Maximum Average Probability*, *Restricted Maximum Sum Probability*, and *Maximum Predicted Class* as examples. The probability values calculated by the classification pipeline for a segment to belong to one of the target classes can be regarded as outcomes of a set of stochastic variables specific to the true class of the document. The class for which the expected value of that variable is highest will have the highest probability value most of the time, although due to variance this will not always be the case. The average values of a document's segments will then be sample means from the same distribution. They will exhibit

[8] As noted in Sect. 2, *Maximum Sum Probability* and *Maximum Average Probability* have the same results.

the same expectation values, but lower variances, so that it becomes less probable for a class with minor expectation to receive the highest probability score.

With a little twist, we can apply the same reasoning to the *Restricted Maximum Sum Probability* algorithm. We introduce a second set of stochastic variables which take the same value as the corresponding variables in the first set when the probability predicted for the class in question is highest and 0 otherwise. The predictions produced by the outcomes of those stochastic variables would be the same. Like before, average values from all of a document's segments will have lower variances, making it more probable for the class with highest expectation to win.[9] The expectation values of these new variables will of course be lower than those of the older variables. However, we cannot know in general how, e.g., the difference between the two highest expectation values will develop, nor how the variances behave. Yet, our results show that for the data considered here, *Restricted Maximum Sum Probability* performs comparably to *Maximum Average Probability*.

Let us introduce a third set of variables which take only values 1 and 0, depending on whether the probability predicted for a class is highest or not. Again, the outcomes on segment level would of course be the same. Aggregating over all of a document's segments leads to *Maximum Predicted Class* in this case, so we can understand why even this method with its exceedingly simple logic consistently improves upon the prediction quality at the segment level. Take note that we refrained from using any additional information such as prediction probabilities for breaking ties[10] and relied on a random choice instead. This rule is consistent in the following sense: For $l \geq 2$ consider the probability α that one of the l highest-valued classes[11] is the true class. It will always be at least as high as the probability of the single highest score, if it exists, coming from the true class. In order to predict a single class, we can always choose from those l classes randomly, achieving a probability of α/l to hit the true one. This is what we do in the case of ties. In the standard untied case, we improve on that outcome by using the additional information contained in the segment level predictions and choosing the single highest-scoring class.[12] By restricting the input to segment level predictions, we demonstrate that it is possible to obtain quite usable predictions with this aggregation method when the scores fueling those predictions are unknown. This algorithm may thus be considered a baseline for aggregating results that provides the minimum level of quality reasonably to be expected from any aggregation method.

---

[9] The fact that *Restricted Maximum Sum Probability* considers the sum rather than the average of these stochastic variables does no harm to this argument, we can simply divide all sums by the number of segments, as long as we treat all classes the same. This is different from *Maximum Average Probability*, where the mean is only calculated over the segments won.

[10] In our experiments, we have a tie in 6.3% till 11.2% of the 1169 documents depending on the classifier and pipeline. The logistic regression results in the lowest proportion comparing to support-vector classifier and random forest.

[11] Disregarding the possibility of a tie for rank l.

[12] It appears plausible to assume that the improvement is positively correlated to the gap between the scores of the leading and the runner-up classes and will approach zero as that gap closes.

In our experiment, *Maximum Predicted Class* occupies an intermediate position among our algorithms, ranking fourth in most cases.

Restricted Maximum Sum Probability can be viewed as an enhanced version of *Maximum Predicted Class* that takes into account the score by which a class wins for one segment.[13] It can thus turn around cases where one class wins a majority of a document's segments by small margins while another class wins fewer segments decisively. This might especially be the case whenever one highly unambiguous marker appears in just one of a document's segments. Like its simpler version, this aggregation method can be applied in some circumstances involving limited information at the segment level, in that it only requires knowledge of the winning classes and probabilities.

On the other hand, *Restricted Maximum Average Probability* goes one step further by considering *only* the winning probability, but not the number of classes won. This strategy leads in all cases to significantly lower scores than those obtained from *Restricted Maximum Sum Probability*. For the support-vector classifier, it performs worse than the baseline formed by the *Maximum Predicted Class* method.

Probability values for the shorter last segments of documents can be expected to exhibit higher variances than other segments. Weighting the average probability on segment length consequently has a smoothing effect that again decreases variance and leads to the often better performance of *Maximum Weighted Average Probability* in comparison with *Maximum Average Probability*. The former improves upon the unweighted version in most cases, except for the second pipeline with the support-vector classifier and for the fourth pipeline.

On the other hand, there is no sign of systematic improvement when weights are introduced into the *Restricted Maximum Average Probability* method, which leads to *Restricted Maximum Weighted Average Probability*. The two algorithms produce similar results, with only slight differences in both directions.

5 Conclusion

Text classification is an important task in industrial applications. Besides finding suitable classification algorithms, the issue of differences in document length has to be faced. One way to handle this is to split the documents into text segments of similar length. Once the segments have been assigned to the different classes, the predictions for the different segments of one document need to be summarized in one class for the document. For this step, we introduced seven different aggregation algorithms which we investigated in combination with four different preprocessing pipelines and three different classifiers.

[13] Alternatively, *Maximum Predicted Class* could be considered a special case of *Restricted Maximum Sum Probability* where the winning class of a segment is always assigned probability 1, while all other classes receive a value of 0.

Regarding the different aggregation methods, we identified three success criteria:

- The algorithm can be interpreted as finding, from a set of stochastic variables, the one with maximal sample average.
- It makes use of the prediction probabilities for the individual segments.
- Averages are weighted on segment length.

While there is no single algorithm which performs best in all of the 12 settings we tested, only 3 aggregation methods always share the leading 3 places. All of these three algorithms adhere to the first two criteria named above, and the one which is additionally based on weighted averages is in first place most often. Of the other two, one is a restricted method, meaning that it uses only information about the winner classes of all segments, while the other one considers all prediction probabilities.

One more restricted algorithm is worth mentioning here which satisfies only the first criterion: *Maximum Predicted Class* is a very simple aggregation algorithm which uses even less information, yet produces quite useable aggregate predictions, ranking in a middle position in our experiment.

We considered three different classification algorithms: logistic regression, random forest, and support-vector machine. Further common classification algorithms like neural networks can be used and will be subject of further investigation. Also the preprocessing pipelines can be modified in multiple ways.

In our study, we focused on splitting the text documents, obtaining segments with similar length. Each text segment was classified separately, and afterwards all segment-based predictions of one document were gathered in one prediction for the document. Another way of dealing with text documents of different lengths would be normalization.

Acknowledgments This work was supported by the Ministry of Education and Research Germany (BMBF, grant number 01IS20061 ("EP-KI")). We are also gratefully for the support by Governikus GmbH & Co. KG who provided the data. Furthermore, we thank Nils Witznick for his supporting work.

Appendix

See Tables 5, 6, 7, 8, 9, 10.

Table 5 Best hyperparameters of logistic regression as classifier

Pipeline	Regularization parameter	Tolerance for stopping criteria
1	1.9308	$1 \cdot 10^{-6}$
2	0.0009	$1 \cdot 10^{-6}$
3	89.0228	$1.163 \cdot 10^{-3}$
4	0.0004	$1 \cdot 10^{-6}$

Table 6 Best hyperparameters of random forest as classifier

Pipeline	Number of trees	Maximum depth of the tree
1	666	52
2	983	526
3	533	588
4	984	821

Table 7 Best hyperparameters of support-vector machine as classifier

Pipeline	Regularization parameter	Kernel	Kernel coefficient	Tolerance for stopping criterion
1	0.3359	linear	$1.5908 \cdot 10^{-6}$	$1 \cdot 10^{-2}$
2	2.7568	rbf	$3.7664 \cdot 10^{-5}$	$1 \cdot 10^{-6}$
3	1.0079	linear	$1 \cdot 10^{-2}$	$1 \cdot 10^{-2}$
4	0.0001	linear	$1.1633 \cdot 10^{-6}$	$5.499 \cdot 10^{-3}$

Table 8 Results in percentage of logistic regression as classifier

Pipeline	Aggregation Algorithm	Accuracy	Precision	Recall	F1-Score
1	Segment Level Values	78.99	79.54	78.99	78.89
	Max. Predicted Class	84.86	85.37	84.86	84.85
	R-Max. Sum	86.48	86.64	86.48	86.32
	R-Max. Average	82.98	83.07	82.98	82.80
	R-Max. Weighted Average	83.06	83.13	83.06	82.90
	Max. Sum	86.40	86.54	86.40	86.21
	Max. Average	86.40	86.54	86.40	86.21
	Max. Weighted Average	86.57	86.69	86.57	86.36
2	Segment Level Values	78.05	78.90	78.05	78.15
	Max. Predicted Class	82.29	83.79	82.29	82.63
	R-Max. Sum	85.54	86.04	85.54	85.47
	R-Max. Average	81.09	81.28	81.09	80.91
	R-Max. Weighted Average	81.01	81.22	81.01	80.82
	Max. Sum	85.12	85.56	85.12	85.04
	Max. Average	85.12	85.56	85.12	85.04
	Max. Weighted Average	85.29	85.67	85.29	85.17
3	Segment Level Values	78.81	79.50	78.81	78.56
	Max. Predicted Class	85.71	86.78	85.71	85.62
	R-Max. Sum	87.08	87.70	87.08	86.83
	R-Max. Average	84.86	85.33	84.86	84.58
	R-Max. Weighted Average	84.69	85.15	84.69	84.42
	Max. Sum	86.91	87.52	86.91	86.64
	Max. Average	86.91	87.52	86.91	86.64
	Max. Weighted Average	87.25	87.89	87.25	86.99
4	Segment Level Values	78.66	79.81	78.66	78.84
	Max. Predicted Class	83.06	84.45	83.06	83.26
	R-Max. Sum	86.57	87.08	86.57	86.45
	R-Max. Average	83.66	84.17	83.66	83.46
	R-Max. Weighted Average	83.49	84.06	83.49	83.29
	Max. Sum	86.31	86.74	86.31	86.14
	Max. Average	86.31	86.74	86.31	86.14
	Max. Weighted Average	86.23	86.64	86.23	86.07

Table 9 Results in percentage of random forest as classifier

Pipeline	Aggregation Algorithm	Accuracy	Precision	Recall	F1-Score
1	Segment Level Values	73.23	73.82	73.23	72.74
	Max. Predicted Class	77.93	79.27	77.93	77.93
	R-Max. Sum	78.61	79.60	78.61	78.36
	R-Max. Average	74.17	75.13	74.17	73.78
	R-Max. Weighted Average	73.91	74.96	73.91	73.54
	Max. Sum	79.30	80.05	79.30	79.03
	Max. Average	79.30	80.05	79.30	79.03
	Max. Weighted Average	79.81	80.72	79.81	79.56
2	Segment Level Values	73.51	74.41	73.51	73.31
	Max. Predicted Class	76.82	77.90	76.82	76.79
	R-Max. Sum	79.04	79.95	79.04	78.95
	R-Max. Average	75.28	76.25	75.28	75.07
	R-Max. Weighted Average	75.02	76.10	75.02	74.87
	Max. Sum	80.07	80.80	80.07	79.99
	Max. Average	80.07	80.80	80.07	79.99
	Max. Weighted Average	80.41	81.13	80.41	80.32
3	Segment Level Values	75.45	75.54	75.45	74.68
	Max. Predicted Class	80.07	81.64	80.07	80.04
	R-Max. Sum	81.52	82.76	81.52	81.28
	R-Max. Average	75.88	78.00	75.88	75.21
	R-Max. Weighted Average	75.71	77.74	75.71	74.99
	Max. Sum	81.95	82.85	81.95	81.70
	Max. Average	81.95	82.85	81.95	81.70
	Max. Weighted Average	82.21	83.07	82.21	82.00
4	Segment Level Values	74.71	75.05	74.71	73.94
	Max. Predicted Class	79.64	81.18	79.64	79.55
	R-Max. Sum	81.01	81.97	81.01	80.59
	R-Max. Average	73.74	75.67	73.74	72.99
	R-Max. Weighted Average	73.74	75.80	73.74	72.99
	Max. Sum	81.27	82.06	81.27	80.90
	Max. Average	81.27	82.06	81.27	80.90
	Max. Weighted Average	80.75	81.66	80.75	80.37

Table 10 Results in percentage of support-vector machine as classifier

Pipeline	Aggregation Algorithm	Accuracy	Precision	Recall	F1-Score
1	Segment Level Values	77.52	77.92	77.52	77.21
	Max. Predicted Class	83.15	84.33	83.15	83.27
	R-Max. Sum	84.43	85.41	84.43	84.45
	R-Max. Average	81.01	82.04	81.01	80.97
	R-Max. Weighted Average	80.75	81.72	80.75	80.66
	Max. Sum	84.43	85.30	84.43	84.42
	Max. Average	84.43	85.30	84.43	84.42
	Max. Weighted Average	84.94	85.85	84.94	84.95
2	Segment Level Values	71.57	72.88	71.57	70.74
	Max. Predicted Class	77.25	78.86	77.25	77.15
	R-Max. Sum	79.64	81.08	79.64	79.33
	R-Max. Average	77.07	78.60	77.07	76.83
	R-Max. Weighted Average	77.25	78.77	77.25	77.01
	Max. Sum	79.98	81.19	79.98	79.68
	Max. Average	79.98	81.19	79.98	79.68
	Max. Weighted Average	79.38	81.13	79.38	79.13
3	Segment Level Values	77.16	78.44	77.16	76.57
	Max. Predicted Class	84.17	85.69	84.17	83.94
	R-Max. Sum	85.63	86.56	85.63	85.26
	R-Max. Average	83.32	84.10	83.32	82.98
	R-Max. Weighted Average	83.49	84.26	83.49	83.15
	Max. Sum	85.80	86.77	85.80	85.35
	Max. Average	85.80	86.77	85.80	85.35
	Max. Weighted Average	85.89	86.82	85.89	85.40
4	Segment Level Values	69.35	72.26	69.35	68.16
	Max. Predicted Class	75.96	79.33	75.96	75.86
	R-Max. Sum	78.44	80.88	78.44	78.07
	R-Max. Average	76.05	78.93	76.05	76.11
	R-Max. Weighted Average	75.96	78.74	75.96	75.96
	Max. Sum	78.44	80.88	78.44	78.02
	Max. Average	78.44	80.88	78.44	78.02
	Max. Weighted Average	77.25	80.19	77.25	76.88

References

1. Bogumil, J., & Holtkamp, L. (2013). *Kommunalpolitik und Selbstverwaltung: Eine praxisorientierte Einführung*. Bonn: Bundeszentrale für politische Bildung.
2. Breiman, L. (1984). *Classification and regression trees*. Routledge.
3. Bund-Länder-Kommission für IT-Standards in der Justiz. (2021). Neue Regelungen für den Elektronischen Rechtsverkehr mit den Gerichten seit 01.01.2018. (Available at https://egvp.justiz.de/behoerdenpostfach/Informationsmaterial_fuer_Behoerden_zu_BeBPo_und_eEB_V1_3_2.pdf).

4. Cortes, C., & Vapnik, V. (1995). Support-vector networks. *Machine Learning, 20*(3), 273–297.
5. Devlin, J., Chang, M., Lee, K., & Toutanova, K. (2019). BERT: Pre-training of Deep Bidirectional Transformers for Language Understanding. (Available via ArXiv http://arxiv.org/abs/1810.04805).
6. Domeniconi, G., Moro, G., Pasolini, R., & Sartori, C. (2015). A study on term weighting for text categorization: a novel supervised variant of tf.idf. In *Proceedings of 4th International Conference on Data Management Technologies and Applications - DATA* (pp. 26–37). France: SciTePress.
7. Fan, R.-E., Chang, K.-W., Hsieh, C.-J., Wang, X.-R., & Lin, C.-J. (2008). LIBLINEAR: A library for large linear classification. *Journal of Machine Learning Research, 9*, 1871–1874. (Available at https://www.csie.ntu.edu.tw/~cjlin/papers/liblinear.pdf).
8. Hastie, T., Tibshirani, R., & Friedman, J. (2009). *The elements of statistical learning* (2nd edn.). Springer.
9. Jurafsky, D., & Martin, J. H. (2008). *Speech and language processing: An introduction to natural language processing, computational linguistics, and speech recognition.* Pearson Prentice Hall.
10. Kowsari, K., Meimandi, K. J., Heidarysafa, M., Mendu, S., Barnes, L., & Brown, D. (1995). Text classification algorithms: A survey. *Information, 10*(4), 321–354.
11. Luhn, H. P. (1957). A statistical approach to mechanized encoding and searching of literary information. *IBM Journal of Research and Development, 1*(4). 309–317.
12. Mikolov, T., Chen, K., Corrado, G., & Dean, J. (2013). Efficient estimation of word representations in vector space. (Available via ArXiv https://arxiv.org/abs/1301.3781).
13. Mockus, J. (1989). Bayesian approach to global optimization. In *Mathematics and its Application* (vol. 37). Springer Dordrecht.
14. Pennington, J., Socher, R., & Manning, C. D. (2014). Glove: Global vectors for word representation. In *Proceedings of the 2014 Conference on Empirical Methods in Natural Language Processing (EMNLP)* (pp. 1532–1543). Qatar: Association for Computational Linguistics.
15. Peters, M. E., Neumann, M., Iyyer, M., Gardner, M., Clark, C., Lee, K., & Zettlemoyer, L. (2018). Deep contextualized word representations. In *Proceedings of the 2018 Conference of the North American Chapter of the Association for Computational Linguistics: Human Language Technologies* (pp. 2227–2237). Louisiana: Association for Computational Linguistics.
16. Skiena, S. S. (2017). *The data science design manual.* Cham: Springer.
17. Sparck Jones, K. (1972). A statistical interpretation of term specificity and its application in retrieval. *Journal of Documentation, 28*(1), 11–21.
18. Weißweiler, L., & Fraser, A. (2017). Developing a stemmer for German based on a comparative analysis of publicly available stemmers. In *Proceedings of the German Society for Computational Linguistics and Language Technology (GSCL).* (Available at http://www.cis.lmu.de/~weissweiler/cistem/).

Hardware-Aware Neural Architecture Search

Dominik Marek Loroch

Abstract The manual search for optimal neural network topologies, like maximizing the accuracy or minimizing the latency, is a time-consuming process that often requires multiple iterations. Depending on the use case, several target objectives are optimized simultaneously. If the underlying hardware platform needs to be considered as well, the complexity of the search grows even further, with a search space too large for manual search. Neural architecture search automatizes the search for optimal topologies. We will outline the implementation of an evolutionary search algorithm for neural networks, which can take several criteria for the search into consideration. As a final step, hardware awareness is introduced for an exemplary hardware architecture.

Glossary

CGP	Cartesian genetic programming
DAG	directed, acyclic graph
DL	deep learning
DNN	deep neural network
EA	evolutionary algorithm
FPGA	field-programmable gate array
GPU	graphics processing unit
HPO	hyperparameter optimization
ML	machine learning
NAS	neural architecture search
RL	reinforcement learning
RNN	recurrent neural network

D. M. Loroch (✉)
Fraunhofer ITWM, Kaiserslautern, Germany
e-mail: dominik.loroch@itwm.fraunhofer.de

J. Franke, A. Schöbel (eds.), *Statistical Machine Learning for Engineering with Applications*, Lecture Notes in Statistics 227,
https://doi.org/10.1007/978-3-031-66253-9_9

1 Introduction

Deep neural networks have proven to have an incredible potential in many application domains, yet at the same time, it is quite difficult to design good neural network models. The difficulty comes from the unknown relation between a concrete problem with its data set and the performance of such a DNN model. Therefore, for example, in the domain of computer vision, it is very common to take a large, off-the-shelf model for a general vision problem and to fine-tune it for one's own application. However, this approach is not feasible if the type and format of the input data is less common (e.g., time series data), where there are no general and well-suited models available. Another problem is that general models are often oversized for the problem at hand, even with postprocessing methods which can reduce the models' computational requirements after training.

If, on the other hand, one chooses to design a model from scratch, the first question that needs to be addressed is what kind of input data is processed, for example, 2D/3D images, time series, nominal, ordinal, etc., and what kind of output should be produced for the classification, regression, translation, etc. task. This influences many design options, for example, convolution layers work very well with image data, softmax activation is good for classification, and so on. Even if the pallet of DNN building blocks is chosen well, still there is an overwhelming amount of design choices left for not only the DNN topology but also for the training procedure (e.g., optimizer, learning rate), the data preprocessing (e.g., augmentation, embeddings), and more. Even after trying out many design options, it is possible that some single but powerful hyperparameter was not considered for adjustment, which would have improved the model considerably.

Another problem of manual design is that the focus often lies on a single performance metric like the model accuracy, ignoring other properties like the model runtime and robustness toward variations of the data set distribution. The situation is aggravated if the model is deployed on a different hardware platform, which spawns yet another set of performance issues which must be considered for the design. Given how difficult it can be to find a model optimized for a single objective, it is clear how intractable manual design is for several objectives. Therefore, some form of automatization is necessary to find designs which are not only well suited for the application performance, but also work well with regard to other aspects like the target hardware or variations in the input quality. This automatization can be provided by NAS. It is a method to navigate the vast space of neural network designs. Section 2 outlines how to implement a NAS based on evolutionary search from scratch. It highlights exemplary implementations of genetic encoding, mutation operations, and selection algorithms. In Sect. 3 the NAS algorithm is used to demonstrate how hardware awareness can be included for an DNN model executed on a FPGA. Before diving into the more practical part, the next section reviews how NAS fits into the larger field of automatic machine learning or auto-ML.

1.1 Overview of Auto-ML and NAS Methods

Automatic ML encompasses all methods which automatically identify the best machine learning algorithms and their parametrization. It does not only include DNN approaches, but also more classic ones like support vector machines, random forest algorithms, etc. There are two categories of parameters that must be distinguished in any ML approach. The first category are the trainable parameters or weights, which are adjusted based on the training data. The second category are hyperparameters, which determine design options for the specific ML algorithm. For DNNs, this can be, for example, the depth and width of the neural network. Usually one would try to find hyperparameters manually while using a training algorithm to optimize the weights for a given training data set. Auto-ML understands this as a bi-level optimization procedure, where both types of parameters are optimized in nested optimization loops. Roughly speaking, the outer loop is the hyperparameter optimization, which passes ML models to the inner loop. The inner loop trains the model weights and returns the evaluation results as feedback to the outer loop, which in turn passes down a better model. Formally this is described as

$$\min_{w,\theta} J(w, \theta) \tag{1}$$

$$\text{s.t.} \quad w \in \arg\min_{w'} L(w', \theta), \tag{2}$$

with w the weights, θ the hyperparameters, J a first-level target function, and L a second-level target function. This type of optimization in the context of auto-ML is known as HPO. Famous algorithms in this field are Hyperbands [17] and their improvements like Bayesian optimization Hyperbands (BOHB) [11] or tree-structured Parzen estimator (TPE) [2]. A good overview on the topic of HPO is provided in [3].

NAS can be seen as a specialized type of HPO, which focuses on finding DNN topologies. Notice that the design of a DNN topology, which in essence is the design of a computational graph, goes beyond simply finding a good set of values. The design choices in a graph are interdependent, meaning the choice of one hyperparameter can determine the existence and type of other hyperparameters. For example, the depth of a DNN determines how many further hyperparameter choices about the layers are available. The choice of a layer type determines which parameters like kernel size and number of output channels are present. So the search space in NAS is rather structured like a tree than a hypercube with a fixed amount of hyperparameters. Dealing with this type of search space is what distinguishes NAS from regular HPO, but it also means it is more of a specialized tool. Therefore, the majority of auto-ML tools support more general HPO algorithms, as indicated by the overview in Fig. 1. The rest of this section introduces some of the more prominent approaches for NAS.

One NAS approach is to grow a neural network from a relatively small seed model into a more complex and capable model. The idea was introduced by Net2Net

Solution	Hyperparameter Tuning	Neural Architecture Search	Open Source
MLJAR	✓		
Google Cloud Auto-ML	✓		
AWS SageMaker Autopilot	✓		
IBM AutoAI	✓		
Azure Automated ML	✓		
H2O.ai	✓		✓
Dataiku	✓		
Deci	✓	✓	
DataRobot	✓		
Hyperopt	✓		✓
Optuna	✓		✓
Auto SciKit-Learn	✓		✓
CATBoost	✓		✓
Auto-Pytorch	✓	✓	✓

Fig. 1 Overview of available auto-ML tools

[7], who used knowledge transfer between DNN models in the form of network morphisms. A network morphism is a modification of the topology and the model, which does not change the output of the model. A simple example is to inline the identity matrix as an additional layer somewhere in the network. The new layer does not change the feature map; therefore, the output remains the same. However, the network has more parameters now; thus, the network's capacity increased. By fine-tuning the modified network, one can hope to get an even better prediction performance from the model. Thus, by successively adding new layers, the model's capabilities grow and new design options are explored. However, the models can never shrink, since removing layers would not preserve the output of the model. Later, [9] combined network morphisms with an evolutionary search algorithm to explore the design space in a multi-objective setting. They also add the notion of approximate network morphisms, which allow to shrink the network by applying knowledge distillation techniques [13].

Another approach to NAS is to learn how to design a good DNN. The idea is to build a DNN that learns to design DNNs. In [34] they use RL in combination with a RNN. The RNN is a DNN, which can produce a sequence of outputs, for example, an LSTM [14]. In [34], the RNN would be trained such that the sequence from the RNN describes a blueprint for a DNN. The problem is that training a RNN that produces blueprints is not directly possible, since a training loss cannot be directly formulated. That is why [34] use RL, which allows to construct a setup that produces a gradient for the RNN. In RL, an agent or actor learns to manipulate an environment by taking actions and receiving a reward from the environment as a consequence of its actions. The goal is to reach a certain state in the environment, which maximizes the reward. In the case of NAS, the RL agent is the RNN; the actions are the sequence of choices in the blueprint, i.e., the hyperparameters, and the reward is the accuracy of the DNN model from the blueprint, evaluated on some data set. The RL update rule is formulated as

$$\nabla_{\theta_c} J(\theta_c) = \frac{1}{m} \sum_{k=1}^{m} \sum_{t=1}^{T} \nabla_{\theta_c} \log P(a_t | a_{(t-1):1}; \theta_c) R_k, \tag{3}$$

where P is the RNN model and a_t are the different design choices, which lead to the reward R_k. θ_c are the weights of the RNN; thus, by minimizing $J(\theta_c)$ via gradient descent, the RNN learns to design better DNNs. Once the RNN learned how to create DNNs, it can design a neural network from scratch very quickly. A disadvantage of RL-based NAS is that the model P, here the RNN, needs training. If the search space is expanded or the training data set is changed, the RL model needs to be trained again.

The RL-based approach is compared against the much simpler EA-based search in [22]. Roughly speaking, the key idea of EAs is that a population of DNNs is subjected to random changes and only the best candidates are kept alive in the population. Therefore, the population keeps improving over time. A more detailed introduction to EAs is given later in Sect. 2. Real et al. [22] found that EAs although not finding better solutions than RL do so in significantly less time and by evaluating less models. EAs are simple to use and very flexible with respect to implementation details and application scenario.

A more recent approach harnesses the power of gradient descent to search for the best topologies. DARTS by Liu et al. [19] is the first paper which uses gradient descent to search for a DNN topology. The idea is to use a big supergraph, which contains all topologies of the search space as subnets. One can imagine the supergraph as a DNN where every layer is replaced by a parallel bundle of different layers, which are combined with a special summation operation. The summation attributes a trainable weight α to every layer in the bundle using the softmax like expression:

$$\bar{o}(x) = \sum_{o \in \mathcal{O}} \frac{\exp(\alpha_o)}{\sum_{o' \in \mathcal{O}} \exp(\alpha_{o'})} o(x), \tag{4}$$

where $o(x)$ are the different layers in the bundle $\mathcal{O}$. Notice that there exists gradient with respect to the α, so gradient descent can be used to train both the DNN weights and the α. Formally, this implies the bi-level optimization procedure:

$$\min_{\alpha} L_{\text{val}}(w^*(\alpha), \alpha) \tag{5}$$

$$\text{s.t.} \quad w^*(\alpha) = \arg\min_{w} L_{\text{train}}(w, \alpha). \tag{6}$$

In practice, both α and the model weights are trained simultaneously, instead of first training all weights and then updating α for one step. This way, a subnet emerges from the supergraph, which is used as the final model from the search. Differentiable NAS is very fast, because it incorporates gradient information into the search, instead of sampling networks randomly. However, the necessity of a supergraph constrains the search space drastically and does not allow to use very large and diverse search spaces. Also, all the optimization targets must be formalized in a loss function, which might not be always possible. However, [32] demonstrates that hardware awareness in the form of execution time on mobile devices can be introduced to DARTS under certain conditions. Another interesting approach is [5], where the supergraph is trained in such a way that any subnet is a well-trained DNN through a form of weight sharing. By employing an EA, the supergraph can be explored with respect to additional hardware-aware objectives to identify the best subnet, without the need to retrain any subnet candidate.

This section was just a brief overview of some of the most prominent works in the field of NAS. Elsken et al. [10] and White et al. [31] are recommended overview papers and serve as a good starting point to dive deeper into the field of NAS. In the following sections, an evolutionary NAS is implemented due to the high flexibility and ease of use. The interchangeable components of the EA allow to mix and match different implementations, so that the basic implementation introduced here can be customized and augmented for a more advanced NAS.

2 Implementation of Evolutionary Neural Architecture Search

2.1 The Basic Evolutionary Algorithm

EAs are iterative procedures, where candidates, or individuals, are successively optimized in a loop as shown in Fig. 2. One iteration of the loop is called a generation. Usually several individuals are considered during every iteration step. The starting set of individuals in a generation is called the parents. The very first candidates are usually sampled randomly from the search space or seeded with some candidate DNNs which are known to perform good. In every step of the iteration, modifications are applied to the parents. In the case of DNNs, the topology is changed by changing the hyperparameters of layers, adding new layers, etc. Often the mutation step in Fig. 2 can be more complex, for example, by using a preceding operation called crossover, which swaps out structural elements between two parents. The individuals created from the mutation are called children. Their properties are unknown, so they have to be evaluated. In the case of DNNs, this means to train the network and to evaluate its performance, measured, for example, by the accuracy. If there is more than one number that indicates the performance of the children, they can be weighted and summarized in the fitness number. The

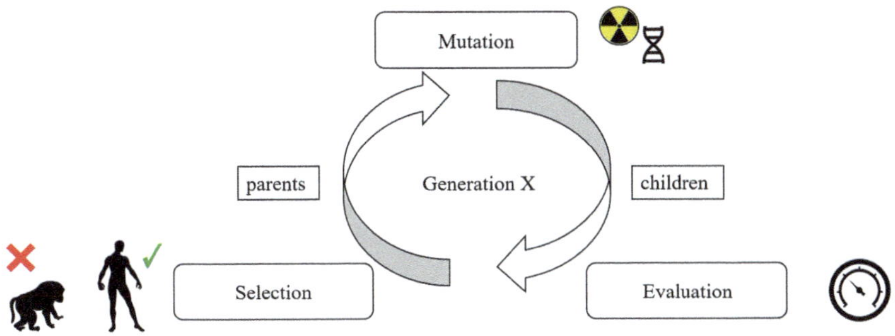

Fig. 2 The basic optimization loop of evolutionary algorithms

goal is to maximize the fitness of the population; therefore, the bad individuals are sorted out in a selection step. The selection decides which individuals from both the children and the parents enter the next generation. In the simplest form, the selection picks the candidates with the highest fitness values. It should be noted that this selection strategy is not good and leads to rather poor optimization results. More details are given in Sect. 2.4. The selection is the last step of an EA iteration. The loop stops after individuals with a certain performance are found or a maximum number of generations are reached. This basic EA description serves as the outline for the NAS implementation. For a broader overview on evolutionary NAS, [20] is recommended.

Another important aspect is the encoding of the DNN structure of an individual, so that the mutation algorithms can work on them. As discussed before in Sect. 1.1, DNNs can be represented by a graph, and very often by a DAG, i.e., a graph without loops. Genetic encoding is an encoding which is naturally compatible with EAs and which can describe DAGs. A genetic encoding scheme for DNNs is described next.

2.2 Genetic Encoding of Neural Networks

A genetic encoding is a sequence of elements or genes, where each gene encodes a building block of the individual. A single gene can be a number, bit pattern, or a more complex data structure. Figure 3 shows how a DAG describing a DNN is represented with a sequence or list of genes. Every gene describes a DNN layer and its input connections. The layer description contains fields for the name, the layer type, and a set of type-dependent hyperparameters. The input connections are a list of indices, which reference genes in the genetic sequence itself. The number of inputs is determined by the layer type, for example, the "concat"-type layer has two and the "input"-type layer has none. This type of genetic encoding is good for building DNN models in DL frameworks, because the list can stepped

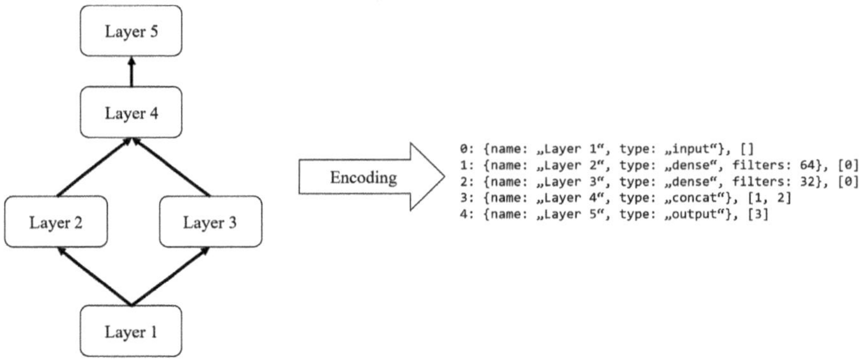

Fig. 3 Genetic encoding of a DNN

through line by line while adding the corresponding layers to the framework-specific representation of the DL model, as shown in Listing 1.

```
# Keep track of all the output tensors from every layer.
layer_outputs = []
# Iterate over the gene list and assemble the DNN.
for layer, gene_inputs in gene_list:
    # Decode inputs.
    inputs = [ layer_outputs[gene_input] for gene_input in
    gene_inputs ]
    # Create layer and add it to the graph.
    layer_output = build_layer(layer, inputs)
    # Add new output to list of layer_outputs.
    layer_outputs.append(layer_output)
```

Listing 1 Converting the gene list into a DNN model

The "build_layer" function call is framework specific and takes care of adding a layer to the model while also returning the output of the layer operation. Since the genes encode a DAG, no indices appear in the connections of a gene which are above the index of the gene itself.

The list encoding allows to build a DNN model from the genetic encoding easily, but modifying the graph structure in the list is cumbersome. For example, if another layer is inserted after "Layer 1" in Fig. 3, all the input indices of the following genes must be incremented by one. There are genetic encodings which allow the graph structure to be modified more easily. An approach called CGP [6, 25] is used here, where the genes are arranged in a matrix; see Fig. 4. There are other ways to encode the genes, for example, code-based approaches as in [26]; also [20] discusses more encoding strategies. The genes are mostly the same, but instead of single-input indices, the indices within the matrix are referenced with tuples. Notice that this representation can be very sparse. Practically, the encoding should not be

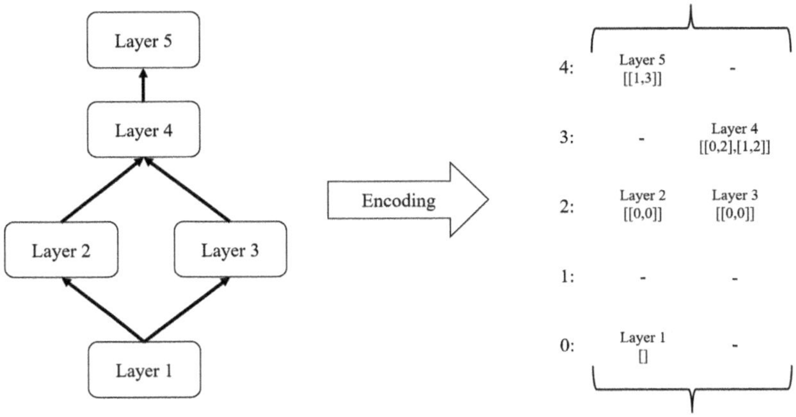

Fig. 4 Genetic encoding of a DNN with Cartesian genetic programming

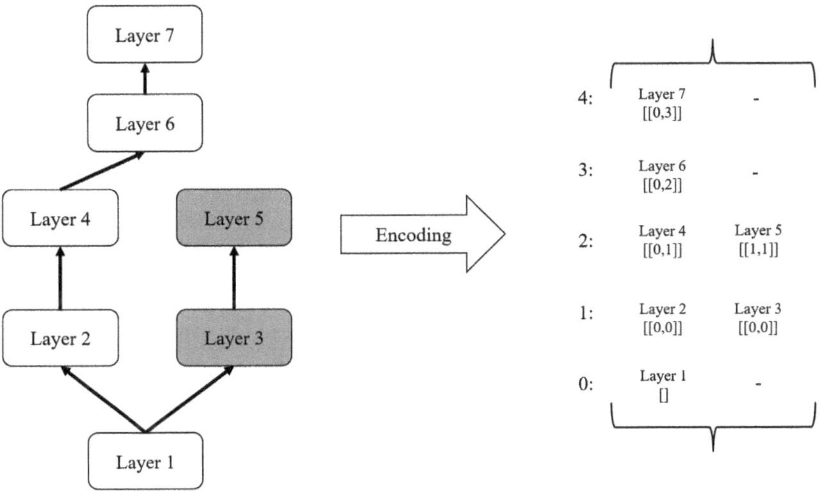

Fig. 5 Genetic encoding of a DNN with dormant genes

stored as an actual matrix, but, for example, as a map or dictionary, which avoids storing empty elements. Another aspect of this representation is that it does not describe the graph uniquely, as many configurations of the matrix can encode the same graph. The CGP allows to make changes to the graph structure more easily, but it also has additional benefits for the implementation of the NAS. First, the height and width of the matrix can be bound, which controls the maximum depth and interlayer parallelism of the topologies in the search space. Acyclic graphs are automatically enforced if only input indices from a row below the own one are allowed. Lastly, the most interesting property coming from CGP are the concepts of genotypes, phenotypes, and dormant genes. Figure 5 visualizes these concepts:

The output of "Layer 5" does not connect to the model's output, so the information processed by this layer is not relevant for the model output. Therefore, these genes are not active, or dormant, which means they do not contribute to the DNN model. There is a difference between the information in the genetic encoding, which is the genotype, and the actual manifestation of the DNN, which is the phenotype. The purpose of dormant genes lies in their potential to be reactivated. In Fig. 5, the input of "Layer 6" can be modified to connect with "Layer 5," reactivating "Layer 5" and "Layer 3." This drastically changes a large portion of the DNN topology, although the actual mutation is on only one connection. Such radical changes, invoked by small mutations, can help to traverse the search space more efficiently.

The CGP representation can be easily transformed into a sequence, which in turn can be used to build a DNN model in a DL framework, as previously demonstrated in Listing 1. The matrix can be sequenced column by column, row by row, until all genes have been visited, while keeping track of the mapping between matrix indices and list indices for the connections. However, inactive or dormant genes must be excluded from the model, since they do not contribute to it. A gene is active if it connects to both the model's input and output, potentially via some other genes. This can be checked by using, for example, breadth-first search, as shown in Listing 2.

```
def has_path_to_input(target_gene):
        # Use 'nodes_to_check' as a fifo
        nodes_to_check = target_gene.get_inputs()
        # Go down the graph, until a node ends in the global
    input.
        node = nodes_to_check.pop(0)
        while node is not None:
            if node in global_inputs:
                # A path to the global input is found.
                return True
            else:
                new_nodes = node.get_inputs()
                # If there are no inputs, then this node
    cannot be connected to the global input.
                if not new_nodes:
                    pass
                else:
                    nodes_to_check.extend(new_nodes)
            # Get the next node
            if nodes_to_check:
                node = nodes_to_check.pop(0)
            else:
                node = None
        # If the global input was not reached, then there is
    no path.
        return False
```

Listing 2 Checking if a gene is connected to the global input

Checking for the model's output connection can be implemented analogously. If both conditions are met, the gene is active and it emerges in the phenotype; therefore, it can be included into the list representation of the DNN.

The first generation of the EA requires randomly generated genotypes. They can be generated by iterating the CGP matrix row by row, starting from the model's input, and randomly inserting layers from the search space. The layer connections have to be chosen from row indices which are smaller than the current row to ensure the acyclic structure. Although the generated graph is valid, the resulting DNN model might not. The reason is that the shapes of the layer outputs are not considered. For example, too much pooling can effectively reduce the input spatial dimensions to zero. Therefore, the phenotypical DNN model must be checked for its validity. This usually can be done with the help of the deep learning framework, for example, by passing a dummy input. If an invalid phenotype is encountered, it can be simply discarded and a new random network can be generated. Another problem arises from layers with several inputs, like adding or concatenation, where the input shapes can mismatch. For example, adding two inputs requires that the shapes match in all dimensions. Suganuma et al. [25] uses auxiliary "glue layers" that help to transform the input shapes. Spatial dimensions can be equalized by using pooling and cropping operations, and the channel dimension can be transformed with a 1×1 convolution layer with a matching number of output channels. Glue layers can increase the probability of a successful DNN generation and modification, if layers with several inputs are introduced into the search space.

A single CGP matrix is already describing a valid DNN. It is possible to stack several matrices in order to concatenate multiple smaller networks into a bigger one. Every matrix can have a different configuration and search space, or they can be copies of a single matrix. The matrix with the search space configuration is referred to as a cell. The use of cells is motivated by the observation that large models are usually made up from repeating, almost identical building blocks, for example, like ResNet [12]. The first to use cells in NAS are [35], where they search for only two different cell types, which they stack into deeper DNN topologies using a fixed stacking order. Cell-based approaches are very popular, since they allow to search for large, high-capacity models by using a relatively small search space. An overview of other cell-based NAS methods is given in [31].

Until this point, the implementation is equal to a generator for random DNN topologies. In principle, a large amount of DNN models can be generated, and the best candidate can be picked. This method is known as random search and despite its simplicity can be quite effective [22]. Nevertheless, a more powerful approach is to use the knowledge gained from previous topologies for the search. Evolutionary NAS does so by creating children from parent neural networks via modifications.

2.3 Modifying Neural Networks

The modification or manipulation of DNN topologies has essentially one goal: to find the best neural network. In order to do so efficiently, the NAS has to balance between two fundamental strategies: exploration and exploitation. Exploration discovers new, unknown features in the topologies. Exploitation recombines already known features in a single individual. Although a search without exploitation is also possible, the search is less efficient, since good features need to be rediscovered for every individual in the population. For both strategies, there are a large number of methods.

For exploration, the size and diversity of the search space determines the effectiveness of all methods. For one part, the search space is a set of combinations of layer parameters, where the parameters can be just a collection of values like in Fig. 3. When mutating, the resulting parameter combinations must be in the search space set. This can be done by simply drawing samples from the search space, for example, with constraint random number generators. A generator function as outlined in Listing 3 can produce random layer parametrizations, which are within a search space.

```
def layer_generator(search_space):
    type = random_number_generator.choice(search_space.types)
    if type == "dense":
        units = random_number_generator.randint(search_space.
    min_units, search_space.max_units)
        layer = { "type": "dense",
                  "units": units,
                }
    elif type == "concat":
        layer = { "type": "concat"}
    return layer
```

Listing 3 A search space generator

The intensity of the mutation is controlled with a parameter called the mutation rate, and most commonly it controls how many parameters are changed in a mutation. For the CGP matrix, the mutation selects random elements in the matrix, where the number of elements can be controlled via the mutation rate. The mutation can be further restricted to select only active, dormant, or both types of genes. Remember that although mutating dormant genes has no immediate impact on the phenotype, another mutation later can reactivate the dormant genes. Mutation can also add new, potentially dormant genes into the CGP matrix. For every selected gene, active or dormant, a new parametrization from the search space is drawn. Any combination from the search space can be allowed, but it is possible to restrict the new parametrization to be of the same layer type as the old one, or to change only a single parameter, or other restrictions.

Besides the parametrization of the layers, mutation can also target the connections between the layers. Changing the input connections demands special attention, because the resulting graph might end up invalid. For example, the layers can be connected such that there is no path from the model's input to the output. Similarly, if the layer type changes to one with more inputs than before, new inputs have to be assigned, filtered for options that connect to the model's input by using Listing 2. Listing 4 shows a basic mutation method, which does not distinguish between active or dormant genes and randomizes the layers.

```
def mutate(individual):
    mutation_rate = individual.get_mutation_rate()
    search_space = individual.get_search_space()
    genotype = individual.get_genotype().copy()
    # Randomly mutate the genotype genes
    for gene in genotype:
        location = genotype.get_location(gene)
        # Mutate the content
        if random_number_generator.rand() <= mutation_rate:
            new_layer = layer_generator(search_space)
            genotype.update_gene(location, new_layer)
        # Mutate the connection
        if random_num_generator.rand() <= mutation_rate:
            candidates = genotype.get_input_candidates
    (location)
            # filter for candidates which are connected to
    inputs
            filtered_candidates = filter(lambda x:
    has_path_to_input(x), candidates)
            new_inputs = random_number_generator.choice
    (filtered_candidates, gene.get_number_inputs())
            gene.set_inputs(new_inputs)
    return genotype
```

Listing 4 A mutation method

More mutation methods can be conceived by applying additional constraints for the mutation locations, parameter combinations, and input candidates. Lastly, mutation can be extended to other hyperparameters, such as learning rate, the used optimizer, and more.

A common method for exploitation is crossover, which is a bioinspired method from genetics. The genes of two individuals are combined with the hope that an even better individual is created, which carries the best features from both parents. Crossover swaps out equivalent genetic sequences between individuals. The problem is that finding equivalent sequences can be challenging, since it requires mutual compatibility. In the worst case, the resulting DNN after the crossover is not valid. Introducing cells to the genetic encoding helps greatly to define equivalent sequences. If all individuals share the same cell structure, it is fair to assume that cells at the same location are equivalent. They can be safely swapped, with a low risk of the resulting individual to be invalid. Crossover can also be done with encodings that do not use cells; see [26] for an example. Crossover can also be used for other hyperparameters, for example, by swapping out the learning rate or training optimizer between two individuals to see if the training setup of one individual is superior to the other.

The mutation rate can be chosen to be constant over the entire search, but a better approach is to have a dynamic rate instead. In the beginning, bigger changes are good to advance quickly to interesting regions in the design space. Later, a smaller rate is preferred, so that the discovered features are optimized, but not destroyed. A mutation rate schedule, where the mutation rate decays with increasing generations, can be implemented easily. Also the distribution of the models' fitness over the generations can be a good indicator to decide which rate should be applied.

The model modifications discussed here only control the degree of change to a model, but the EA implementation still lacks the intelligent component of an evolutionary approach, that is, the selection algorithm.

2.4 Selection Algorithms

2.4.1 Fitness-Based Selection

Evolution is often explained with the term "survival of the fittest." So the concept of fitness is apparently central to EA and how it selects the best candidates. Generally, fitness is a scalar value that describes the goodness of an individual as an aggregation of several performance indicators. Depending on the prediction task, the application performance can be measured with accuracy, area under curve, intersection over union, etc. Additional to the prediction performance, other objectives or targets can be defined like the model size in number of parameters, the inference speed in milliseconds, robustness to adversarial attacks, and many more. The goal of the NAS is to minimize or maximize these target functions. Without loss of generality, all target functions used here are formulated as minimization problems. The optimization problem is multidimensional, where the targets are possibly contradicting each other. For example, the minimization of the number of parameters contradicts the minimization of the prediction error, since a very few parameters will suffer in expressive capacity and therefore in the prediction error. As mentioned before, the fitness is the aggregation of the target values. One option is to use a weighted sum over all objectives, which effectively turns the optimization problem one-dimensional. The weights are usually chosen a priori to the NAS, which encodes the user's preferences, but also poses a strong bias. Formally, the fitness can be described as

$$f_m := \sum_{i=1}^{n} w_i \cdot \tilde{f}_i \tag{7}$$

$$\text{s.t.} \quad \sum_{i=1}^{n} w_i = 1, \quad \tilde{f}_i := \frac{f_i - l_i}{u_i - l_i} \in [0, 1] \tag{8}$$

with f_i the target functions and w_i their respective weights. The weights are normalized to sum up to one, but also the target functions are normalized in the range $[l_i, u_i]$. Finding a good range $[l_i, u_i]$ can be difficult, since it determines the impact of an objective on the fitness. A dynamic range determined by the maximum and minimum values of the current set of individuals can work quite well, although it does not allow to compare the fitness between two sets of individuals, only the individuals within the current set. A fixed range allows global comparability between different sets of individuals, but can introduce another bias to the fitness, just like the weights. Notice that since all targets f_i are posed as minimization problems, f_m is also a minimization problem, which is contradictory to the regular use of the term fitness as a property to maximize ("survival of the fittest"). Therefore, f_m will be referred to as the fitness measure in contrast to the fitness. The optimization problem can be extended with the introduction of constraints. Hard constraints pose a limit above which the individual is considered to be infeasible, so the fitness assumes the worst value. Soft constraints pose a limit above which the individual is punished, where the penalty can be, for example, the distance to the soft constraint limit. On the other hand, if the objective value is below the limit, all values are seen as equally good and the fitness is the same. This allows to prevent an over-optimization for objectives, which stop to have technical relevance at some point (overengineering).

The fitness allows to implement a very basic selection algorithm for the NAS. After the individuals are evaluated with respect to the target functions, their fitness measure f_m is calculated, and a certain number of individuals with the lowest measure are selected for the parent generation. The children are expected to improve over the generations, since the fitness can only improve. However, this naive selection algorithm does not work well often, because it suffers from a collapse in model diversity. By selecting and modifying only the best individuals, the individuals quickly cluster around a single "champion" DNN early on. The population consists mostly of very similar individuals. There is no mechanism which pushes the population apart and allows for novel but slightly worse individuals to emerge. A possible solution is to introduce aging as in [22], which removes individuals after a certain amount of generations, irrespective of their fitness. This prevents that an "immortal champion" dominates the search until the end. Further approaches are discussed in [1, 20].

A major problem with fitness-based selection is the large bias which is introduced by aggregating all target functions into a single value. It possible to search for multiple objectives at once using the notion of Pareto optimality. In [8] this concept was introduced in the context of EA. It has also been adopted in the case of NAS, for example, by Elsken et al. [9]. The next section describes a selection method that is based on Pareto optimality.

2.4.2 Pareto Optimality Based Selection

The definition of Pareto optimality is as follows.

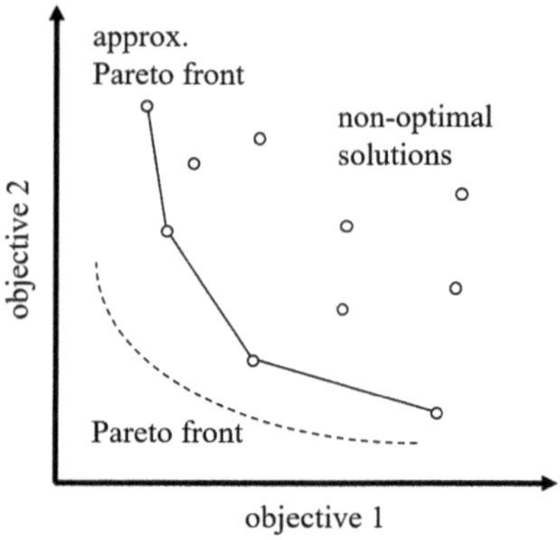

Fig. 6 Visualization of the Pareto front for an optimization problem with two objectives

Definition 1 (Pareto Optimality) An element $\overline{u}$, which is part of a set U, is called Pareto optimal if there exists no other element $u \in U$ with $f_i(u) \leq f_i(\overline{u})$ for all $i \in \{1, .., n\}$ and $f_j(u) < f_j(\overline{u})$ for at least one $j \in \{1, .., n\}$. The set of all Pareto optimal elements is called the Pareto set. The mapping of the Pareto set to the objective space is called the Pareto frontier.

In more gentle terms, the elements in a Pareto optimal set mutually outperform each other in at least one objective. It is the set of the best trade-offs with respect to the objectives. Figure 6 demonstrates the Pareto front for an optimization problem for two objectives. Every point is a valid element in the set of known solutions, but only those towards the bottom left corner are optimal in the Pareto sense. As plotted here in the objective space, the Pareto elements form the Pareto frontier. Strictly speaking, the elements are only the approximation of the Pareto frontier. The real Pareto frontier, i.e., the set of globally Pareto optimal solutions, is usually unknown. For the sake of brevity, the terms Pareto set and frontier will always refer to the approximation. Pareto optimality accounts for the fact that one single best individual does not exist. For example, let there be two objectives, one for the model size and one for the error of the DNN model. A model with less parameters tends to be less accurate than a big model, so the two objectives will oppose each other at some point, because the reduction of model size will result in a larger error. It is impossible to further improve one objective without hurting the other. The Pareto frontier in high-dimensional spaces can be imagined as a hyperplane. This hyperplane does not need to be contiguous, which depends on whether or not there exist feasible solutions in areas of the objective space. These "holes" can be a problem for search algorithms, because the search might focus on exploring candidates in these areas to "close the gap," which is simply not impossible. For a more in-depth survey on Pareto fronts in the context of EA, see [15].

Coming back to the implementation of the selection in the EA, the set of parents shall be equal to the Pareto set. The selection algorithm is to identify the Pareto set in the joint set of the former parents and the new children, following Definition 1. The goal of the selection is to push the Pareto frontier forward, that is, to beat former elements with respect to all objectives and replace them. As a second goal, the Pareto frontier should spread out far and be sampled uniformly, i.e., to "close the gaps." So if a new individual cannot beat the parents with respect to all objectives, it still can be added to the frontier and expand the number of elements in it. Thus, the number of parent individuals is not constant and will fluctuate over time. The parents to generate children can be picked randomly in the hope to contribute to the Pareto front. To improve the Pareto front more efficiently, the first step is to find regions with little or no samples, i.e., the "gaps" in the frontier. A vicinity with many known elements has a lower chance to improve the frontier, because it obviously has already been explored rather thoroughly. The sparse regions can be identified by looking at the density distribution. The distribution of the elements can be approximated with a kernel density estimator p_{KDE} defined as

$$p_{\text{KDE}}(y) = \sum_{i=1}^{N} K(y - x_i; h),$$

where x_i are the N Pareto set elements and h is the bandwidth parameter of the kernel K. The kernel K is a positive function which describes the contribution of an element to the overall density function, e.g., a Gaussian distribution, and the bandwidth h describes the zone of influence. The bandwidth h can be estimated based on the current elements, for example, using a simple rule of thumb like Scott's rule [23] or more advanced methods like [24]. A simple one-dimensional example using a Gaussian distribution as the kernel function is shown in Fig. 7. The variance of the Gaussian distributions is determined by h. A high value of p_{KDE} means that there are many individuals in that region. In order to find the sparse regions in the Pareto front, the inverse distribution

$$p(u) = \frac{c}{p_{\text{KDE}}(f(u))} \tag{9}$$

must be used instead, where c is a normalization factor for the distribution and f is the vector of objectives $(f_1(u), \ldots, f_n(u))$ for an individual u. This distribution is used to sample parents for the mutation. Parents in or close to a sparsely populated area of the Pareto frontier have a higher probability to be selected for mutation.

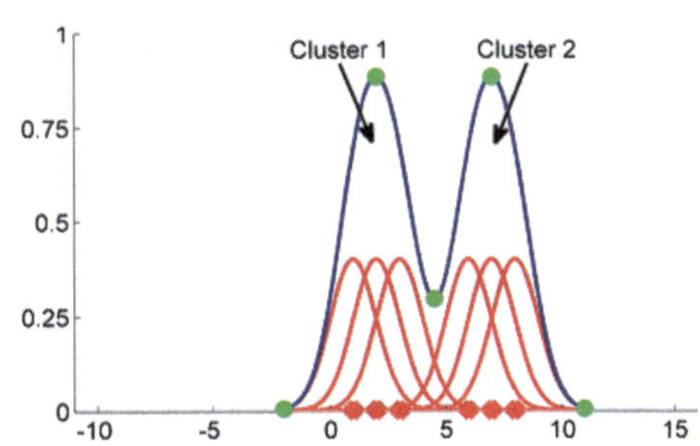

Fig. 7 The density function based on the six data points (red) is approximated through the sum of six Gaussian distributions. The sum of the density functions (blue) is greater in densely populated regions

The assumption is that the mutation applied to the parents produces children which are in the vicinity of the respective parents in objective space. The children thus have a higher chance to contribute to the Pareto front. This loops back into the importance of mutation methods for the search algorithm, especially the control over exploration vs. exploitation and mutation rates, to ensure sufficient proximity of children to their parents.

A major advantage of the Pareto-based selection is that there is no bias in the objectives due to a priori weighting. The Pareto front contains many, usually diverse solutions, from which one can choose the best trade-offs to use for one's application. This is especially important in the case of multiple objectives. The next section introduces hardware awareness as a class of optimization objectives, which makes the outlined multi-objective NAS introduced in this section truly shine.

3 Hardware-Aware Neural Architecture Search

There is an increasing need to run DNNs efficiently on the underlying hardware, often for reasons like energy costs or constraint computational resources. The same DNN topology can behave quite differently on different hardware platforms, as illustrated in the example given in Table 1. The target platform for the DNN is an FPGA, which is a device where the circuitry can be configured electronically to compute any algorithm in a highly efficient way. The same DNN is brought onto two different FPGA platforms, one of them is smaller and consumes 2 W power, the other, bigger one consumes 8 W. The goal, which was not explicitly considered during the design of the DNN, is to find the solution with the least energy consumed per processed input sample. One would think that the smaller, low-power solution would require less energy. However, as can be seen in Table 1, the energy consumed by the 2 W platform is 100 times bigger than by the 8 W platform. So the seemingly good ad hoc decision to choose the smaller platform results in a solution which uses up more energy. In this case, the reason is that the neural network cannot be mapped onto the smaller platform well enough because of hardware resource limitations. The resulting execution on the smaller device is inefficient and results in a much worse execution time, followed by a higher energy consumption. Only if the algorithm can be mapped properly onto the hardware the algorithm can be executed efficiently. For the bigger platform with a lot of available resources, the DNN can be mapped nicely onto the FPGA, which results in a low runtime and

Table 1 Danger of poor ad hoc decisions. Choosing the platform with less power leads to a much higher energy consumption

FPGA Platform	Power (W)	Runtime (μs)	Energy (μJ)
Small FPGA	2.0	756	1513
Large FPGA	8.1	2.1	16.9

low energy consumption. This example demonstrates that hardware awareness is important in the scope of performance, especially if the target platform introduces additional constraints for the DNN.

Another point why hardware awareness leads to better implementations are hardware-software synergies. Hardware accelerators are usually optimized to perform specific operations very efficiently, but they are suboptimal for others. A GPU, for example, is very good in performing matrix-matrix multiplications. But it is very bad in diverging execution paths, like in decision trees. Knowing which operations run best on the hardware and including this knowledge into the search, for example, by favoring the use of specific layer types in the search space, leads to a better performance on the target platform. Efficient hardware usage means that more computational performance can be gained from the platform, which in turn allows to implement more complex algorithms, thus reaching solutions which have not been feasible before.

3.1 Hardware Awareness

Hardware awareness is a very broad term and describes the modeling of any influence of hardware specifics on the performance of the algorithm. The modeling can be done in many ways, which are subdivided here into four classes: proxy values, an analytical estimation, simulators or statistical models, and lookup tables from hardware measurements.

A proxy value is a rough estimate or tendency, which often does not even describe the modeled quantity directly. A popular proxy value is the model size measured by the number of trainable parameters for predicting the latency. Such a proxy can make sense, because a bigger model requires more computations, which means more time. However in the worst case, a proxy value can point in the opposite direction of an optimal solution, thus leading the search away from good designs. For example, [33] demonstrates that the number of floating point operations as a proxy value for the latency does not correlate well enough on different hardware platforms for a reliable prediction. Nevertheless, the big advantage of a proxy value is that it can be evaluated quickly.

A more precise model for hardware awareness is an analytical estimation between the hardware features and the performance. In [16] this is done, for example, by calculating the processing time on an FPGA system based on the DNN hyperparameters. Ney et al. [21] use a similar approach for the energy consumption. However, analytical descriptions normally are not able to encompass all aspects of the hardware platform in a set of formulas. A closed form solution might not even exist. But as an approximation based on design parameters, this approach is evaluated quickly and better than just a proxy value.

Moving further beyond a closed form solution leads to hardware simulators or statistical models. For hardware simulators, such an estimation can be quite costly. A good example is the energy consumption of a system. It requires to model not only the entire system with all of its components and their power but also how it performs in action, where often several use case scenarios need to be evaluated to gain a reasonable estimate for the energy. It should be considered carefully if the inclusion of a simulator into the search process is necessary. In fact, a statistical model can shortcut computations and still provide a good enough estimate of the performance. The statistical model can be trained on simulator or real-world results and thus learn to replace the simulator component as a less precise but faster predictor. For example, ProxylessNAS by Cai et al. [4] trains a model to predict the latency of DNNs based on real-world measurements. During search, they query only the model for latency prediction, which is fast and sufficiently precise. Nevertheless, this approach needs to train a prediction model, which requires a large enough data set for training.

A more radical option for hardware awareness is to measure the performance of the concrete hardware directly. In MnasNet [28], they directly measure the real-world inference latency of every DNN on a mobile device. Although extremely precise, this approach is very expensive in the technical setup. Alternatively in FBNet [32], a small subset of measurements is performed prior to the search and stored in a lookup table. Measurements include the inference latency for the used layer types with some hyperparameter combinations. The major assumption in FBNet is that the overall latency for a DNN is simply the sum over all the layers, since the DNN is assumed to be computed sequentially. This lookup-based approach can be very accurate yet fast, but can imply strong assumptions on the computational model of the hardware. A sufficiently large database with measurements might not be technically feasible in some cases.

Once the modeling of a specific hardware metric is done, it can be included into the search as an optimization target. Objectives are one way to introduce hardware awareness to the NAS. They can be seen as a posteriori awareness, since the DNN is evaluated on the optimization target after the DNN has been created. Hardware awareness also can be enforced a priori, using proper search space design. The search space can be designed to comprise only layers and hyperparameter combinations, which are favorable for the target hardware platform. For example, FBNet [32] only has layers and hyperparameter combinations in the search space that exist in the lookup table, so that the inference latency can be estimated. Layers that are not supported by the hardware or the implementation framework naturally would be excluded from the search space, for example, [16] concentrates on convolutional neural networks and therefore does not include RNN building blocks into the search space. Instead of forcefully mapping specific layers and hyperparameter combinations of an off-the-shelf DNN onto the hardware, the NAS designs a DNN with the best hardware compatibility while achieving the same prediction performance. Thus, search space design is as important as hardware modeling and is a very powerful tool to enforce hardware awareness.

For further reading on the topic of hardware awareness, the survey [1] is recommended. The next sections will demonstrate how hardware awareness can be introduced to the previously implemented NAS, given an FPGA chip and a custom implementation of the hardware architecture and the convolution layers.

3.2 The Target Platform: An FPGA Pipeline

The target platform is an FPGA, which allows to freely describe and map a custom hardware accelerator onto hardware. The capabilities of an FPGA will be utilized to design an application-specific, deep pipeline for DNNs, which uses massive hardware-level parallelism to compute the output as quickly as possible. The unique property of the FPGA is that the hardware architecture can be reconfigured; therefore, hardware-software synergies can be leveraged to a greater extent than with a fixed architecture. This makes FPGAs interesting candidates for hardware-aware NAS. In the brevity of this article, different hardware approaches for running DNNs cannot be explained here, but a good introduction into the field of hardware-efficient DNN design is given by Sze et al. [27].

The goal is to map a DNN, as found by the NAS in the form of some DL framework-specific representation, into the FPGA configuration. There are plenty of tools which are available for this purpose and a good overview is given by Venieris et al. [30]. Moreover, these tools can optimize the execution of the DNN by applying various optimization steps. For example, quantization can reduce the memory requirements of all operands at the cost of reduced numerical precision. Layer fusion can join several smaller operations into a bigger, more efficient one on the hardware accelerator. Although powerful, these optimization methods are limited, because they cannot apply fundamental changes to the DNN topology, which would require retraining the network.

In this article, a variant of the Xilinx FINN framework [29] is used to implement the FPGA designs. This tool will be referred to as the hardware generator. The hardware generator implements the entire DNN as a pipeline on the FPGA. Although the implementation performed by the generator is automatic, choosing a good DNN is still difficult, because hardware resources on the FPGA are limited, so there are solutions which can be implemented better than others. This is why hardware-aware NAS is needed, so that solutions which meet both the application and hardware requirements optimally can be found.

Before the objectives for the NAS can be formulated, the FPGA implementation of the DNN must be understood. Figure 8 shows the differences between a DNN as it is executed on, for example, a processor or GPU, and the specific FPGA implementation. Typically, the output of a DNN is calculated layer by layer, loading the parameters and weights from memory when needed. Although the computations within a layer are parallelized, the entire DNN is usually computed sequentially. For a better utilization of the hardware resources, batches of several independent inputs are computed simultaneously. The FPGA, on the other hand, computes inputs in a pipelined fashion. Instead of computing an entire output feature map of a layer, single elements (e.g., pixels) of the output are computed and streamed to the next layer as soon as possible. This allows to execute all layers in parallel. The weights of all the layers are always on-chip, so no memory transfers apart from loading the input and writing the output are needed. The computations within a layer are also parallelized. So an advantage of the FPGA pipeline implementation is that all layers

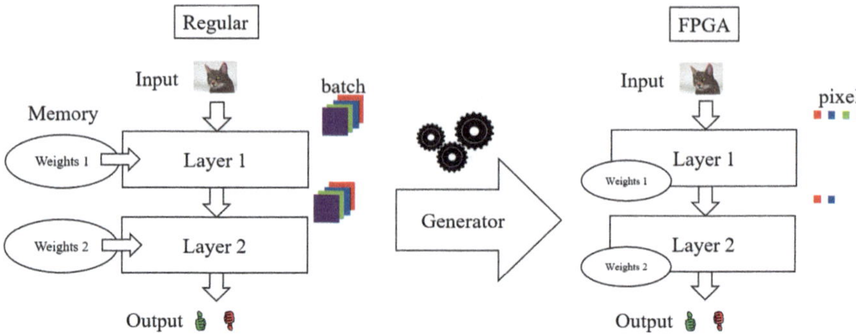

Fig. 8 Conversion from a software DNN model into a hardware model for the FPGA

compute in parallel, instead of sequentially. For more details on how to implement an efficient pipeline on FPGA, see [18]. The hardware generator maps the DNN from the "regular" to the FPGA side by using a one-to-one correspondence between the layers, i.e., every layer in the DNN has its own implementation in the FPGA pipeline.

The runtimes between the regular and FPGA implementation for the DNN are expected to be high. The regular execution is sequential, so the overall runtime is the sum of the latencies for each layer. For the FPGA pipeline, the runtime is dominated by a single layer, the bottleneck. Every pipeline has a bottleneck, that is, a single pipeline stage which effectively determines the speed of the entire pipeline. The bottleneck can be easily identified, because it is the slowest stage, i.e., it needs the most time to produce an output for the next stage. So the runtime of the hardware DNN is the maximum of all layer latencies, in contrast to the sum of all layer latencies in the case of software DNNs. Notice that due to the max-term, optimizing any other layer but the bottleneck results in no speedup at all. Also, when optimizing the bottleneck, it can become so fast that another layer becomes the bottleneck, rendering further optimization pointless. The next section explains how to include the pipeline speed as an optimization target into the NAS.

3.3 A Hardware-Aware Objective: The FPGA Pipeline Speed

As mentioned, the runtime of the FPGA pipeline is dominated by a single bottleneck layer, which is the max-term over all layer latencies. So in order to identify it, the latencies of all layers must be known. This section concentrates on analyzing the 1D convolution layer, but other layer types can be derived similarly. The algorithm for the computation of a 1D convolution layer is visualized in Fig. 9. The input is a one-dimensional sequence of length L and it has multiple channels C. A collection of F filters is applied to this input, each with a kernel size of K and C channels, the same as the input. Visualized in Fig. 9, a filter is overlapped with the input sequence.

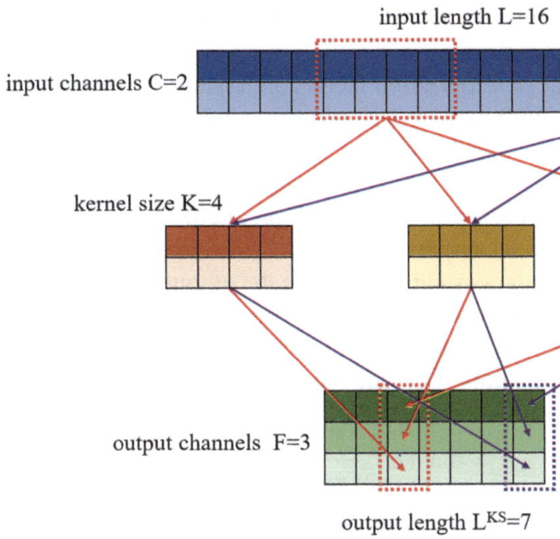

Fig. 9 Visualization of a 1D convolution layer

The overlapping elements are multiplied and summed into a single scalar value, which produces the output in the output channel corresponding to the filter index. This happens with all F filters, each contributing an element to the corresponding output channel. Next, the filters slide down the input sequence by the stride factor S, skipping over $S - 1$ input elements. In Fig. 9, the stride is two. This process is repeated, until the input sequence is completely processed. The output length is L^{KS} with F output channels, where the output length L^{KS} is a function of the kernel size K and the stride S. Formally, the 1D convolution operation is expressed as

$$O_{l_o,c_o} = \sum_{k}^{K} \sum_{c_i}^{C} I_{l_o S+k, c_i} W_{k,c_i,c_o}, \tag{10}$$

where l_o goes up to the output length

$$L^{KS} = \lfloor \frac{L - K}{S} \rfloor + 1 \tag{11}$$

and c_o goes to F. The formula for the output length assumes that the input is not padded, i.e., there are no elements added to the beginning or end of the input sequence.

In principle, all elements of the output can be computed simultaneously, since there are no dependencies. However, in the pipelined setup, the elements are streamed column by column between the pipeline stages, so it is not possible to compute the entire output in one step. What can be parallelized is the computation over all indices k, c_i, and c_o in Eq. (10). The parallelization factors p_k, p_{c_i}, and p_{c_o}

determine how many operations in the respective filter dimension are carried out in parallel. The parallelization factors are naturally limited by the filter sizes K, C and F. For simplicity's sake, the parallelization factors are assumed to always divide the corresponding filter sizes. Without parallelization, i.e., full sequential computation within the layer, there are $K \cdot C$ operations for every output element, of which there are $L^{KS} \cdot F$. This means $L^{KS} \cdot K \cdot C \cdot F$ computational steps or clock cycles are needed if one operation needs one cycle. In the case of full parallelization, i.e., $p_k = K$ and so on, there are only L^{KS} cycles required, where L^{KS} has to be smaller or equal to the input length L. But it is not possible to calculate the output sequence in less than L cycles, since the output depends on the input sequence, which is streamed into the layer with a maximum rate of one input column of C elements per clock cycle. Therefore, the latency of the parallelized convolution layer is lower bound by

$$lat_{1DCNN} = \max(L^{KS} \frac{K}{p_k} \frac{C}{p_{c_i}} \frac{F}{p_{c_o}}, L\sigma), \tag{12}$$

with σ the rate at which the inputs arrive in cycles per input, which is the previous layer's output rate. In particular, if a stride of bigger one is used, i.e., skipping over some input elements, the layer still needs to wait until the corresponding inputs arrive in sequential order from the previous layer. On the other hand, striding can relax the parallelization factors, since more time passes before a new input arrives. The optimal parallelization factors result in a pipeline which produces outputs roughly at the same rate as inputs arrive, or

$$L^{KS} \frac{K}{\hat{p}_k} \frac{C}{\hat{p}_{c_i}} \frac{F}{\hat{p}_{c_o}} \approx L\sigma. \tag{13}$$

The bottleneck of the entire pipeline can be identified now; it is the stage where Eq. (12) is the highest. Assuming all stages are parallelized optimally according to Eq. (13), the bottleneck is the stage with the highest $L\sigma$. Notice that a stride does not affect $L\sigma$, since a decrease in L means the same increase in σ. Since σ cannot be increased otherwise, because of the optimal parallelization, only L is important. But L can only decrease, so the bottleneck is the first 1D convolution layer of the pipeline, irrespective of the rest of the entire pipeline! Again, this is only true if all stages of the hardware DNN can be parallelized optimally, which might not always be possible, since the hardware resources are limited. So in a setup, where the layers are not parallelized optimally, the bottleneck could be any layer.

Finally, the execution speed is formulated as an objective. Equation (12) computes the latencies of all convolution layers in the DNN, and similar derivations can be made for other layer types. The overall latency is still the maximum over all layers, irrespective of the fact that there are multiple layer types. The parallelization factors, if they cannot be chosen to be optimal, remain to be an additional degree of freedom. This is an optimization problem embedded into the NAS, where the best trade-off between hardware resources and latency must be found for each DNN candidate. For example, an FPGA design tool can be used to determine if the model would fit on the target platform for a given set of parallelization factors. In order to keep the following NAS simple, the optimal parallelization factors are used, assuming that they always fit onto the hardware. Therefore, the pipeline speed is simply the input sequence length for the first convolution layer. There does not seem to be much potential for the optimization of the DNN runtime here; however, it is indeed possible. The idea is to skip elements from the raw input sequence in main memory, trading off the sequence length against the information content of the input data.

3.4 Hardware-Aware NAS with a Pipelined DNN Accelerator

In this section, the NAS and the FPGA-based DNN accelerator are put together, using the hardware-aware optimization objectives as a bridge. The data set in this example comprises 1D time series data. The task is binary classification, i.e., identifying if a specific feature appears in the input sample. The goal is to achieve a minimum accuracy of 90 % and at the same time to speed up the hardware accelerator as much as possible. The search space for this example mainly consists of convolution and pooling layers. Additionally, a special subsampling layer is included. The subsampling layer is similar to a pooling layer, but it blindly selects the first element of its pooling window. If placed as the first layer of the network, it skips accordingly in main memory when loading the input sequence. This is the key to improving the pipeline speed with respect to Eq. (13), simply by reducing the input sequence length for the pipeline. However, the decision whether or not to use the subsampling layer and where to put it is left completely to the NAS. Thus, the effectiveness of the NAS can be checked, since the only way to achieve speedup is to use the subsampling layer as the first layer. Additionally, the number of trainable parameters is another optimization target to be minimized as well, as a simply proxy to ensure the DNN can fit onto the hardware.

In order to see the effect of the hardware-aware objective, two setups for the NAS are compared. The first "normal" NAS setup contains search criteria only for the accuracy and the number of parameters. The setup "HW-aware" additionally includes the minimization of the pipeline latency. First, the effect of both setups on the layer placement is studied in Fig. 10. It shows the percentage of all candidates in a generation, which had convolution, pooling, or subsampling as the first layer. With the additional objective, there is a trend for choosing subsampling layers in favor of convolution layers, as expected. Note that it is not strictly the best option to have a subsampling layer at the front, since the accuracy is an additional objective, which can benefit from having a convolution layer in the beginning.

Next, the speed of the candidates in both setups is compared. Figure 11 shows the average speed of all individuals within a generation, normalized to some upper estimate for the maximum runtime. Indeed, the average speed within the population is lower in the case where there is the additional objective.

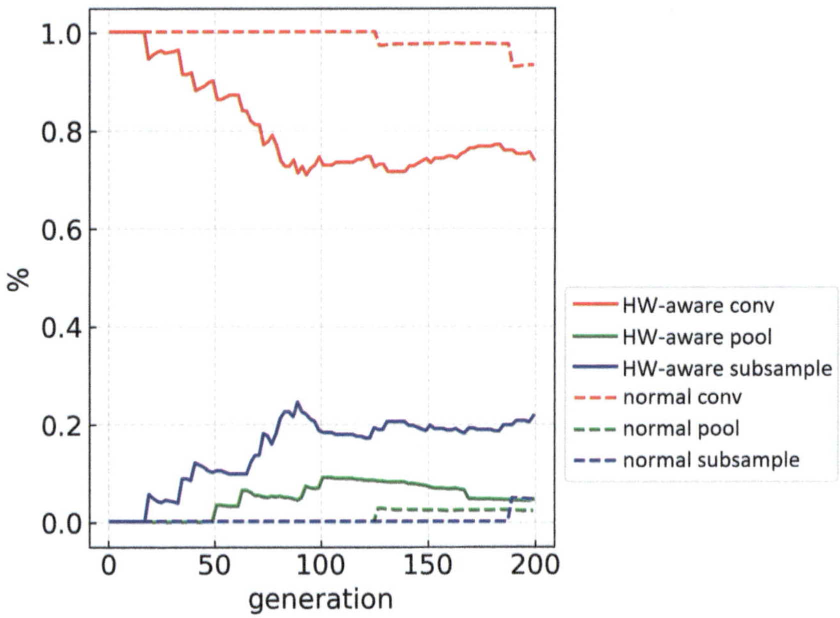

Fig. 10 Comparison of a NAS run without specific constraints (normal) and including the formula for the latency as a search metric (clocks)

The bottom line of this small experiment is that the additional objective helps to guide the search in the desired direction. But although the chance to find a fast solution increases, there is no guarantee to find "the fastest" solution due to the stochastic nature of the search algorithm. For example, in another scenario, where it could not be assumed that the parallelization factors are always optimal, the hardware resources need to be considered. This leads to an optimization problem within an optimization problem. One possibility is to let the NAS optimize the additional parallelization parameters. However, this would increase the search space dramatically. It is advisable not to employ stochasticity on every level of the search, but to find a solution, even if approximate, to such embedded optimization problems. On the other hand, if an embedded optimization problem is complex, it will cost time, so it can be more important to have a higher rate of evaluated models with low fidelity than evaluating fewer models with high fidelity. Therefore, a proper design of not only the NAS selection algorithms but also the search space and the evaluation methods is the key to successful neural architecture search.

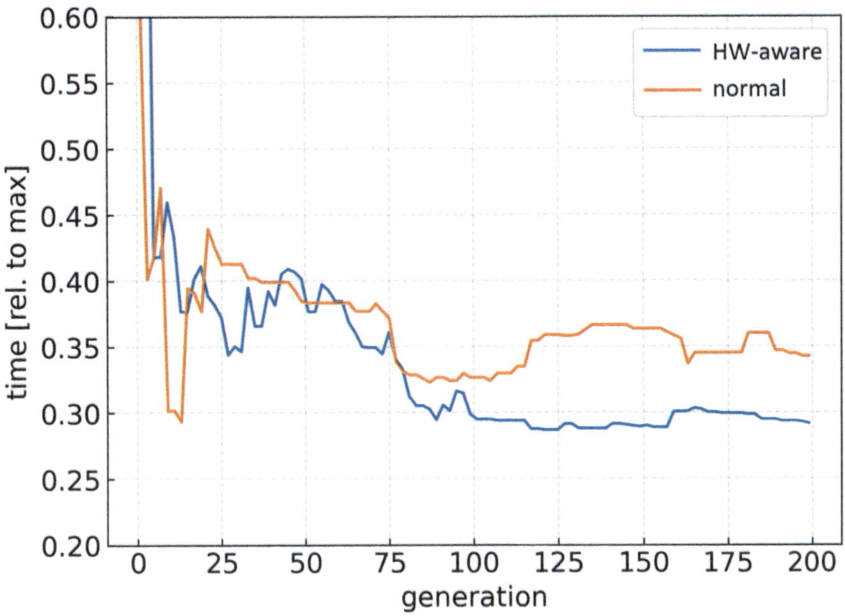

Fig. 11 Normalized number of clock cycles per generation, averaged over all individuals in a generation

References

1. Benmeziane, H., Maghraoui, K. E., Ouarnoughi, H., Niar, S., Wistuba, M., & Wang, N. (2021). A comprehensive survey on hardware-aware neural architecture search. arXiv preprint arXiv:2101.09336.
2. Bergstra, J., Bardenet, R., Bengio, Y., & Kégl, B. (2011). Algorithms for hyper-parameter optimization. *Advances in Neural Information Processing Systems, 24*.
3. Bischl, B., Binder, M., Lang, M., Pielok, T., Richter, J., Coors, S., ... & Lindauer, M. (2023). Hyperparameter optimization: Foundations, algorithms, best practices, and open challenges. *Wiley Interdisciplinary Reviews: Data Mining and Knowledge Discovery, 13*(2), e1484.
4. Cai, H., Zhu, L., & Han, S. (2018). Proxylessnas: Direct neural architecture search on target task and hardware. arXiv preprint arXiv:1812.00332.
5. Cai, H., Gan, C., Wang, T., Zhang, Z., & Han, S. (2019). Once-for-all: Train one network and specialize it for efficient deployment. arXiv preprint arXiv:1908.09791.
6. Cartesian Genetic Programming. https://www.cartesiangp.com/
7. Chen, T., Goodfellow, I., & Shlens, J. (2015). Net2net: Accelerating learning via knowledge transfer. arXiv preprint arXiv:1511.05641.
8. Deb, K., Pratap, A., Agarwal, S., & Meyarivan, T. A. M. T. (2002). A fast and elitist multiobjective genetic algorithm: NSGA-II. *IEEE Transactions on Evolutionary Computation, 6*(2), 182–197.
9. Elsken, T., Metzen, J. H., & Hutter, F. (2018). Efficient multi-objective neural architecture search via lamarckian evolution. arXiv preprint arXiv:1804.09081.
10. Elsken, T., Metzen, J. H., & Hutter, F. (2018). Neural architecture search: A survey. arXiv 2018. arXiv preprint arXiv:1808.05377.

11. Falkner, S., Klein, A., & Hutter, F. (2018). BOHB: Robust and efficient hyperparameter optimization at scale. In International Conference on Machine Learning (pp. 1437–1446). PMLR.
12. He, K., Zhang, X., Ren, S., & Sun, J. (2016). Deep residual learning for image recognition. In Proceedings of the IEEE Conference on Computer Vision and Pattern Recognition (pp. 770–778).
13. Hinton, G., Vinyals, O., & Dean, J. (2015). Distilling the knowledge in a neural network. arXiv preprint arXiv:1503.02531.
14. Hochreiter, S., & Schmidhuber, J. (1997). Long short-term memory. *Neural Computation, 9*(8), 1735–1780.
15. Hua, Y., Liu, Q., Hao, K., & Jin, Y. (2021). A survey of evolutionary algorithms for multi-objective optimization problems with irregular Pareto fronts. *IEEE/CAA Journal of Automatica Sinica, 8*(2), 303–318.
16. Jiang, W., Zhang, X., Sha, E. H. M., Yang, L., Zhuge, Q., Shi, Y., & Hu, J. (2019). Accuracy vs. efficiency: Achieving both through fpga-implementation aware neural architecture search. In Proceedings of the 56th Annual Design Automation Conference 2019 (pp. 1–6).
17. Li, L., Jamieson, K., DeSalvo, G., Rostamizadeh, A., & Talwalkar, A. (2016). Hyperband: A Novel Bandit-Based Approach to Hyperparameter Optimization. arXiv 2018. arXiv preprint arXiv:1603.06560.
18. de Fine Licht, J., Besta, M., Meierhans, S., & Hoefler, T. (2020). Transformations of high-level synthesis codes for high-performance computing. *IEEE Transactions on Parallel and Distributed Systems, 32*(5), 1014–1029.
19. Liu, H., Simonyan, K., & Yang, Y. (2018). DARTS: Differentiable architecture search. In *International Conference on Learning Representations.*
20. Liu, Y., Sun, Y., Xue, B., Zhang, M., Yen, G. G., & Tan, K. C. (2021). A survey on evolutionary neural architecture search. *IEEE Transactions on Neural Networks and Learning Systems, 34*(2), 550–570.
21. Ney, J., Loroch, D., Rybalkin, V., Weber, N., Krüger, J., & Wehn, N. (2021). HALF: Holistic auto machine learning for FPGAs. In *2021 31st International Conference on Field-Programmable Logic and Applications (FPL)* (pp. 363–368). IEEE.
22. Real, E., Aggarwal, A., Huang, Y., & Le, Q. V. (2019). Regularized evolution for image classifier architecture search. In *Proceedings of the AAAI Conference on Artificial Intelligence* (Vol. 33(01), pp. 4780–4789).
23. Scott, D. W. (1992). *Multivariate density estimation: Theory, practice, and visualization.* Wiley.
24. Sheather, S. J., & Jones, M. C. (1991). A reliable data-based bandwidth selection method for kernel density estimation. *Journal of the Royal Statistical Society Series B, 53*, 683–690.
25. Suganuma, M., Shirakawa, S., & Nagao, T. (2017). A genetic programming approach to designing convolutional neural network architectures. In *Proceedings of the Genetic and Evolutionary Computation Conference* (pp. 497–504).
26. Sun, Y., Xue, B., Zhang, M., Yen, G. G., & Lv, J. (2020). Automatically designing CNN architectures using the genetic algorithm for image classification. *IEEE Transactions on Cybernetics, 50*(9), 3840–3854.
27. Sze, V., Chen, Y. H., Yang, T. J., & Emer, J. S. (2017). Efficient processing of deep neural networks: A tutorial and survey. *Proceedings of the IEEE, 105*(12), 2295–2329.
28. Tan, M., Chen, B., Pang, R., Vasudevan, V., Sandler, M., Howard, A., & Le, Q. V. (2019). Mnasnet: Platform-aware neural architecture search for mobile. In *Proceedings of the IEEE/CVF Conference on Computer Vision and Pattern Recognition* (pp. 2820–2828).
29. Umuroglu, Y., Fraser, N. J., Gambardella, G., Blott, M., Leong, P., Jahre, M., & Vissers, K. (2017). Finn: A framework for fast, scalable binarized neural network inference. In *Proceedings of the 2017 ACM/SIGDA international symposium on field-programmable gate arrays* (pp. 65–74).

30. Venieris, S. I., Kouris, A., & Bouganis, C. S. (2018). Toolflows for mapping convolutional neural networks on FPGAs: A survey and future directions. *ACM Computing Surveys (CSUR), 51*(3), 1–39.
31. White, C., Safari, M., Sukthanker, R., Ru, B., Elsken, T., Zela, A., ... & Hutter, F. (2023). Neural architecture search: Insights from 1000 papers. arXiv preprint arXiv:2301.08727.
32. Wu, B., Dai, X., Zhang, P., Wang, Y., Sun, F., Wu, Y., ... & Keutzer, K. (2019). Fbnet: Hardware-aware efficient convnet design via differentiable neural architecture search. In *Proceedings of the IEEE/CVF Conference on Computer Vision and Pattern Recognition* (pp. 10734–10742).
33. Zhang, L. L., Yang, Y., Jiang, Y., Zhu, W., & Liu, Y. (2020). Fast hardware-aware neural architecture search. In *Proceedings of the IEEE/CVF Conference on Computer Vision and Pattern Recognition Workshops* (pp. 692–693).
34. Zoph, B., & Le, Q. V. (2016). Neural architecture search with reinforcement learning. arXiv preprint arXiv:1611.01578.
35. Zoph, B., Vasudevan, V., Shlens, J., & Le, Q. V. (2018). Learning transferable architectures for scalable image recognition. In *Proceedings of the IEEE Conference on Computer Vision and Pattern Recognition* (pp. 8697–8710).

Optimal Experimental Design Supported by Machine Learning Regression Models

Philipp Seufert, Jochen Schmid, Michael Bortz, Jan Schwientek, Tobias Seidel, David Mogalle, Martin Bubel, and Karl-Heinz Küfer

Abstract Modern industry heavily relies on accurate mathematical models to optimize processes. Models are obtained by performing experiments and adapting the model parameters to the measured data. Optimal experimental design (OED) provides methods to obtain precise parameter estimates for mathematical models with as few experiments as possible, which in turn saves costs and time. Classical algorithms for OED require pre-computations on a finite design grid or repeated solutions of a nonlinear program. Both methods need a large amount of model evaluations, which can be computationally expensive and make classical OED unviable for complex and time-dependent models. Utilizing Gaussian process-based regression surrogate models we can approximate the sensitivity function needed in the OED algorithms. This results in a Bayes-like sampling and solution approach for the OED. We show the effectiveness of the machine learning approach by applying the algorithm to several models from chemical process engineering. For small-scale models the method obtains similar results, and for large-scale models the method drastically outperforms classical OED algorithms.

1 Introduction

A lot of real-world applications heavily rely on mathematical models $x \mapsto f(x, \theta)$ of the input-output relationships to be modelled. In most cases, these models are not completely determined, but come with a set of unknown model parameters θ which have to be estimated. It is important that the estimates for these unknown parameters θ are precise enough so that the model is an adequate representation of reality. In order to obtain reliable estimates for the model parameters, one usually performs a set of experiments $x_1, \ldots, x_n$ and takes corresponding measurements

P. Seufert (✉) · J. Schmid · M. Bortz · J. Schwientek · T. Seidel · D. Mogalle · M. Bubel · K.-H. Küfer
Fraunhofer Institute of Industrial Mathematics ITWM, Kaiserslautern, Germany
e-mail: philipp.seufert@itwm.fraunhofer.de

J. Franke, A. Schöbel (eds.), *Statistical Machine Learning for Engineering with Applications*, Lecture Notes in Statistics 227,
https://doi.org/10.1007/978-3-031-66253-9_10

$y_1, \ldots, y_n$ of the output quantity to be modelled. Subsequently, one adjusts the model parameters θ such that the discrepancy between the measurements y_i and the predictions $f(x_i, \theta)$ of the model $x \mapsto f(x, \theta)$ becomes minimal.

As conducting experiments costs time and money, it is usually important to reduce the number of experiments required to obtain reliable models f. This task is the subject of optimal experimental design (OED), which is often also called model-based design of experiments. So, in other words, in OED one computes experiments that lead to an optimal model precision. A mathematical overview of OED can be found in [1–4], for instance, and we also give a brief introduction in Sect. 2.

In many cases, the determination of the best model parameters is an iterative process, where one repeatedly computes optimal experiments, performs the experiments and takes measurements, and then adjusts the model parameters to fit the measurements. This iterative OED workflow is depicted in Fig. 1 and is described, for example, in [2, 4–6].

A variety of algorithms have been developed to compute optimal experimental designs. These include classical methods like a vertex-direction method introduced by Federov and Wynn [1, 8], a vertex-exchange method introduced by Böhning [9], or a multiplicative algorithm by Tritterington et al. [10]. More state-of-the-art methods include a cocktail algorithm by Yu [11], an algorithm by Yang et al. [12], or a random-exchange method by Harman et al. [13].

What most of these methods have in common is that they scale badly in the model dimension, which is given by the number of factors the experimenter can set, and in the model complexity. In this respect, dynamical models can be particularly

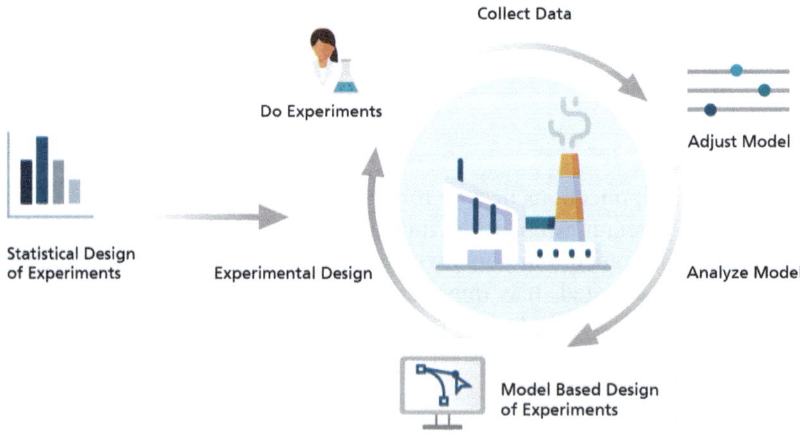

Fig. 1 Iterative model determination workflow with optimal experimental design. This picture is taken from the Fraunhofer Institute for Industrial Mathematics home page [7] and is reprinted with kind permission of the copyright owner, the Fraunhofer Institute for Industrial Mathematics (ITWM)

problematic, as they may contain time-dependent control variables $u(t)$, which are infinite-dimensional.

In this contribution, we present an adaptive discretization method for OED. Via the use of Gaussian process regression, this method is viable to compute optimal designs even for high-dimensional models and dynamical models.

2 Optimal Experimental Design

In this section, we give a brief mathematical introduction to the theory of optimal experimental design (OED). We follow the ideas and notation of [1, 2], where more details can also be found. We begin by introducing the Fisher information matrix of an experimental design for static and dynamic models. We then formulate the optimization problem for the log-D-criterion, state results on the existence of optimal designs, and give necessary and sufficient optimality conditions. These results are mainly due to the well-known Kiefer-Wolfowitz equivalence theorem [14].

2.1 *Fisher Information Matrix for Stationary Models*

In optimal experimental design theory, one considers parametrized models $x \mapsto f(x, \theta)$ of an input-output relationship of interest in a given application. Such models predict an output value $f(x, \theta) \in \mathbb{R}^{d_y}$ for every value $x \in X$ from the input space $X \subset \mathbb{R}^{d_x}$ and for every possible value $\theta \in \Theta$ of the model parameters from a set $\Theta \subset \mathbb{R}^{d_\theta}$ of admissible model parameters. In order to estimate the true but unknown value θ^* of these model parameters, one performs experiments $x_1, \dots, x_n$ to obtain corresponding measurements $y_1, \dots, y_n$. We assume these measurements to be given by the model predictions $f(x_i, \theta^*)$ plus some independent and identically distributed measurement errors $\varepsilon_i \sim \mathcal{N}(0, \Sigma)$, with $\Sigma \in \mathbb{R}^{d_y \times d_y}$ denoting the covariance matrix of each of the measurement errors. In formulas

$$y_i = f(x_i, \theta^*) + \varepsilon_i \qquad (i \in \{1, \dots, n\}). \tag{1}$$

After performing experiments x_i and taking measurements y_i, we can estimate the unknown value of the model parameters. A standard estimate for θ^* is the least-squares estimator θ_{LSE} which is obtained by minimizing the least-squared error. In formulas

$$\theta_{LSE} = \theta_{LSE}(\xi, y) = \operatorname*{argmin}_{\theta \in \Theta} \sum_{i=1}^{n} (y_i - f(x_i, \theta))^T \cdot \Sigma^{-1} \cdot (y_i - f(x_i, \theta)),$$

where $\xi := (x_1, \ldots, x_n)$ and $y := (y_1, \ldots, y_n)$ denote the combination of the performed experiments and of the corresponding measurement values, respectively.

As the estimator depends on the random errors ε_i, it itself is also a random variable. Via a linearization of the model f about the parameter $\overline{\theta}$, given by

$$f(x, \theta) \approx f(x, \overline{\theta}) + D_\theta f(x, \overline{\theta})(\theta - \overline{\theta}), \tag{2}$$

we obtain the distribution of the estimator θ_{LSE}. In particular, we can also approximate the covariance matrix as

$$\operatorname{Cov}[\theta_{LSE}] \approx \left(\sum_{i=1}^{n} D_\theta f(x_i, \overline{\theta})^T \cdot \Sigma^{-1} \cdot D_\theta f(x_i, \overline{\theta}) \right)^{-1}, \tag{3}$$

where $D_\theta f(x, \overline{\theta}) = \left(\frac{\partial f_j}{\partial \theta_k}(x, \overline{\theta}) \right)$ is the $d_y \times d_\theta$ Jacobi matrix of the model f at $(x, \overline{\theta})$ with respect to the model parameters θ. In the special case where the model is already linear in θ, this approximate relation becomes exact and the right-hand side does not depend on $\overline{\theta}$. In the general case of nonlinear models, however, the above approximation depends on the value of $\overline{\theta}$, of course.

In the following, we formalize the notions of experimental plans by defining design measures and introduce the Fisher information matrix for these design measures. As we will see below, the Fisher information matrix is closely related to the approximated covariance matrix of the estimator θ_{LSE} and is thus of particular interest.

A *(continuous) design* is an experimental plan consisting of experiments (design points) $x_1, \ldots, x_n$ with weights $\omega_1, \ldots, \omega_n$. The weights of the design points are positive, $\omega_i \geq 0$, and fulfill $\sum \omega_i = 1$. The points x_i with strictly positive weights ω_i are called the *support points (of the design)*. We can extend the notion of a design to the *design measures*, which are probability measures ξ on the design space X endowed with its Borel-σ-algebra $B(X)$. In the following we denote the space of all design measures by

$$\Xi(X) = \{\xi \mid \xi \text{ is a probability measure on } (X, B(X))\}.$$

The *Fisher information matrix* $m(x, \overline{\theta}) \in \mathbb{R}^{d_\theta \times d_\theta}$ of a design point x is defined as

$$m(x, \overline{\theta}) := D_\theta f(x, \overline{\theta})^T \cdot \Sigma^{-1} \cdot D_\theta f(x, \overline{\theta}).$$

If the matrix-valued function $x \mapsto m(x, \overline{\theta})$ is integrable with respect to the design measure ξ, the Fisher information matrix $M(\xi, \overline{\theta})$ of ξ is defined by

$$M(\xi, \overline{\theta}) := \int_X m(x, \overline{\theta}) \xi(dx). \tag{4}$$

We end this section by illustrating how the notion of designs and the Fisher information matrix translate to our original setting. For a given set of experiments $x_1, \ldots, x_n$, we obtain a design measure by assigning the design points x_i the weights $\omega_i = 1/n$. The Fisher information matrix of this continuous design is given as

$$M(\xi, \overline{\theta}) = \sum_{i=1}^{n} \frac{1}{n} \cdot D_\theta f(x_i, \overline{\theta})^T \cdot \Sigma^{-1} \cdot D_\theta f(x_i, \overline{\theta}).$$

With this observation at hand, we can rewrite the approximation (3) as

$$\text{Cov}\,[\theta_{LSE}] \approx 1/n \cdot M(\xi, \overline{\theta})^{-1}. \tag{5}$$

This shows the relation between the covariance of the estimator θ_{LSE} and the Fisher information matrix of the considered design.

2.2 Fisher Information Matrix for Dynamic Models

In many process engineering applications, stationary models are inadequate, and dynamic models are needed instead. In this contribution, we consider models given by ordinary differential equations of the form:

$$\begin{aligned} \frac{ds(t)}{dt} &= g(s(t), u(t), p, \theta, t) \quad \big(t \in [0, t_f]\big) \quad \text{and} \quad s(0) = s_0, \\ y(t) &= h(s(t)), \end{aligned} \tag{6}$$

with t from an interval $[0, t_f]$. This model description follows ideas presented in [15, Chapter 10] or [3, 16].

Here $s(t) \in \mathbb{R}^{d_s}$ denotes the state of the system of differential equations at a given time t, $s_0 \in \mathbb{R}^{d_s}$ denotes the initial state, $u(t) \in \mathbb{R}^{d_u}$ denotes the time-dependent control variables of the system, $p \in \mathbb{R}^{d_p}$ denotes the static control parameters $\theta \in \mathbb{R}^{d_\theta}$ denotes the unknown model parameters, and $y(t) \in \mathbb{R}^{d_y}$ the measurable outputs of the system. The variable $t \in [0, t_f]$ corresponds to the time, and t_f denotes the final time of the system.

In the following we derive a formula for the Fisher information matrix of the dynamic model. For this purpose we adapt the dynamic setting to obtain a structure similar to that of stationary models. Under corresponding assumptions, we then define the Fisher information matrix analogously to Eq. (4).

As in the static OED setting, we assume the measurements $y(t)$ to be perturbed by independent and identically distributed measurement errors $\varepsilon \sim N(0, \Sigma)$. We consider the design variables of the dynamic model. These consist in particular of the control variables $u(t)$ and p, the initial state s_0 and the measurement time $\overline{t}$.

Typically, we replace the controls $u(t)$ by a discretized version $\overline{u} \in \mathbb{R}^{d_{\overline{u}}}$. This can be, for example, a piecewise constant or a piecewise linear function, represented by a finite number of values $u_1, \ldots, u_m \in \mathbb{R}^{d_u}$ at fixed time points $t_1, \ldots, t_m \in [0, t_f]$. The vector $\overline{u}$ then consists of the concatenated values $u_1, \ldots, u_m$. With $\overline{u}$ we can then construct the time-dependent control $u(t)$.

Additionally, by $\overline{t} \in [0, t_f]$, we denote a new variable corresponding to the time point at which we measure the solution $y(t)$ of the system.

We denote by $S_0 \subset \mathbb{R}^{d_s}$ the set of admissible initial states s_0, via $P \subset \mathbb{R}^{d_p}$ the set of admissible stationary control parameters p, and via $\overline{U} \subset \mathbb{R}^{d_{\overline{u}}}$ the set of admissible vectors $\overline{u}$ for the discretized version of the controls $u(t)$. Similarly to the stationary OED setting, we define the design variables as $x = (s_0, \overline{u}, p, \overline{t})$ and the design space as $X = S_0 \times \overline{U} \times P \times [0, t_f] \subset \mathbb{R}^{d_x}$. The dimension d_x of the design space can vary depending on the considered discretization of $u(t)$.

With these adjustments, we can consider the dynamic model as a black box function parametrized by θ. This function maps the inputs $x = (s_0, \overline{u}, p, \overline{t})$ to outputs $y(\overline{t})$, where y is a solution to (6). We have now obtained a setup analogous to the stationary OED setup, given by

$$\overline{y} = f(x, \theta) + \varepsilon.$$

In this context, a design measure ξ is a probability measure over the space X endowed with its Borel-σ-algebra $B(X)$. Assuming the model f is differentiable in θ, the Fisher information matrix for a design point x is given as

$$m(x, \overline{\theta}) := D_\theta f(x, \overline{\theta})^T \cdot \Sigma^{-1} \cdot D_\theta f(x, \overline{\theta}).$$

If the matrix-valued function $x \mapsto m(x, \overline{\theta})$ is integrable with respect to the design measure ξ, the Fisher information matrix $M(\xi, \overline{\theta})$ of ξ is defined by

$$M(\xi, \overline{\theta}) = \int_X m(x, \overline{\theta}) \xi\,(dx)\,.$$

2.3 Characterization of Optimal Experimental Designs

After having introduced the Fisher information matrix for both static and dynamic models, we now recall what makes an experimental design optimal and how such optimal experimental designs can be characterized.

In optimal experimental design, one aims at finding a design measure ξ that results in a precise model, that is, in a precise parameter estimate. In order to achieve this, one tries to minimize the covariance matrix of the least-squares estimator $\theta_{LSE}(\xi)$ with respect to ξ. As it is difficult to minimize matrix-valued functions, one usually minimizes suitable scalar-valued functions of the covariance matrix.

A particularly common choice for such a scalar-valued function is the volume of the ellipsoid of concentration corresponding to the covariance matrix [1]. Up to a constant prefactor, this volume is given by $\det(\mathrm{Cov}[\theta_{LSE}])^{1/2}$. Applying a logarithmic scaling and inserting the approximation (5), we obtain the optimization problem:

$$\min_{\xi \in \Xi(X)} \log\left(\det\left(M(\xi, \overline{\theta})^{-1}\right)\right). \qquad (P_{\overline{\theta}})$$

In view of the logarithmic determinant, this optimization problem is often referred to as the *log-D-criterion*. Instead of minimizing the volume of the ellipsoid of concentration, one can also minimize the sum of its eigenvalues or its maximal eigenvalue w.r.t. ξ. In this manner, one arrives at the so-called A-criterion and E-criterion, respectively. We will not use these other design criteria here and refer to [1, 2, 17] for more information.

It is clear that for nonlinear models in θ, the information matrix $M(\xi, \overline{\theta})$, and thus also the optimization problem $(P_{\overline{\theta}})$, depend on the specific reference value $\overline{\theta}$. A typical approach in OED is to insert the current best estimate of θ for $\overline{\theta}$. Then, an optimal design for this fixed value of $\overline{\theta}$ is computed. Such designs are called *locally optimal designs (w.r.t. the log-D-criterion)*, as the linearization applied only holds locally in $\overline{\theta}$. After performing the new experiments, a new estimate for θ is computed. This cycle of performing experiments, updating the estimator and computing optimal designs, is repeated until the estimator has reached a required precision [5, 6].

In the remainder of this contribution, we consider locally optimal designs and thus assume to have a fixed value $\overline{\theta}$ given. To simplify notation, we omit $\overline{\theta}$ from the Fisher information matrices and from the optimization problem, and instead write $m(x)$, $M(\xi)$ and

$$\min_{\xi \in \Xi(X)} \log\left(\det\left(M(\xi)^{-1}\right)\right). \qquad (P)$$

In order to compute optimal experimental designs, it is important to have good necessary and sufficient optimality conditions. Such characterizations of optimal experimental designs were first established by Kiefer and Wolfowitz [14] for several design criteria in the form of the Kiefer–Wolfowitz equivalence theorem. In particular, these results hold for the log-D-criterion. We state a modified version of this theorem, which is based on [1]. It expresses the optimality conditions in terms of the so-called sensitivity function ψ of the log-D-criterion, which is given by

$$\psi : \Xi_{fin}(X) \times X \to \mathbb{R}, \qquad (\xi, x) \mapsto d_\theta - \mathrm{tr}\left[M(\xi)^{-1} m(x)\right]. \qquad (7)$$

Here, $\Xi_{fin}(X)$ denotes the subset of all design measures $\xi \in \Xi(X)$ with finite objective value. For the log-D-criterion, these correspond to all design measures with non-singular—or, equivalently, with positive definite—Fisher information matrix $M(\xi)$. Also, d_θ, as before, denotes the dimension of the considered model parameter space $\Theta \subset \mathbb{R}^{d_\theta}$.

Condition 1

(i) *The design space $X \subset \mathbb{R}^{d_x}$ is compact:*
(ii) *The Fisher information matrix $m(x, \overline{\theta})$ exists for each $x \in X$ and depends continuously on $x \in X$.*
(iii) *There exists a design measure ξ^0, such that the Fisher information matrix $M(\xi^0, \overline{\theta})$ is positive definite.*

With these basic assumptions, one obtains the following characterization of locally optimal experimental designs.

Theorem 2 (Kiefer–Wolfowitz) *Suppose that Condition 1 is satisfied and that ψ is the sensitivity function defined in* (7)*. Then, the following assertions hold true:*

(i) *A locally optimal design ξ^* with finite support consisting of at most $\frac{d_\theta \cdot (d_\theta+1)}{2}$ support points exists.*
(ii) *The set of locally optimal designs is convex.*
(iii) *A design ξ^* is locally optimal if and only if $\xi^* \in \Xi_{fin}(X)$ and $\psi(\xi^*, x) \geq 0$ for all $x \in X$.*
(iv) *If ξ^* is a locally optimal design, then $\psi(\xi^*, x) = 0$ for ξ^*-almost every $x \in X$.*

In particular, this theorem reveals that solving the OED problem (P) is equivalent to finding a design $\xi^* \in \Xi_{fin}(X)$ such that $\psi(\xi^*, x) \geq 0$ for all $x \in X$.

We close this section by relating the sensitivity function (7) of the log-D-criterion to the uncertainty in the model prediction. In order to do so, we approximately compute the covariance matrix of the estimated model. Inserting the estimator θ_{LSE} into the linearization (2) and using the approximation (5), we obtain:

$$\begin{aligned}\operatorname{Cov}[f(x, \theta_{LSE})] &\approx D_\theta f(x, \overline{\theta}) \cdot \operatorname{Cov}[\theta_{LSE}] \cdot D_\theta f(x, \overline{\theta})^T \\ &\approx 1/n \cdot D_\theta f(x, \overline{\theta}) \cdot M(\xi, \overline{\theta})^{-1} \cdot D_\theta f(x, \overline{\theta})^T\end{aligned}$$

for the covariance of the estimated model at x. In particular, it follows that

$$n \cdot \operatorname{tr}\left[\operatorname{Cov}[f(x, \theta_{LSE})] \cdot \Sigma^{-1}\right] \approx \operatorname{tr}\left[m(x, \overline{\theta}) \cdot M(\xi, \overline{\theta})^{-1}\right]. \tag{8}$$

In the particularly simple case where the model is already linear in θ and maps to 1-dimensional outputs ($d_y = 1$), all the above approximations become exact and (8) simplifies to

$$\frac{n}{\Sigma} \cdot \operatorname{Cov}[f(x, \theta_{LSE})] = \operatorname{tr}\left[m(x) \cdot M(\xi)^{-1}\right]. \tag{9}$$

So, in view of (8) and (9), the value

$$U^{f,\xi}(x, \overline{\theta}) := \operatorname{tr}\left[m(x, \overline{\theta}) \cdot M(\xi, \overline{\theta})^{-1}\right] \tag{10}$$

is a measure for the uncertainty of the estimated model's prediction at a given input point $x \in X$. We will refer to this value $U^{f,\xi}(x, \overline{\theta})$ as the *prediction uncertainty of the estimated model (given design ξ) at the point x*. And, accordingly, we refer to

$$U^{f,\xi}(\overline{\theta}) := \max_{x \in X} U^{f,\xi}(x, \overline{\theta}) \tag{11}$$

as the *maximal prediction uncertainty of the model (given design ξ)*. As with the Fisher information matrices, we typically assume a value $\overline{\theta}$ to be given and omit this value from the notation. Instead we only write $U^{f,\xi}(x)$ and $U^{f,\xi}$ for simplicity. In view of (7) and (10), the sensitivity function of the log-D-criterion and the prediction uncertainty are related as

$$\psi(\xi, x) = d_\theta - U^{f,\xi}(x).$$

And therefore, the optimality conditions for locally optimal designs can be expressed in a particularly neat form in terms of the maximal prediction uncertainty.

Corollary 3 *Suppose that Condition 1 is satisfied and that ψ is the sensitivity function defined in (7). A design $\xi^* \in \Xi(X)$ is then locally optimal w.r.t. the log-D-criterion if and only if $\xi^* \in \Xi_{fin}(X)$ and*

$$U^{f,\xi^*} = \max_{x \in X} U^{f,\xi^*}(x) = d_\theta.$$

3 Adaptive Discretization Algorithms for Optimal Experimental Design

In this section, we introduce a class of algorithms to solve the OED problem (P). The optimization problem (P) is an infinite-dimensional problem in the design measure ξ. In order to solve this problem, we use an approach similar to the Blankenship and Falk algorithm [18] from semi-infinite programming, resulting in an adaptive discretization of the design space X.

We consider a finite subset $\dot{X} = \{x_1, \ldots, x_k\}$ of the design space X, consisting of k points. Any design measure over the space $\dot{X}$ can be described by a vector $\omega \in \mathbb{R}^k$ from the standard simplex. The vector ω then satisfies $\omega_i \geq 0$ for $i = 1, \ldots, k$ and

$\sum_{i=1}^{k} \omega_i = 1$. By assigning each design point $x_i \in \dot{X}$ the weight ω_i, we obtain a design measure:

$$\xi := \sum_{i=1}^{k} \omega_i \delta_{x_i}, \tag{12}$$

where δ_x denotes the point measure at $x \in X$. We check the optimality conditions from Theorem 2 for the design ξ. If the design is optimal, we are done. Otherwise we refine the discretization and update the vector ω. We repeat this process with the new discretization and weight vector.

We state the resulting iterative scheme. Denote the initial discretization by $\dot{X}_{k_0} = \{x_1, \dots, x_{k_0}\}$ and the initial weight vector by $\omega^{k_0} \in \mathbb{R}^{k_0}$. The resulting design measure is denoted by ξ^{k_0}. We then iterate over $k = k_0$ and perform the following steps:

1. Solve $x_{k+1} = \operatorname{argmin}_{x \in X} \psi(\xi^k, x)$ and refine $\dot{X}_{k+1} = \dot{X}_k \cup \{x_{k+1}\}$.
2. Update the weight vector to obtain $\omega^{k+1} \in \mathbb{R}^{k+1}$ and assign the weights ω^{k+1} to the design points $\dot{X}^{k+1}$ as in Eq. (12) to get ξ^{k+1}.

3.1 *Adaptive Discretization with Optimal Weights*

A variety of algorithms based on this scheme exist, mostly differing in the way the weight vector ω^k is updated. A first form with convex weight update was introduced by Federov [1] and Wynn [8]. A more recent version was introduced by Yang, Biedermann, and Tang [12], which computes optimal weights ω^k over the discrete set $\dot{X}_k$ in each iteration. We call this algorithm the *YBT algorithm*.

The weight optimization is given by

$$\begin{aligned} \min_{\omega} \ & \log \left(\det \left(\sum_{i=1}^{k} \omega_i \cdot m(x_i) \right)^{-1} \right) \\ \text{s.t.} & \sum_{i=1}^{k} \omega_i = 1, \quad \omega_i \geq 0. \end{aligned} \tag{$P(\dot{X}_k)$}$$

Since $(P(\dot{X}_k))$ is a convex optimization problem, it can be solved to global optimality by nonlinear programming algorithms. Since, on the other hand, $(P(\dot{X}_k))$ is not strictly convex in the weights ω, the optimal solution need not be unique, and a set of optimal solutions can also exist. The combined algorithm is given in Algorithm 1.

Algorithm 1 Yang-Biedermann-Tang (YBT) algorithm

Require: Tolerance $\varepsilon > 0$, a finite reference grid $\dot{X}_{grid}$ and an initial subset $\dot{X}_{k_0} \subset X_{grid}$, such that a design $\xi^0 \in \Xi_{fin}(\dot{X}_{k_0})$ exists.
Set $tol = \infty$.
while $tol > \varepsilon$ **do**
  Compute a solution ω^k of the problem $(P(\dot{X}_k))$.
  Set ξ^k as the design obtained by assigning the weights ω^k to the points $\dot{X}_k$.
  Solve $x_{k+1} = \text{argmin}_{x \in \dot{X}_{grid}} \psi(\xi^k, x)$.
  Set $tol = -\psi(\xi^k, x_{k+1})$ and $\dot{X}_{k+1} = X_k \cup \{x_{k+1}\}$.
  Iterate $k \to k+1$.
end while
return The design ξ^k.

As the YBT algorithm computes optimal weights in each iteration, it converges significantly faster than other versions of the adaptive discretization algorithm. The convex weight optimization can also be solved to global optimality with little computational effort.

However, in each iteration of the iterative scheme, a global optimizer of $x \mapsto \psi(\xi^k, x)$ has to be computed. In general, this is a non-convex nonlinear program. Additionally, ψ depends on the derivatives of the model f with respect to θ, which are often only given implicitly. The computation of x_{k+1} can thus be difficult. A typical approach to solve $x_{k+1} = \text{argmin}_{x \in X} \psi(\xi^k, x)$ is a brute-force optimization. In such an approach, the design space X is replaced with a reference grid $\dot{X}_{grid}$, and the function ψ is evaluated on all grid points $x \in \dot{X}_{grid}$. We also apply this approach in our implementation of the YBT algorithm.

For fine grids consisting of many points per dimension, this approach leads to good results. However, for each grid point x, the derivative $D_\theta f(x, \overline{\theta})$ of the model has to be evaluated. For computationally expensive models like dynamic models or models with a high-dimensional design space, this can be very time-consuming.

3.2 *Adaptive Discretization via Gaussian Process Regression*

We have seen that while adaptive discretization algorithms in theory provide solutions to the OED problem (P), they can perform poorly for computationally expensive models and high-dimensional design spaces. We now incorporate machine learning methods into the adaptive discretization scheme to improve the performance in these cases. This approach and the corresponding notation and ideas can be found in [2].

For the new approach, we focus in particular on the subproblem:

$$x_{k+1} = \underset{x \in X}{\text{argmin}}\, \psi(\xi^k, x). \tag{13}$$

This function can be costly to evaluate and is often only given implicitly. Therefore, we consider ψ as a black box model. Then, we utilize a Bayesian optimization [19, 20] approach in order to approximate the solution x_{k+1}.

Based on a finite set of design points $\dot{X}_k = \{x_1, \ldots, x_k\}$ and corresponding function values $Y_k = \{\psi(\xi^k, x_i),\ i = 1, \ldots, k\}$, we compute a Gaussian process regression model (GPR) for the data $\left(\dot{X}_k, Y_k\right)$. We then obtain a Gaussian predictive a posteriori distribution $y^* \mapsto \widehat{p}(y^* \mid x, \dot{X}_k, Y_k)$, with mean $\widehat{\mu}_k(x)$ and variance $\widehat{\sigma}_k^2(x)$ (see, e.g., [21, Chapter 15.2] for details). As the data Y_k depends on the current design ξ^k, so does the predictive distribution $\widehat{p}$. Using this distribution, we replace the problem (13) of globally minimizing the sensitivity function $x \mapsto \psi(\xi^k, x)$ by the problem:

$$x_{k+1} = \underset{x \in X}{\operatorname{argmin}} \left(\widehat{\mu}_k(x) - c \cdot \widehat{\sigma}_k^2(x)\right), \tag{14}$$

of minimizing the so-called lower confidence bound (LCB) of $x \mapsto \psi(\xi^k, x)$, where $c > 0$ is a scaling factor of the LCB. The LCB presents a combination of exploration, via the variance term, and exploitation, via the expectation. The scaling factor c regulates the trade-off between both goals.

The resulting adaptive discretization scheme is given as follows. We start with an initial finite set $X_{k_0} = \{x_1, \ldots, x_{k_0}\}$ and then perform the following steps:

1. Compute optimal weights ω^k as a solution to $(P(\dot{X}_k))$. Set ξ^k as the design obtained by assigning the weights ω^k to the points $\dot{X}_k$.
2. Compute the values $Y_k = \{\psi(\xi^k, x_i),\ i = 1, \ldots, k\}$.
3. Fit a Gaussian process regression model to the data $(\dot{X}_k, Y_k)$.
4. Solve $x_{k+1} = \operatorname{argmin}_X \left(\widehat{\mu}_k(x) - c \cdot \widehat{\sigma}_k^2(x)\right)$ and refine $\dot{X}_{k+1} = \dot{X}_k \cup \{x_{k+1}\}$.

In contrast to Algorithm 1, we cannot simply use the values $\psi(\xi^k, x_{k+1})$ for the stopping tolerance. Indeed, as x_{k+1} now is only a minimizer of the LCB surrogate of $x \mapsto \psi(\xi^k, x)$, and thus there may be points $x' \in X$ with $\psi(\xi^k, x') < \psi(\xi^k, x_{k+1})$. Consequently, the value of $\psi(\xi^k, x_{k+1})$ should not be used in the stopping criterion of the algorithm.

Instead we consider the relative progress of the algorithm over a specified number of iterations k_1. For a current iteration $k > k_1$, we consider the value:

$$tol\,(k, k_1) = \frac{\log\left(\det\left(M(\xi^{k-k_1})\right)^{-1}\right) - \log\left(\det\left(M(\xi^k)\right)^{-1}\right)}{\log\left(\det\left(M(\xi^k)\right)^{-1}\right)}.$$

As soon as this tolerance is less than or equal to a threshold ε, we terminate the algorithm. Additionally, we set a number of initial iterations k_{init}, in which we do not terminate no matter the tolerance, and a number of maximum iterations k_{max}, at which we always terminate if reached.

The complete algorithm is given in Algorithm 2.

Algorithm 2 Adaptive discretization algorithm with Gaussian process regression

---

Require: Tolerance $\varepsilon > 0$, values $k_1, k_{init}, k_{max}, c$ and an initial subset $\dot{X}_{k_0} \subset X$, such that a design $\xi^0 \in \Xi(\dot{X}_{k_0})$ exists with non-singular Fisher information matrix $M(\xi^0)$.
 Set $tol = \infty$ and $k = k_0$.
 while $tol > \varepsilon$ **do**
  Compute a solution ω^k of the problem $(P(\dot{X}_k))$.
  Set ξ^k as the design obtained by assigning the weights ω^k to the points $\dot{X}_k$.
  Compute the values $Y_k = \{\psi(\xi^k, x_i),\ i = 1, \ldots, k\}$.
  Fit a Gaussian process regression model to the data $(\dot{X}_k, Y_k)$.
  Solve $x_{k+1} = \text{argmin}_{x \in X} \left(\widehat{\mu}_k(x) - c \cdot \widehat{\sigma}_k^2(x)\right)$.
  if $k - k_0 < k_{init}$ **then**
   Set $tol = \infty$ and $\dot{X}_{k+1} = \dot{X}_k \cup \{x_{k+1}\}$.
  else if $k - k_0 < k_{max}$ **then**
   Set $tol = tol(k, k_1)$ and $\dot{X}_{k+1} = \dot{X}_k \cup \{x_{k+1}\}$.
  else
   Set $tol = 0$.
  end if
  Iterate $k \to k + 1$.
 end while
 return The design ξ^k.

---

4 Application Examples

In this section, we introduce a variety of examples from process engineering applications. For these examples we compute optimal design measures. First, we evaluate the Fisher information matrices on a reference grid in the design space and use the YBT algorithm to find the optimal designs on these grids. Next, we compute an optimal design via the ADAGPR, the adaptive algorithm supported by Gaussian process regression. We compare the performance of the algorithms, as well as the quality of the results.

We have implemented both algorithms in *Python*. For the optimization of the weights, which is required as a step in both algorithms, we have used the *SLSQP* [22] method from the *scipy.optimize* package. The Gaussian process regression for the ADAGPR was implemented via the *scikit-learn* package [23] and with a radial basis function kernel.

In order to compare the computational performance of the algorithms, we compare the runtimes of the YBT algorithm (Algorithm 1) and the adaptive discretization algorithm with Gaussian process regression (ADAGPR, Algorithm 2). When stating the runtime in the provided examples, we distinguish between two phases, a pre-computational phase and a main algorithm phase. The pre-computational phase consists of the runtime to pre-compute the Jacobian matrices $D_\theta f(x, \overline{\theta})$ for the reference grid points $x \in X_{grid}$. It is denoted by *runtime (pre-computations)*. This phase is only required for the YBT algorithm in which we utilize a brute-force approach to find a minimizer x^{k+1} of the sensitivity function $x \mapsto \psi(\xi^k, x)$. The main algorithm phase consists of the runtime for the adaptive discretization scheme of each algorithm, as described in Algorithms 1 and 2. We refer to this time as *runtime (algorithm)*.

For some of the examples, we have used parallelization of code in the pre-computational phase to evaluate the matrices $D_\theta f(x, \overline{\theta})$. As we have not optimized the algorithms with respect to parallelization, we compare the runtime with the cumulative times (without parallelization). For completeness, we do list both times, however.

In order to compare the quality of the computed designs, we mainly consider the value $\log(\det(M(\xi)^{-1}))$ of the objective function at the design ξ returned by the algorithms. This value is listed as *objective value* in the following sections. We also want to compare the value of the maximum prediction uncertainty $U^{f,\xi}$. In view of Corollary 3, this value is closely related to the optimality conditions of the OED problem. As we cannot efficiently compute the exact value $U^{f,\xi}$, we instead evaluate the maximal prediction uncertainty on the reference grid $\dot{X}_{grid}$, given by

$$U^{f,\xi}_{grid} = \max_{x \in \dot{X}_{grid}} U^{f,\xi}(x).$$

As inputs for the YBT algorithm, we use equidistant grids and a random subset of k_0 initial points. The design spaces and the number of grid points used in each dimension are given in Tables 3, 5, and 8. The number k_0 and the considered tolerances $\varepsilon > 0$ used for each example are given in Table 1.

For the ADAGPR algorithm, we sample k_0 initial points via a uniform distribution from the specified design space. The tolerance ε and the numbers k_{init}, k_1 and k_{max} differ for each example and are given in Table 2.

Table 1 Algorithm settings for the YBT algorithm

	Flash	Cinnamaldehyde Hydrogenation	Yeast Fermentation
Tolerance ε	10^{-4}	10^{-4}	10^{-4}
Initial points k_0	9	21	9

Table 2 Algorithm settings for the ADAGPR

	Flash	Cinnamaldehyde Hydrogenation	Yeast Fermentation
Tolerance ε	10^{-2}	10^{-2}	10^{-2}
Initial points k_0	69	451	729
Initial iterations k_{init}	20	200	600
Iterations k_1	20	200	600
Maximum iterations k_{max}	200	1000	1500

4.1 Flash

The first example we consider is a flash, which is also used as an example in [2, 4]. A flash, as illustrated in Fig. 2, is a pressurized container with a liquid input stream (feed) $\dot{F}$ with molar composition x_F, an evaporator with heat duty $\dot{Q}$, a liquid output stream $\dot{L}$ with molar composition x, and a vapor output stream $\dot{V}$ with molar composition y. Within the flash unit, we assume thermodynamic equilibrium between the vapor and liquid phase and that no chemical reactions take place. The mixture considered in this example consists of water and methanol. For more details on the flash example, we refer to [2, 4].

In this example, the following OED setting is chosen:

- As inputs x_1 and x_2, we specify the pressure p at which the flash operates and the molar fraction of methanol $x_{F,m}$ in the feed stream.
- As the unknown model parameters θ_1 and θ_4, we choose the parameters of the activity coefficient model, which is used to describe the thermodynamic equilibrium between the vapor and liquid phases in the flash. We denote these parameters as a_{12}, a_{21}, b_{12} and b_{21}.
- As measurement outputs y_1 and y_2 of the flash example, we take the equilibrium temperature T inside the flash and the molar fraction of methanol y_m in the vapor output stream.

We consider our model f as a function mapping the inputs (x_1, x_2) to a solution (y_1, y_2) of the implicit system of *MESH* equations; see [2] or [4] for details. To construct the design space, pressures are varied between 0.5 bar and 5 bar and the molar fraction of methanol and water between 0 and 1 mol/mol. The feed stream flow rate $\dot{F}$ is kept constant at 1 kmol/h for this example and thus is not contained in the design space. Furthermore, as in [2, 4], we choose the heat duty $\dot{Q}$ such that the vapor output stream flow rate $\dot{V} = 10^{-6}\frac{\text{kmol}}{\text{h}}$, i.e., we choose a minimum heat duty so the entire feed stream leaves the flash in liquid state. As we compute locally

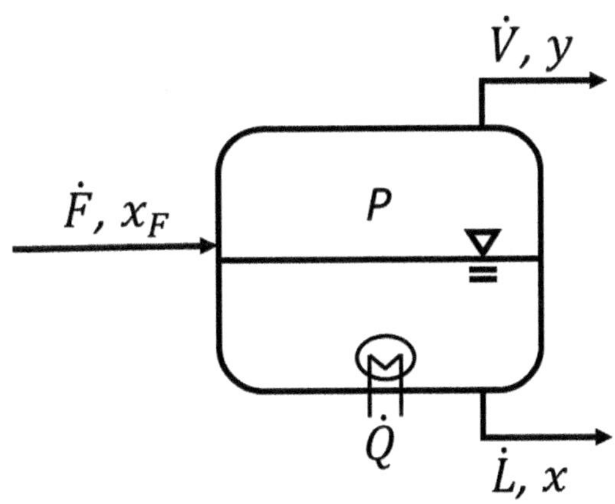

Fig. 2 A sketch of the flash with feed $\dot{F}$, equilibrium at pressure p, heat duty $\dot{Q}$, and the liquid and vapor output streams $\dot{L}$ and $\dot{V}$

Table 3 Design space for the flash example

	$x_{F,m}$	p
Lower bound	0.0	0.5
Upper bound	1.0	5.0
Discrete grid points	101	91

Table 4 Algorithm performance for the flash example

Algorithm	YBT	ADAGPR
Objective value	23.6109	23.7830
Maximal prediction uncertainty $U^{f,\xi}_{grid}$	4.0001	4.3614
Runtime (pre-computations)	3467.14 s	–
Runtime (algorithm)	3.51 s	159.47 s
Number of Jacobian evaluations	9191	130

optimal designs, we have to choose a reference value $\overline{\theta}$ for the model parameters. We take the reference values to be $\overline{a}_{12} = -3.9402, \overline{a}_{21} = 6.3037, \overline{b}_{12} = 1337.558$, and $\overline{b}_{21} = -1891.945$.

As a reference grid $\dot{X}_{grid}$, we consider an equidistant grid with 9191 grid points, based on the lower and upper bounds on the pressure and feed stream molar fractions as specified above. The grid is obtained by (equidistantly) discretizing the molar fraction of methanol in the $x_{F,m}$ dimension with 101 points and the p dimension by 91 points. The resulting design space is also given in Table 3.

In Table 4 the results for both algorithms are given. We see that the YBT algorithm efficiently computes an optimal design on the reference grid. However, if we take the pre-computation of the Jacobians $D_\theta f$ on the reference grid into account, the runtime drastically increases.

The ADAGPR also computes an approximately optimal design. The objective value is worse than that of the YBT. This is to be expected, as we have a low-dimensional design space with a very fine reference grid. However, the ADAGPR computes the design significantly faster. We see that this is mainly due to the number of evaluations of the Jacobian $D_\theta f(x, \overline{\theta})$. The YBT requires a Jacobian computation on all 9191 grid points, whereas the ADAGPR only requires the Jacobian at 130 design points. Also the maximal prediction uncertainty on the reference grid is of comparable size.

This example illustrates that the ADAGPR can compete in terms of result quality and of performance with state-of-the-art methods, even for low-dimensional stationary models.

4.2 Cinnamaldehyde Hydrogenation

Next we consider the cinnamaldehyde hydrogenation, which we model as described in [24, 25]. This is a dynamic model given by the reaction rate expression:

$$\frac{d}{dt}C(t) = \frac{c_{H_2}\left(p_{H_2}, T\right)}{(LHHW_*\left(C(t)\right))^3} \cdot A \cdot C(t) \quad (t \in [0, 700]) \quad \text{and} \quad C(0) = C^0,$$

where $C(t)$ is the vector of mass flows, $c_{H_2}\left(p_{H_2}, T\right)$ is the Henry correlation, $LHHW_*(C(t))$ is the Langmuir-Hinshelwood-Hougen-Watson expression and A the reaction matrix. The time $t \in [0, 700]$ is given in minutes.

In general, the model describes a reaction of seven components. However, in the specific Model 2 from [24, Chapter 7], a variety of reaction rates are set to $k = 0$, and thus the model reduces to four relevant and active components. In Fig. 3, taken from [24], the reaction scheme of the cinnamaldehyde hydrogenation is depicted.

For the optimal experimental design, we take the following setup:

- We consider seven model inputs x_1 to x_7. These consist of the initial mass flows $C^0_{CAL}, C^0_{COL}, C^0_{HCAL}$ and C^0_{HCOL} in $\frac{\text{mol}}{\text{l}}$ of cinnamaldehyde, cinnamyl alcohol, hydrocinnamaldehyde, and 3-phenyl-1-propanol. Additionally, we also set the temperature T in K, the partial pressure p_{H_2} in MPa, and the measurement time $\bar{t}$ in minutes as model inputs.
- As unknown model parameters $\theta_1, \ldots, \theta_{10}$, we consider the reaction rates $k_1, \ldots, k_5$ and the activation energies $E_{a1}, \ldots, E_{a5}$ of the five active reactions.
- As outputs $y_1, \ldots, y_4$, we take the mass flows of $C_{CAL}(\bar{t}), C_{COL}(\bar{t}), C_{HCAL}(\bar{t})$ and $C_{HCOL}(\bar{t})$ at the measurement time $\bar{t}$.

In Table 5 the bounds for the seven design variables are given, as well as the number of grid points in each dimension for the equidistant reference grid. As we have to solve a system of differential equations, computing the Jacobian matrix $D_\theta f(x, \overline{\theta})$ is computationally expensive. We thus cannot sample finely in each dimension and are restricted to 28672 design points.

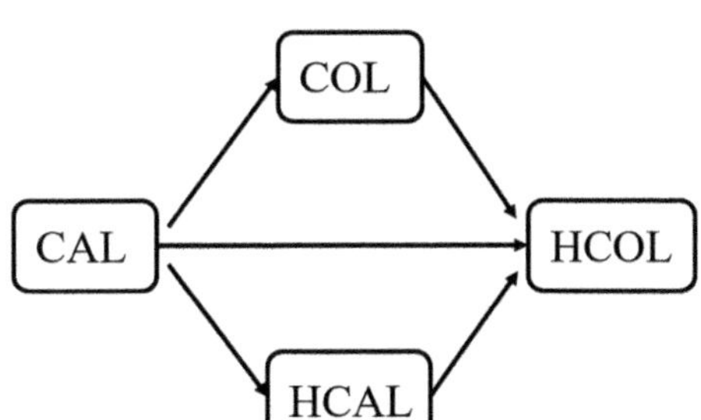

Fig. 3 Reaction scheme of the cinnamaldehyde hydrogenation with Model 2. In comparison with [24], the inactive reactions are not depicted here

Table 5 Design space for the cinnamaldehyde hydrogenation example

	C^0_{CAL}	C^0_{COL}	C^0_{HCAL}	C^0_{HCOL}	T	p_{H_2}	$\bar{t}$
Lower bound	0.5	0.002	0.01	0.01	250.0	1.0	100
Upper bound	4.0	0.5	0.5	0.5	350.0	5.0	700
Discrete grid points	4	4	4	4	4	4	7

Table 6 Reference values of the model parameters for the cinnamaldehyde hydrogenation example

j	1	2	3	4	5
k_j	0.0721	0.0914	0.0541	0.3992	$1.29 \cdot 10^{-8}$
E_{aj}	45000	55000	45000	35000	50000

Table 7 Algorithm performance for the cinnamaldehyde hydrogenation example

Algorithm	YBT	ADAGPR
Objective value	123.11	121.50
Maximal prediction uncertainty $U^{f,\xi}_{grid}$	10.0057	11.3140
Runtime (pre-computations)[a]	168297.34 s (24595.10 s)	–
Runtime algorithm	6.87 s	7512.26 s
Number of Jacobian evaluations	28672	687

[a]Time with parallelization given in brackets

The reference values $\overline{\theta}$ for the model parameters are given in Table 6. These are used for the computation of the Jacobian matrices and result in locally optimal designs.

The results for the YBT algorithm and the ADAGPR are given in Table 7. As for the flash example, we see that the YBT algorithm efficiently computes an optimal design on the reference grid, if we ignore the pre-computations of the Jacobians.

However, even for the coarse grid in the 7-dimensional space, these pre-computations require over 46 hours in total. In comparison, the ADAGPR requires only 687 evaluations of the Jacobian matrix and is thus significantly faster. Despite utilizing fewer Jacobian matrices, the ADAGPR obtains a better objective value than the YBT algorithm.

This example shows that the ADAGPR can outperform the YBT for dynamical models. This is because we adaptively evaluate the Jacobian matrix $D_\theta f(x, \overline{\theta})$ at interesting design points, instead of performing many evaluations on a coarse uniform grid.

4.3 *Yeast Fermentation*

Last we consider a batch reaction, in the form of yeast fermentation, which is a dynamic model with time-dependent control variables. The example is taken from [2, 3, 26].

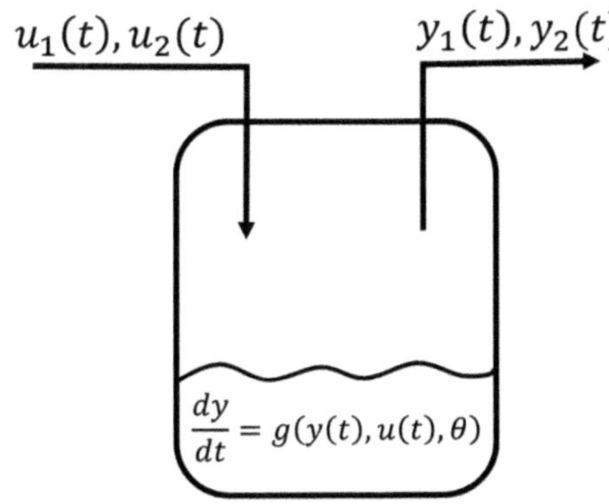

Fig. 4 A sketch of a batch reaction with controls u_1 and u_2

In a reactor we have a biomass (given by a mass flow y_1 in g/l) and a substrate (given by a mass flow y_2 in g/l) which react (fermentation). During the fermentation process, we continuously add substrate via an input feed governed by a dilution factor $u_1(t)$ in h^{-1} and a substrate mass flow $u_2(t)$ in g/l, both of which are time-dependent controls. The mass flows $y_1(t)$ and $y_2(t)$ can be measured during the reaction. This process is depicted in Fig. 4, which is taken from [2].

The yeast fermentation is modeled by the following system of differential equations:

$$\frac{dy_1}{dt}(t) = (r(t) - u_1(t) - \theta_4)y_1(t) \qquad (t \in [0, 20]) \quad \text{and} \quad y_1(0) = y_1^0,$$
$$\frac{dy_2}{dt}(t) = -\frac{r(t)y_1(t)}{\theta_3} + u_1(t)(u_2(t) - y_2(t)) \quad (t \in [0, 20]) \quad \text{and} \quad y_2(0) = y_2^0,$$
$$r(t) = \frac{\theta_1 \cdot y_2(t)}{\theta_2 + y_2(t)}$$

where the parameters $\theta_1 - \theta_4$ are $four$ unknown model parameters. The parameters θ_1, θ_2, and θ_3 are bounded by the range $[0.05, 0.98]$, and θ_4 is bounded by the range $[0.01, 0.98]$. These ranges span the parameter space Θ.

For the OED settings, we proceed as described in Sect. 2.2. We consider the controls $u_1(t)$ and $u_2(t)$ and replace them with a continuous linear interpolation. As support points we set a time point every 4 hours via $t_0 = 0,\ t_1 = 4,\ \ldots,\ t_5 = 20$, and corresponding values $u_{1,0}, \ldots, u_{1,5}$ and $u_{2,0}, \ldots, u_{2,5}$. The values $u_{1,j}$ are assumed to be in the range $[0.05, 0.2]$ and the values $u_{2,j}$ in the range $[5.0, 35.0]$. In Fig. 5 an example of the interpolated controls is shown.

In addition to the 12 values $u_{1,j}$ and $u_{2,j}$ defining the controls, we also take the initial biomass mass flow y_1^0, from the range $[1.0, 10.0]$, as an additional input parameter. This results in a 13-dimensional design space given by

$$X = [0.05, 0.2]^6 \times [5.0, 35.0]^6 \times [1.0, 10.0].$$

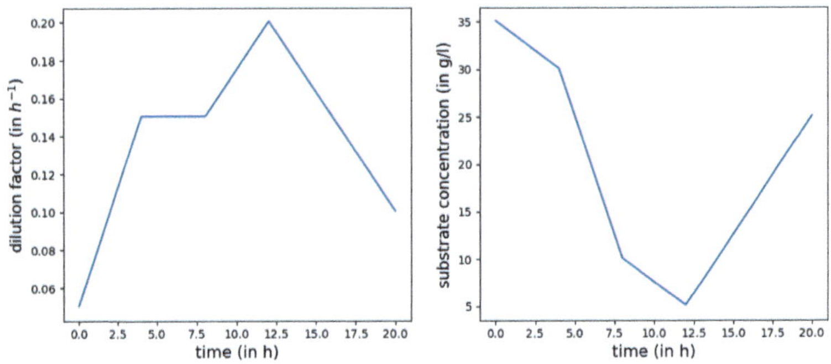

Fig. 5 Example figure of the control functions u_1 and u_2, obtained via continuous linear interpolation

Table 8 Design space for the yeast fermentation example

	$u_{1,j}$	$u_{2,j}$	y_1^0
Lower bound	0.05	5.0	0.1
Upper bound	0.2	35.0	10.0
Discrete grid points	3	3	3

In contrast to Sect. 2.2 however, we do not add the measurement time to the design variables. Instead we consider a fixed measurement every 2 hours at the time points $\bar{t}_0 = 0, \bar{t}_1 = 2, \ldots, \bar{t}_{11} = 20$. As model outputs we then have the 11 measurements of the mass flows $(y_1(t), y_2(t))$, which results in a 22-dimensional model output:

$$\bar{y} = \left(y_1(\bar{t}_1), \ldots, y_1(\bar{t}_{11}), y_2(\bar{t}_1), \ldots, y_2(\bar{t}_{11})\right)^T.$$

The reference values of the four model parameters are taken to be $\overline{\theta}_1 = \ldots = \overline{\theta}_4 = 0.5$.

For the YBT algorithm, we require a reference grid in the design space. As we have 13 dimensions, we use a coarse equidistant grid with only 2 or 3 points per dimension. By taking the corresponding cross product as our reference grid in the design space, we obtain $3^{13} = 1\,594\,323$ grid points. The design space and the reference grid are given in Table 8.

The results for both algorithms are given in Table 9. We observe that the computation of the Jacobian matrices $D_\theta f(x, \overline{\theta})$ over the reference grid takes over 350 hours, even for the coarse discretization with only 3 points per dimension. This illustrates why the YBT algorithm may be unreasonable for dynamic models with time-dependent control variables. The ADAGPR only evaluates 1471 Jacobian matrices and is thus significantly faster, with approximately 14 hours of runtime.

With respect to the runtime, we also note that the ADAGPR trains a Gaussian process regression model every iteration. For large data sets, this training can

Table 9 Algorithm performance for the yeast fermentation example

Algorithm	YBT	ADAGPR
Objective value	−127.34	−120.6812
Maximal prediction uncertainty $U_{grid}^{f,\xi}$	4.0003	43.9805
Runtime (pre-computations)[a]	1270400.84 s (161130.12 s)	–
Runtime algorithm	191.57 s	49327.23 s
Number of Jacobian evaluations	1594323	1471

[a]Time with parallelization given in brackets

also be very time-consuming. Utilizing more efficient implementations or different regression models can improve the performance of the ADAGPR in this example.

Comparing the objective values both algorithms obtain, it appears that the YBT algorithm outperforms the ADAGPR. However, if we revert the scaling via the logarithm, we see that the algorithms obtain practically the same values for the determinants of the Fisher information matrices at the computed designs, namely, $\exp(-127.34) \approx 0 \approx \exp(-120.6812)$. From these values we observe that the OED problem may be ill-conditioned for the yeast fermentation model. We expect this behavior to carry over to the regression models with radial basis function kernel.

5 Conclusion

In this contribution, we have presented a class of adaptive discretization algorithms which can be used to solve optimal experimental design problems. By incorporating a Bayesian optimization step into these algorithms, we have derived the ADAGPR, an adaptive discretization algorithm with Gaussian process regression. Compared to most OED algorithms, this method does not rely on pre-computations over a reference grid. For dynamic models, high-dimensional models, and for models which are time-consuming to evaluate in general, this method can lead to a performance increase by reducing the number of Jacobian evaluations required.

Via three examples from process engineering, we have compared the algorithm with Gaussian process regression to the state-of-the-art YBT algorithm. For a small-scale model (the flash), the algorithm can compute near-optimal designs faster than the YBT algorithm. As we use approximations of the sensitivity function in each iteration, we cannot expect to achieve the same tolerance as the exact method obtains. However, even here we can get a performance increase.

For higher-dimensional dynamic models, it is not viable to pre-compute Jacobian matrices on a fine grid. Thus, the ADAGPR can compute better designs in a shorter time (e.g., the cinnamaldehyde hydrogenation). This is achieved by only evaluating the Jacobian matrix of interesting design points instead of on a predetermined grid.

However, we also observe that there is still room for further improvement. So far we have only considered GPR as our regression model. These models also become expensive to compute and evaluate for a large number of training data points. A GPR may also perform badly when approximating irregular functions. Utilizing further machine learning models or considering more efficient implementations can thus lead to a further performance increase. As we utilize Bayesian optimization techniques, we can also consider additional acquisition functions like the expected improvement in new implementations.

Acknowledgments We gratefully acknowledge funding from the DFG within the SPP 2331 "Machine Learning in Chemical Engineering." We would also like to thank EUROKIN for providing an application example used in this contribution.

References

1. Fedorov, V. V., & Leonov, S. L. (2013). *Optimal design for nonlinear response models*. CRC Press.
2. Seufert, P., Schwientek, J., & Bortz, M. (2021). Model-based design of experiments for high-dimensional inputs supported by machine-learning methods. *Processes, 9*(3), 508.
3. Asprey, S. P., & Macchietto, S. (2002). Designing robust optimal dynamic experiments. *Journal of Process Control, 12*, 545–556.
4. Vanaret, C., Seufert, P., Schwientek, J., Karpov, G., Ryzhakov, G., Oseledets, I., Asprion, N., & Bortz, M. (2020). Two-phase approaches to optimal model-based design of experiments: How many experiments and which ones? *Computers & Chemical Engineering, 146*, 107218.
5. Galvanin, F., Macchietto, S., & Bezzo, F. (2007). Model-based design of parallel experiments. *Industrial & Engineering Chemistry Research, 46*(3), 871–882.
6. Franceschini, G., & Macchietto, S. (2008). Model-based design of experiments for parameter precision: State of the art. *Chemical Engineering Science, 63*(19), 4846–4872.
7. Rinck, A. (2022). Fraunhofer Institute for Industrial Mathematics ITWM homepage, department OPT. https://www.itwm.fraunhofer.de/de/abteilungen/optimierung/verfahrenstechnik-prozesstechnik/versuchsplanung-chemische-verfahrenstechnik.html. Cited 20December2022.
8. Wynn, H. P. (1970). The sequential generation of d-optimal experimental designs. *The Annals of Mathematical Statistics, 41*, 1655.
9. Böhning, D. (1986). A vertex-exchange-method in D-optimal design theory. *Metrika: International Journal for Theoretical and Applied Statistics, 33*(1), 337–347.
10. Silvey, S. D., Titterington, D. H., & Torsney, B. (1978). An algorithm for optimal designs on a design space. *Communications in Statistics—Theory and Methods, 7*(14), 1379–1389.
11. Yu, Y. (2011). D-optimal designs via a cocktail algorithm. *Statistics and Computing, 21*, 475–481.
12. Yang, M., Biedermann, S., & Tang, E. (2013). On optimal designs for nonlinear models: A general and efficient algorithm. *Journal of the American Statistical Association, 108*(504), 1411–1420.
13. Harman, R., Filová, L., & Richtárik, P. (2020). A randomized exchange algorithm for computing optimal approximate designs of experiments. *Journal of the American Statistical Association, 115*(529), 348–361.
14. Kiefer, J., & Wolfowitz, J. (1960). The equivalence of two extremum problems. *Canadian Journal of Mathematics, 12*, 363–366.
15. Bortz, M., & Asprion, N. (2022). *Simulation and optimization in process engineering: The benefit of mathematical methods in applications of the chemical industry / edited by Michael Bortz, Norbert Asprion*. Elsevier.

16. Körkel, S., Kostina, E., Bock, H. G., & Schlöder, J. P. (2004). Numerical methods for optimal control problems in design of robust optimal experiments for nonlinear dynamic processes. *Optimization Methods and Software, 19*(3-4), 327–338.
17. Atkinson, A. (2008). Examples of the use of an equivalence theorem in constructing optimum experimental designs for random-effects nonlinear regression models. *Journal of Statistical Planning and Inference, 138*, 2595–2606.
18. Blankenship, J. W., & Falk, J. E. (1976). Infinitely constrained optimization problems. *Journal of Optimization Theory and Applications, 19*(2), 261–281.
19. Frazier, P. I. (2018). A tutorial on bayesian optimization. *ArXiv:1807.02811* https://arxiv.org/pdf/1807.02811.pdf. Cited 04August2023.
20. Snoek, J., Larochelle, H., & Adams, R. P. (2012). Practical bayesian optimization of machine learning algorithms. In F. Pereira, C. J. Burges, L. Bottou, & K. Q. Weinberger (eds.). *Advances in neural information processing systems* (Vol. 25). Curran Associates, Inc.
21. Murphy, K. P. (2012). Machine learning: A probabilistic perspective / Kevin P. Murphy. In *Adaptive computation and machine learning series*. MIT Press.
22. Kraft, D. (1988). *A Software Package for Sequential Quadratic Programming*. Deutsche Forschungs- und Versuchsanstalt für Luft- und Raumfahrt Köln: Forschungsbericht. Wiss. Berichtswesen d. DFVLR.
23. Pedregosa, F., Varoquaux, G., Gramfort, A., Michel, V., Thirion, B., Grisel, O., Blondel, M., Prettenhofer, P., Weiss, R., Dubourg, V., Vanderplas, J., Passos, A., Cournapeau, D., Brucher, M., Perrot, M., & Duchesnay, E. (2011). Scikit-learn: Machine learning in Python. *Journal of Machine Learning Research, 12*, 2825–2830.
24. Vergunst, T. (1999). *Carbon coated monolithic catalysts: Preparation aspects and testing in the three-phase hydrogenation of cinnamaldehyde: Zugleich: Delft, Technical University, Professor-Schr., 1999*. Ponson & Looijen.
25. Vergunst, T., Kapteijn, F., & Moulijn, J. A. (2001). Kinetics of cinnamaldehyde hydrogenation–concentration dependent selectivity. *Catalysis Today, 66*(2–4), 381–387.
26. Barz, T., Arellano-Garcia, H., & Wozny, G. (2010). Handling uncertainty in model-based optimal experimental design. *Industrial & Engineering Chemistry Research—IND ENG CHEM RES, 49*(12), 5702–5713.

Data Analytics, Artificial Intelligence, and Machine Learning in Mobility and Vehicle Engineering

Michael Burger

Abstract In recent years, the availability of data is growing rapidly and drastically in almost all areas of life. The fields of vehicle engineering, vehicle operation, traffic, and mobility are included here quite naturally. Modern vehicles are recording a huge amount of data, collections of historic measurement campaigns are available as well, and today, there is technology available that helps recording, administrating, processing, and communicating large data sets in a very efficient way. All this opens the door for data-based mathematics and tools from artificial intelligence to efficiently and effectively derive essential information from data and to combine with classic techniques from modeling and simulation. This leads to large potential for applications, especially in the field of vehicle engineering and mobility. This article gives an overview of possible application and throws some spotlights on specifically chosen examples and case studies.

1 Introduction

In the vehicle and mobility sector, the availability of data has been increasing rapidly for many years; in addition to data and information measured directly on the vehicle, among others, this also applies to environmental data. In particular, there are historical measurements from campaigns with dedicated test vehicles as well as from fleet operation. In addition, modern vehicles are increasingly recording data continuously during operation, so that more and more field data from real customer use is available, often even in real time, e.g., via telematics systems. Technologies for data acquisition, data management and computational data processing have also steadily improved and evolved, especially during the last decade. Both make it possible, with suitable methods and tools of data-based mathematics, artificial intelligence (AI), and machine learning (ML), to gain

M. Burger (✉)
Fraunhofer Institute for Industrial Mathematics ITWM, Kaiserslautern, Germany
e-mail: michael.burger@itwm.fraunhofer.de

J. Franke, A. Schöbel (eds.), *Statistical Machine Learning for Engineering with Applications*, Lecture Notes in Statistics 227,
https://doi.org/10.1007/978-3-031-66253-9_11

valuable insights and information from existing data volumes and to use these, possibly already at an early stage in the development process, during the operational lifetime of a single vehicle or for (optimally) controlling vehicle traffic.

The specific application field of data-driven methods and techniques is rich and diverse. Methods of data analysis and machine learning, for example, are suitable for deriving data-based dynamic prediction models or even dynamic controllers, or for identifying structures, patterns, and correlations in existing data sets. These include long-understood structures such as neural networks, whose applicability and usefulness have, however, again been significantly increased by modern hardware in recent years (cf. "deep learning"). For example, trained ML models can complement existing simulation tools and calculation chains or make them significantly more efficient; see Sect. 3. Other important areas are systems for machine monitoring (monitoring and predictive maintenance) as well as control systems directly on the vehicle (e.g., advanced driver assistance systems and automated driving functions (ADAS/AD)), on a vehicle test rig, e.g., for tracking specified reference variables, or even for controlling traffic lights.

In addition to the vehicle and customer usage data just mentioned, the quantity and quality of the available *environmental* data is also steadily increasing. This includes, for example, digital map data, climate data, topographical data, and also socio-economic information. With the help of suitable mathematical methods, it is possible to derive vehicle-independent environment data from certain vehicle measurements. Here, we refer to our work on the identification of road profiles and road roughness based on simple vehicle measurements; details and detailed investigations can be found in [9]. At Fraunhofer ITWM, a comprehensive collection of methods and procedures has been developed in recent years, including the Virtual Measurement Campaign (VMC; cf. [8]) software suite, whose fundamental goal is to compile and maintain the aforementioned environmental data in the form of a geo-referenced database covering as much area as possible worldwide and to combine it with mathematical analysis and simulation tools. However, a profound benefit for the entire design, development, test, and validation process in vehicle engineering often only arises from a combination of the two types of data mentioned, vehicle or customer data on the one hand and environment data on the other hand, in connection with appropriate models and analysis tools.

Existing data of both basic types can be processed and analyzed very efficiently using mathematical methods in order to draw insights and conclusions for future development and testing steps. In addition, data can be combined with simulation models in many ways: obviously for parametrization and validation of the models, as input variables for simulation processes (e.g., for overall vehicle simulation), up to the use of data-based (sub-) models in existing simulation models, so-called graybox approaches. In the context of the vehicle development process, a typical goal is to obtain realistic information about vehicle loads and customer usage behavior, which can then be used, for example, to define design targets and test criteria (cf. Sect. 2). This naturally concerns the field of durability and fatigue strength of vehicle components, but also extends to the attribute of energy efficiency and thus in particular to the development of modern and innovative driving technologies, as well as the design and testing of ADAS/AD systems.

Today's vehicle development process addresses global markets, and, therefore, it is important to collect knowledge about usage and the corresponding vehicle loads and energy requirements in almost any region of the world. The conventional, classic solution approach is to equip specially tailored measurement vehicles with numerous complex sensors and measurement devices to collect a comprehensive data set describing vehicle loads and energy consumption. However, a number of challenges arise in doing so: First, fully equipping such a vehicle is a costly and time-consuming task. Some sensors may not be sufficiently robust and fail, and, typically, the measurement campaign must always be accompanied by at least one test engineer. In addition, a measurement campaign must be planned in such a way that it is both statistically representative and also records relevant events, a difficult task, if at all completely solvable. Furthermore, vehicle loads and energy requirements depend heavily on driving behavior and the traffic situation; in a classic campaign, there are usually a few specially trained test drivers on the road, which is a significant difference from real customer use.

Taking all the above aspects into account, it would be much more desirable to have measured vehicle data from as large a fleet of real customers and users as possible. If the fleet is sufficiently large and diverse and covers the typical usage variations in specific regions, a reliable estimate of the actual usage variability in terms of vehicle loads and energy demand can be obtained. Endurance data, fleet data, and data from actual customer operations can be used for this purpose. It must be noted, however, that when data is collected from the sources just mentioned, it is usually not possible to install high-quality sensor technology; rather, quantities that are easy to measure are collected, often moreover in aggregated and classified form; this is a first place where data-based techniques can help. Data collected in this way must be statistically analyzed and evaluated and, as indicated above, ideally related to geo-referenced environmental data.

To sum up, the popularity of data-driven approaches, AI and ML naturally covers the sectors vehicle engineering, vehicle operation, and mobility. As in all other field, in general, the main reason for the ongoing hype is basically the *availability* of data, which includes the pure existence of many different data sources (i.e., modern equipment and sensors are available at affordable costs), but also, and foremost, the possibilities of communicating, recording, and processing the available (large) data sets. This comes along with the naturally improving computing hardware options, which contributes at a second level as well.

The goal of this paper is to give an overview of the application of data-driven and AI-based methods in mobility and vehicle engineering. With the help of application and project examples from Fraunhofer ITWM's work in the mentioned fields, we aim at throwing spotlights on specific examples. The choice of examples and success stories presented here shall demonstrate and highlight the utility and potential of AI-based methods. However, at this point, it is worth stressing our view that a sustainable success of AI in those sectors is only achieved by combining these methods with physics-based modeling and system knowledge and not by system-agnostic pure black-box application. The remaining part of this contribution is organized as follows: In the next three sections, we present exemplary successful applications of AI-based method in different situations. In Sect. 2, we show how machine and vehicle data can be used for recognizing what the machine is doing; this is important information both for future developments and during operation of the considered machine. In Sect. 3, we demonstrate the usage of AI-methods as surrogate models that help accelerating classic and established simulation workflows; thus, AI can contribute here significantly in the process of virtual product development. Last, not least, we consider many vehicles and vehicle flows, in short: traffic. We present two different scenarios, in which AI helps in the control of traffic flows and, thereby, may contribute to solve current challenges in mobility.

2 AI-Based Application Detection for Vehicles and Machines

Modern vehicles already record many state variables as standard, and commercial vehicles are also often equipped with telematics systems that send data to a data cloud at regular intervals. In the area of commercial vehicles in particular, especially agricultural and construction machinery, the variability of use is typically very high. Possible applications and usage scenarios of such machines are very diverse and also often depend on special customer groups and their regions of use. For the development process, it is therefore of particularly essential interest to have the most precise knowledge possible about the actual use of a vehicle type, e.g., in order to define specifically tailored design targets and testing criteria. Classical sources of such information are historical knowledge as well as estimates based on sales and marketing; however, this information can be very uncertain and is usually also not directly based on actual and real machine operation.

At this point, we propose a ML-based detection algorithms, e.g., ensemble-based methods like random forest classifiers, that identify the type of use (e.g., "digging" in the case of an excavator) based on machine or vehicle state information; see also Fig. 1. Data recorded on the vehicle or machine is either stored in aggregated,

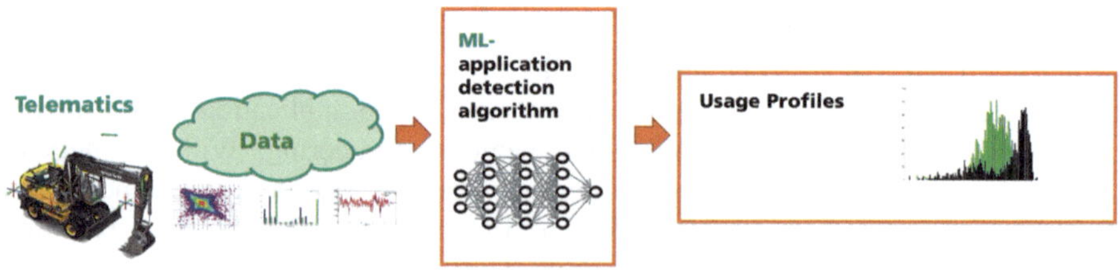

Fig. 1 Process overview: Detection of machine usage

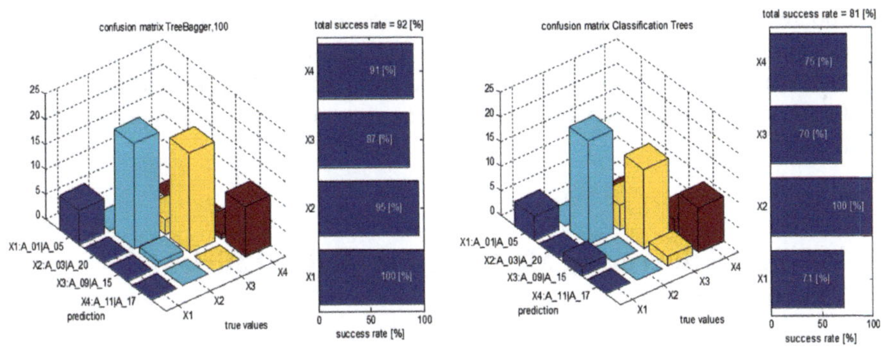

Fig. 2 Test results with a classification tree (right) and a bagging method (left). The figures show the confusion matrices (absolute) and the accuracy for each application to be detected, X_1, X_2, X_3, X_4

coarse, or classified form, respectively, or sent to a cloud in a suitable form at regular intervals.

Once a corresponding ML model has been trained, very accurate region- and customer group-specific usage profiles can be generated when applied to regularly incoming machine data.

The procedure is demonstrated using the example of a mobile excavator; for further details, please refer to [6, 7]. Here, such a detection algorithm is trained based on data from a measurement campaign. Inputs of the algorithm are so-called damage values (a scalar quantity motivated from durability analysis, which reflects the load level of a time series) for certain channels, temporally normalized to a suitable time window, for example, half an hour. Time-normalized mean values would also be conceivable here, for example. Based on a data analysis of the measurement campaign, four different applications have been chosen. The application assignment as a reference in the campaign has taken place by reviewing corresponding video material. A total of approx. 200 h of data material has been available. With this reference data, first of all a problem of "supervised learning" is to solve; more precisely it is a classification problem, since the output variable, i.e., the application, can only take finitely many values. As models for the detection algorithm, a number of methods from the field of machine learning can be used: nearest neighbor methods, decision/classification trees, random forests, boosting, bagging, and in general ensemble methods. In addition, neural networks, support-vector machines, and Gaussian processes could also be used for this problem. Figure 2 shows test results (70% of the data is used to train the method; the remaining data is used for testing) for a classification tree and a bagging (bootstrap aggregation) method for classification trees. Bagging is a technique to increase the accuracy by clever combination of simple classification methods. Already the simple first procedure shows good results in the example, the bagging procedure building on it with a success recognition rate of 92% even very good results. A model trained in this way can now in principle be operated in real time, during vehicle's operation (on-board). If the necessary input channels are available in the form of damage values normalized to a suitable time unit in this specific case, the respective current application, i.e., the one executed in the time unit, can be determined directly. If this is stored or transmitted as part of a telematics system, precise usage profiles can be derived.

3 AI-Based Surrogate Modeling for Highly Efficient Simulation

In this section, an example is shown how data-based models can be used to accelerate classical simulations. Further details and detailed studies can be found in [10]. The analyses and approaches presented here were initiated and motivated by the work of the ITWM within the framework of the EU project LORRY, in which tires with reduced rolling resistance were developed for the commercial vehicle sector.

Based on the training data generated with the high-quality ITWM tire model CDTire, a ML model is trained, which can substantially accelerate the tire simulation in the concrete application scenario. Specifically, the output variables rolling resistance and tread wear are considered for a specific tire. For the relevant input variables for a single tire simulation with CDTire, i.e., wheel load, camber and slip angles, drive torque, and speed, so-called base events are generated as comprehensively as possible in all meaningful value ranges. A single tire simulation is then carried out with these basic events, and rolling resistance and tread wear are numerically determined. These data pairs represent the training data with which a neural network is trained.

In the subsequent application, the corresponding input data can then be generated for a given route, using suitable simulation software, and these are then fed to the trained data-based model rather than the highly accurate tire simulation model. In this way, simulation times can be drastically reduced for a specific tire, for which the model has been trained. Due to the combination of physics-based simulation and data-based submodel, this is a typical graybox model.

In a study (cf. [10]), several approaches are analyzed and compared. Very good results are achieved with a shallow, 3-layer neural network; each layer has 100 neurons; the so-called rectified linear unit (ReLU) is chosen as activation function. Thus, training accuracies of almost 100% can be achieved. Thereby, no typical overfitting problem arises, since here the training data is generated synthetically in such a way that on the input side, the entire relevant value range is covered as comprehensively as possible.

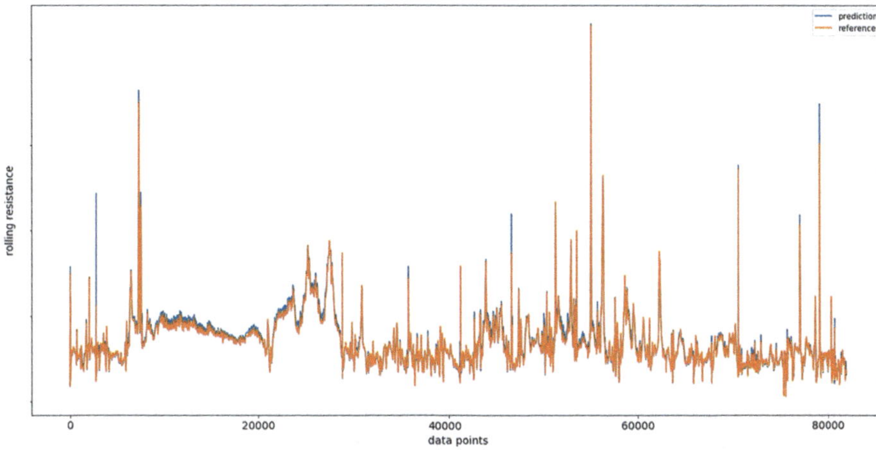

Fig. 3 Test results for a reference route around Kaiserslautern. Exemplarily, results for the rolling resistance prediction are shown: The reference output of the high-fidelity model (orange) and the output of the data-driven surrogate (blue) are plotted. We observe a very high accuracy of 98%

The accuracy results mentioned here and below are based on a similarity measure μ that defined for prediction and reference signals, ρ^{pred}, ρ^{ref}, as follows:

$$\mu(\rho^{pred}, \rho^{ref}) := 100 \cdot \left(1 - \frac{\|\rho^{pred} - \rho^{ref}\|_2^2}{\|\rho^{ref} - \overline{\rho^{ref}}\|^2}\right)$$

where $\bar{\rho}$ denotes the mean value of the quantity. For a selected test route around Kaiserslautern of about 80 km length, the corresponding input signals are generated, and a prediction of rolling resistance and tread wear is performed with the data-based model as well as with the full-scale, highly accurate physical tire model (cf. Fig. 3). In turn, geo-referenced environmental data of the route are also included in the generation of the input data. Very high accuracy is achieved: 98% for rolling resistance and 92% for abrasion. At the same time, there is a very drastic *computation time reduction*: The single tire simulation needs more than 5 days for this test route, while the neural network evaluation is finished in 0.32 s for the whole route. Of course, the neural network is only valid for this one specific tire and by far does not have the physically based extrapolation properties of the high-quality model. Nevertheless, an enormous increase in efficiency can be achieved for targeted analyses and studies with a data-based model and a one-time training process (effort: approx. 10 h). With the procedure presented here, studies can therefore now be carried out for the first time, in which tens of thousands of kilometers of a specific usage scenario are simulated in order to be able to derive usage profiles with regard to the attributes of rolling resistance and tread wear.

4 AI-Based Optimal Traffic Control

Today, the amount of traffic increases strongly nearly all over the world. This results in more accidents, congested roads, and traffic jams, and, all this contributes to more emissions of different types. Thus, there is a need of intelligent mobility solutions that should guarantee efficiency, safety, and sustainability. Intelligent vehicles that communicate with other road users may contribute to these solutions if they not only maximize their own goals (e.g., minimizing travel time) but, additionally, aim to increase and stabilize traffic flow for all vehicles in certain traffic situations, for instance, in stop-and-go waves and congestions that emerge because of inefficient human driving behavior. With intelligent vehicle control, these inefficiencies may be outbalanced, if other vehicles' driving behavior can be directly affected. A crucial point is the design of such intelligent controllers. Models that describe traffic dynamics and human driving behavior have been studied since the last century; nowadays, vehicle technology and mobility infrastructure allow to collect and to transmit large and comprehensive data, especially in the light of emerging and maturing communication technologies, such as 5G. This data may be used by complex driver assistance systems or (semi-) autonomous vehicles. Also, intelligent traffic control systems may use this data to enable communication and cooperation between road users and infrastructure. To develop and implement such solutions, we need intelligent approaches to control traffic flows. These approaches should take into account established modeling as well as available data, with the goal to improve traffic control constantly.

In the sequel, we present two different approaches for optimally controlling two different traffic scenarios. The first one relies substantially on real-time data of the considered traffic situation and aims for an optimal control of traffic lights in a complex roundabout using model-free reinforcement learning (RL) as purely data-driven method. The second example presents a hybrid control approach: It is based on specific modeling approaches for traffic dynamics and uses machine learning to approximate a model-predictive controller.

4.1 Traffic Light Control with Reinforcement Learning

Reinforcement learning belongs to the AI and ML toolbox, it is a data-driven control technique, and, roughly speaking, it can be seen as an observation-based trial-and-error scheme. In particular, RL is based on the idea that an agent (the controller) interacts with its environment (the dynamical system to be controlled) and aims at controlling the system optimally w.r.t certain criteria. In the so-called model-free RL framework, no system description needs to be unknown. The agent, however, is allowed to repeatedly apply different actions (controls) depending on different states of the considered system and to observe the system's response, typically in terms of measurements. With the help of the observed responses and

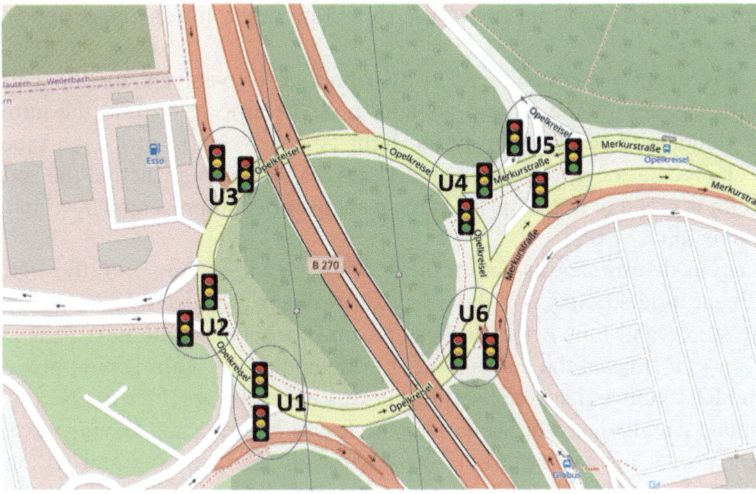

Fig. 4 Opel-roundabout in Kaiserslautern, Germany [2, 12]

using a mathematical formulation of the desired criteria, it is assessed how good those criteria are satisfied. Next, this assessment is used to update the agent, such that controls leading to desirable outputs are encouraged or *reinforced* and disadvantageous outputs are discouraged. In recent years, RL achieved impressive successes, especially in playing games [14] or in robotics [13]. There is a huge potential of the technique, especially in situations, in which model descriptions are hard to derive. For a general introduction of RL, we refer to [17]; for RL introduction with a focus on traffic control, we refer to our work [3] and the references therein.

In the remaining part of this section, we present an example, in which an RL controller is applied to a complex road network, the Opel-roundabout in Kaiserslautern; see Fig. 4. The experiments and analysis are carried out in virtual setup using a digital twin of the roundabout based on OpenStreetMap data and a microscopic traffic model that allows to model realistic traffic behavior running through the roundabout. The results shown here are based on our work in [2, 3], to which we also refer for more details of this analysis; moreover, we refer to [1, 18, 19] for more applications of RL in the context of traffic control.

The considered network is as a roundabout with six incoming and outgoing lanes, and the traffic flow is controlled by several traffic lights as shown in Fig. 4.

In the virtual setup, these traffic lights are controlled by switching through different phases with fixed durations. Such static behavior leaves room for improvement, and in the last decades, several solutions for optimal traffic light control have been proposed. We apply an RL approach to this traffic scenario by aggregating the traffic lights to six different traffic control units, which can switch between two phases showing either green for one or the other direction; see Fig. 4. Every time step the RL agent decides for all units whether to stay at the current phase or to switch to the next one. Thus, following the RL procedure as presented in [2], the traffic flow is optimized iteratively from run to run by optimizing a linear model policy in the RL agent.

The RL agent observes the density (number of vehicles on the lane), the minimum distance of the closest vehicle to next traffic light unit, the number of waiting vehicles in front of the traffic light (speed $v < 0.1\,\text{m/s}$), the time since the last change has occurred, and a binary variable indicating whether the upcoming traffic light is green or not. This leads to an environment represented by a state space with dimension $n = 65$.

The agent determines its action $u_t \in \mathbb{R}^6$ for each of the six traffic light units with a linear model policy. We define a fixed value u_{switch} such that, if $u_t^i > u_{\text{switch}}$ for time step t and unit i, the traffic light unit switches to the next phase and, if $u_t^i \leq u_{\text{switch}}$, it stays at the current one. For details of the optimization process that trains the RL agent and specific settings of this example, we refer to [2].

To obtain a realistic traffic scenario, vehicles are entering and leaving the network persistently over a fixed time horizon. There are six incoming and six outgoing lanes, and we determined realistic frequencies f_{ij} indicating the probability of a vehicle entering the network on lane i and leaving it on lane j. For each time step, we draw samples z_{ij} of a uniformly distributed random variable Z on the interval [0,1], and, if $z_{ij} \leq f_{ij}$, a vehicle starts driving from lane i to lane j. We use randomly chosen frequencies for the training of the controller to model different traffic scenarios and make it less prone to changes of the environment. Additionally, it seems more realistic to assume that vehicles are not entering the system in equal time periods. The entries f_{ij} are summarized in a frequency matrix F. The RL controller is trained in stochastic scenarios based on frequency matrix F, and we compare it with a typical static traffic light switching program. This static controller is based on a realistic fixed cycle time of 90 s and aims at avoiding collisions and achieving high values of traffic flow, see [2].

In Fig. 5, results are shown from a scenario both with the static controller and with the RL-based controller. Most of the time steps, the RL controller achieves a higher average speed over all vehicles which indicates that it can outperform the static controller on a frequency that it is trained on.

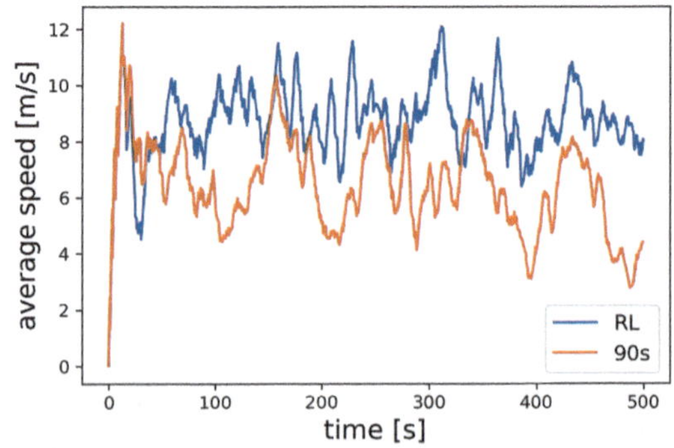

Fig. 5 RL agent and static 90 s controller in terms of the average speed [m/s] over all vehicles for one scenario that was created with frequency matrix F and fixed stochastic speed

These results highlight the strong potential of a RL-based control algorithm, especially in this traffic control context.

4.2 Ring Road Control with Extended Imitation Learning

In the second example, we consider a hybrid approach: a combination of model-based and a data-driven method. The control task, considered here, differs substantially from the traffic light control scenario. Here, we control a single vehicle that aims at stabilizing traffic flow on a ring road. This is a summarized version of our work [4].

In real-world experiments of the ring road [15, 16], it has been shown that human-driving behavior can be inefficient and that connected autonomous vehicles (AVs) may be able to improve traffic flow for all vehicles in the system. As it is a closed scenario, the number of vehicles is fixed which facilitates the mathematical description of the system. However, despite being an artificial scenario, similar inefficiencies (congestion and stop-and-go waves) as in real-world scenarios occur. Therefore, the ring road experiment has proven to be an effective benchmark scenario for developing and testing new traffic control strategies.

In this work, we focus on scenarios with a total number of N vehicles including one AV and $N - 1$ human drivers (HDs). We assume that the dynamics of the HDs are given by a car-following model and cannot be controlled directly. But for the AV, we can set the acceleration behavior. The considered control task consists in finding such controls for the AV that the traffic flow for all vehicles in the system is maximized. One approach to measure the traffic flow in the system is the average speed over all vehicles (Fig. 6).

To find optimal controls u for the AV, i.e., to derive a suitable control algorithm, extensive studies and investigations have been performed (cf. [4]). A model predictive control (MPC) approach, which iteratively computes optimal controls for certain time periods by solving an optimal control problem that uses a mathematical model of the traffic flow, has turned out to be very successful; see Fig. 7.

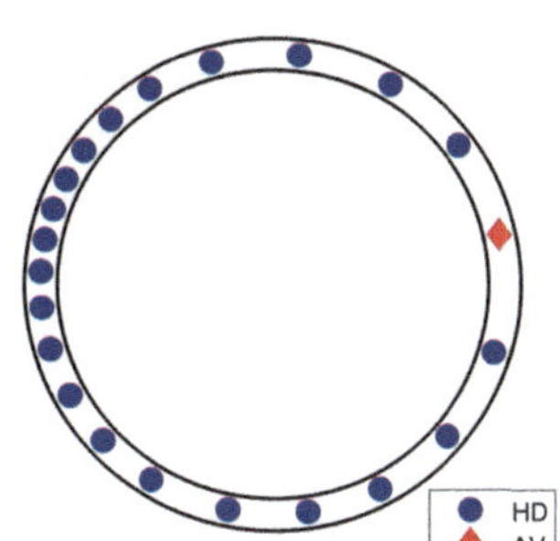

Fig. 6 Visualization of the ring road in *Matlab*

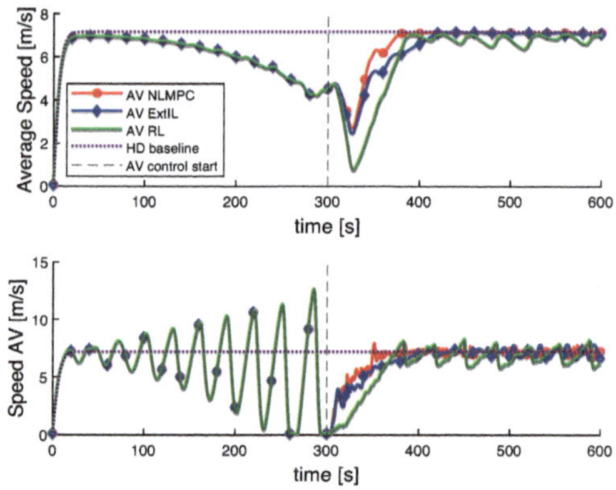

Fig. 7 Comparison of the NLMPC approach with extIL and RL in terms of the average speed over all vehicles and the AV's speed

However, this MPC approach has a severe drawback: Any time, a new control has to be set; the numerical solution of an optimal control problem is required, which is potentially very time-consuming. Even in this case, in which the considered traffic model is of moderate complexity, the average computation time for a 2s control step is approximately 8s, thus, far from real time, which is necessary for a realistic application.

Consequently, there is a need to accelerate the computation. Following the idea presented in 3, we apply an *imitation learning (IL)* technique.

The goal of an imitation learner is to keep the high performance of the nonlinear MPC approach, but to decrease the required computing time. The general idea of IL is to *imitate* (or mimic) another controller at a certain task. This controller, for the ring road scenario, could be either an intelligent human driver or, in our applications, a controller that has been realized using MPC.

In particular, by solving the control problem over a certain timer period $[t_0, t_f]$ numerically, we obtain for each time step t of a time grid $[t_0, \dots, t_f]$ a state $x_t \in \mathcal{X}$ of state space $\mathcal{X}$ and a corresponding optimal control $u_t \in \mathcal{U}$ of control space $\mathcal{U}$. Now, we aim to find a function:

$$\pi_\theta \colon \mathcal{X} \to \mathcal{U}$$

with parameters θ that maps *optimally* from state space to control space. Optimally, in this context, means as similar as possible to the other controller. This can, e.g., be achieved by optimizing θ such that

$$\hat{\theta} = \underset{\theta}{\operatorname{argmin}} \sum_{t=t_0}^{t_f} \|\pi_\theta(x_t) - u_t\|^2 . \tag{1}$$

A typical choice for π_θ are multilayer feed-forward neural nets (NNs) because they are well known for their capability of approximating nonlinear functions [5, 11].

So far, we have introduced an IL controller that shall mimic the MPC controller for *one* specific scenario. However, we aim to find a controller that performs well in different situations. Whence, we have solved L different problems leading to a data set of states and controls $\mathscr{D} = \{(x_t^l, u_t^l)\}$ over time steps $t \in \{t_0, \ldots, t_f\}$ and trajectories $l = 1, \ldots, L$. These problems may differ in their initial values and setups of the dynamical system. Then, analogously to the previous situation, we optimize the parameters of a function π_θ such that

$$\hat{\theta} = \underset{\theta}{\operatorname{argmin}} \sum_{l=1}^{L} \sum_{t=t_0}^{t_f} \left\| \pi_\theta(x_t^l) - u_t^l \right\|^2 . \tag{2}$$

This covering procedure takes into account a variety of controllers that are optimal for different setups, thus leading to a much higher robustness and generalization quality. More details can be found in [4].

For all scenarios, we let the vehicles start with equal leads and slightly vary each HD's driving behavior such that over time a stop-and-go wave occurs. At the beginning, all vehicles are controlled by HDs to observe the emerging stop-and-go wave caused by heterogeneous driving behavior. Then, at $t = 300$ s the AV control for vehicle $i = N$ is switched on with the goal to stabilize the traffic flow (cf. 7). The plot shows that all controllers succeed in stabilizing the flow. Though the MPC-based approach performs best (red curve), the imitation learning controller (blue curve) reaches the same quality. For reasons of comparison, an RL-based control scheme (cf. 4.1) has been studied here as well (green curve).

Table 1 summarizes the results in terms of the reached average speed after stabilizing success; its standard deviation (SD); the time, until this stable state is reached; and, last not least, the needed average computing time. The table includes additionally as reference a control algorithm that is taken from literature ("PI Sat"); it is a controller that is based on heuristics and rules which especially is tailored to that specific control task (cf. [15]).

Table 1 Comparison of different controllers at the ring road scenario in terms of average values over test scenarios. The average time to stabilize the system is defined as the time until all speeds $v_1, \ldots, v_N$ are within ± 0.3m/s of the equilibrium speed

Control-	Av. speed [m/s]	SD speed [m/s]	Av. time for stab. [s]	Av. comp. time [s]
HD	4.60	3.98	–	–
NLMPC	6.61	1.35	72.1	8.1
extIL	6.57	1.40	131.2	0.0041
RL	5.65	2.10	191.1	0.0026
PI Sat	6.41	1.60	181.5	0.0007

5 Summary and Conclusion

As in almost all areas of life, the availability of data is also increasing substantially in the field of vehicle development and mobility. In this article, current developments as well as methods of data analysis, AI, and ML in the vehicle development and engineering process and for mobility challenges have been considered and discussed. For years, the technological possibilities for collecting, processing, and managing data have been growing and improving in this area. All this leads to the fact that it is possible to extract valuable information and insights from existing data volumes by using suitable methods and procedures from data-based mathematics. Projected to the vehicle development process, this concerns both vehicle and customer data and environmental data, and it extends to all areas and components of the overall process, i.e., design, development, layout, validation, testing and verification, and operation; and it additionally extends to the field of many vehicles, traffic, and mobility. The availability of data relating to the vehicle and its environment will still continue to increase in the near future, and the task here is to extract as much knowledge as possible from both historically existing data sets and newly added information; mathematics is able to make a decisive contribution at this point. Especially, the task of testing, validating, and verifying AI-based methods themselves is a crucial challenge that still needs to be solved in a satisfying and an acceptable way. Today, there is much research going on in this area, but there are still many questions and tasks open and unsolved. This is especially important for the application of AI-based methods in safety-critical applications: for instance, when applying the traffic controllers presented in Sect. 4 to the real word, or for AI-based component in ADAS/AD systems as well. The combination with methods of physics-based modeling and classical system simulation is of crucial importance: Here, existing knowledge and proven methodological approaches must be combined with AI- and data-driven techniques in the best possible way to achieve the greatest benefit. As a last point, it is important to emphasize that there is enormous potential, now and in the future, in the current and still growing availability of data, which can be raised with data-driven techniques and AI methods, as the presented spotlight examples underline.

References

1. Arel, I., Liu, C., Urbanik, T., & Kohls, A. (2010). Reinforcement learning-based multi-agent system for network traffic signal control. *IET Intelligent Transport Systems, 4*, 128–135.
2. Baumgart, U., & Burger, M. (2021). A reinforcement learning approach for traffic control. In *Proceedings of the 7th International Conference on Vehicle Technology and Intelligent Transport Systems—VEHITS*, (pp. 133–141). INSTICC, SciTePress.
3. Baumgart, U., & Burger, M. (2022). Optimal control of traffic flow based on reinforcement learning. In C. Klein, M. Jarke, M. Helfert, K. Berns, & O. Gusikhin (eds.). *Smart cities, green technologies, and intelligent transport systems* (pp. 313–329). Springer International Publishing.

4. Baumgart, U., & Burger, M. (2023). Hybrid optimal traffic control: Combining model-based and data-driven approaches. In *Proceedings of the 9th International Conference on Vehicle Technology and Intelligent Transport Systems—VEHITS*. INSTICC, SciTePress.
5. Bishop, C. (2006). *Pattern recognition and machine learning*. Springer.
6. Burger, M., Dressler, K., & Speckert, M. (2019). Datenbasierte Schätzung beanspruchungsrelevanter Fahrzeuggrößen. In *46. Tagung des DVM-Arbeitskreises Betriebsfestigkeit, 09. und 10. Oktober 2019, Wolfsburg. DVM-Bericht 146.*
7. Burger, M., Dressler, K., & Speckert, M. (2021). Load assumption process for durability design using new data sources and data analytics. *International Journal of Fatigue, 145*(106116), 106116.
8. Burger, M., Speckert, M., & Dressler, K. (2018). Nutzungsspezifische Vorhersage von Beanspruchung und Energieverbrauch. In *VDI-Bericht 2333* (pp. 781–793). VDI.
9. Burger, M., Speckert, M., Müller, R., & Weiberle, D. (2018). Model-based identification of road profiles and road roughness indicators using vehicle measurements. In *Proceedings of commercial vehicle technology symposium 2018* (pp. 276–287). Springer.
10. Burger, M., & Steidel, S. (2019). A grey box model approach for the prediction of tire energy loss. In M. K. et al., (ed.). *Advances in dynamics of vehicles on roads and tracks, IAVSD 2019.* Lecture Notes in Mechanical Engineering.
11. Hornik, K., Stinchcombe, M., & White, H. (1989). Multilayer feedforward networks are universal approximators. *Neural Networks, 2*, 359–366.
12. OpenStreetMap (2020). https://www.openstreetmap.org. Accessed Dec.18, 2020.
13. Rajeswaran, A., Kumar, V., Gupta, A., Vezzani, G., Schulman, J., Todorov, E., & Levine, S. (2018). Learning complex dexterous manipulation with deep reinforcement learning and demonstrations. In *Proceedings of Robotics: Science and Systems (RSS).*
14. Silver, D., Hubert, T., Schrittwieser, J., Antonoglou, I., Lai, M., Guez, A., Lanctot, M., Sifre, L., Kumaran, D., Graepel, T., Lillicrap, T., Simonyan, K., & Hassabis, D. (2018). A general reinforcement learning algorithm that masters chess, shogi, and Go through self-play. *Science, 362*(6419), 1140–1144.
15. Stern, R. E., Cui, S., Delle Monache, M. L., Bhadani, R., Bunting, M., Churchill, M., Hamilton, N., Haulcy, R., Pohlmann, H., Wu, F., Piccoli, B., Seibold, B., Sprinkle, J., & Work, D. B. (2018). Dissipation of stop-and-go waves via control of autonomous vehicles: Field experiments. *Transportation Research Part C: Emerging Technologies, 89*, 205–221.
16. Sugiyama, Y., Fukui, M., Kikuchi, M., Hasebe, K., Nakayama, A., Nishinari, K., Tadaki, S.-i., & Yukawa, S. (2008). Traffic jams without bottlenecks—experimental evidence for the physical mechanism of the formation of a jam. *New Journal of Physics, 10*(3), 033001.
17. Sutton, R. S., & Barto, A. G. (2018). *Reinforcement Learning: An Introduction.* Adaptive Computation and Machine Learning (2nd ed.). MIT Press.
18. Vinitsky, E., Kreidieh, A., Flem, L. L., Kheterpal, N., Jang, K., Wu, C., Wu, F., Liaw, R., Liang, E., & Bayen, A. M. (2018). Benchmarks for reinforcement learning in mixed-autonomy traffic. In *Proceedings of the 2nd Conference on Robot Learning. Proceedings of Machine Learning Research* (Vol. 87), pp. 399–409.
19. Wiering, M. (2000). Multi-agent reinforcement leraning for traffic light control. In *Proceedings of the Seventeenth International Conference on Machine Learning (ICML '00)* (pp. 1151–1158).

Index

J. Franke, A. Schöbel (eds.), *Statistical Machine Learning for Engineering with Applications*, Lecture Notes in Statistics 227,
https://doi.org/10.1007/978-3-031-66253-9

Printed by Printforce, the Netherlands